Climate Change and Climate Modeling

The modeling of past, present and future climates is of fundamental importance to the issue of climate change and variability. *Climate Change and Climate Modeling* provides a solid foundation for science students in all disciplines for our current understanding of global warming and important natural climate variations such as El Niño, and lays out the essentials of how climate models are constructed.

As issues of climate change and impacts of climate variability become increasingly important, climate scientists must reach out to science students from a range of disciplines. Climate models represent one of our primary tools for predicting and adapting to climate change. An understanding of their strengths and limitations – and of what aspects of climate science are well understood and where quantitative uncertainties arise – can be communicated very effectively to students from a broad range of the sciences. This book will provide a basis for students to make informed decisions concerning climate change, whether they go on to study atmospheric science at a higher level or not. The book has been developed over a number of years from the course that the author teaches at UCLA. It has been extensively class-tested by hundreds of students, and assumes no previous background in atmospheric science except basic calculus and physics.

This book:

- provides a solid understanding of the physical climate system and the underpinnings of current climate assessments
- provides a bridge between introductory textbooks and popular science books on climate change, and advanced textbooks on atmospheric science
- is supported by a range of internet resources.

J. David Neelin is a professor and chair of the Department of Atmospheric and Oceanic Sciences, and member of the Institute of Geophysics and Planetary Physics at the University of California, Los Angeles. He has published over 100 scientific papers, including contributions to understanding and predictability of the El Niño/Southern Oscillation phenomenon, decadal variability, vegetation interaction with climate variability, how rainfall interacts with natural climate variability and anthropogenic change, and methods of improving representation of rainfall processes in climate models. He has taught courses in climate science from introductory undergraduate to advanced graduate level. He is a fellow of the John Simon Guggenheim Memorial Foundation, the Royal Meteorological Society and the American Meteorological Society, and the recipient of a Presidential Young Investigator Award, National Science Foundation Special Creativity Award and the American Meteorological Society Meisinger Award.

"This is a timely and important book that lucidly and engagingly covers topics related to climate change, topics that currently receive enormous attention and that unfortunately cause polarization."

Professor S. George Philander, *Princeton University*

"David Neelin's book is a very valuable and accessible textbook for students of climate science, and all those with an interest in climate modeling. It is a thorough and highly readable book that neatly spans the gap between general interest climate change texts and higher-level books for specialists."

Dr Drew Shindell, *NASA Goddard Institute for Space Studies*

"This book presents the diverse subjects of climate modeling and climate variability in a way that is clear and understandable to students from different backgrounds. The author is a world-famous climate scientist who has been highly successful both in research and teaching, covering all of the theoretical, modeling and data analysis aspects of climate science. The book is based on a course he has been teaching at UCLA for many years, which has been extremely popular and highly valued by students from a variety of disciplines. I am sure that the book will soon become the standard textbook on climate modeling and climate change."

Professor Akio Arakawa, *University of California, Los Angeles*

"If you're looking for an up-to-date text that deals with the science of climate change and climate modelling in a way that is both rigorous and accessible, then this book is for you. This timely treatment of a vitally important topic presents a novel integration of climate system science, including variability and change, with the fundamental principles of climate modelling and its applications that is accurate, informative and useful in a range of contexts. The book is structured to provide engaging material for both those interested in engaging with the complex science of climate change, and those whose focus is on developing a broader understanding to apply in areas such as ecology, engineering or policy. Neelin's book will be a valuable addition to my library and mandatory reading for my students."

Dr Janette Lindesay, *Australian National University*

"With the looming prospect of serious climate change at hand, it is ever more important to interest the best and brightest minds in the challenging problems of climate science. But those of us who teach climate science have been handicapped by the lack of a comprehensive and engaging text. With his masterful Climate Change and Climate Modeling, David Neelin has answered our prayers."

Professor Kerry A. Emanuel, *Massachusetts Institute of Technology*

Climate Change and Climate Modeling

J. DAVID NEELIN

University of California, Los Angeles

CAMBRIDGE
UNIVERSITY PRESS

CAMBRIDGE
UNIVERSITY PRESS

University Printing House, Cambridge CB2 8BS, United Kingdom

One Liberty Plaza, 20th Floor, New York, NY 10006, USA

477 Williamstown Road, Port Melbourne, VIC 3207, Australia

4843/24, 2nd Floor, Ansari Road, Daryaganj, Delhi - 110002, India

79 Anson Road, #06-04/06, Singapore 079906

Cambridge University Press is part of the University of Cambridge.

It furthers the University's mission by disseminating knowledge in the pursuit of
education, learning and research at the highest international levels of excellence.

www.cambridge.org
Information on this title: www.cambridge.org/9780521602433

First published 2011
4th printing 2015

A catalogue record for this publication is available from the British Library

Library of Congress Cataloging in Publication data
Neelin, J. David.
Climate change and climate modeling / J. David Neelin.
p. cm.
Includes bibliographical references and index.
ISBN 978-0-521-84157-3 (hardback) – ISBN 978-0-521-60243-3 (pb)
1. Climatology–Textbooks. 2. Climatic changes–Textbooks. I. Title.
QC861.3.N44 2010
551.6–dc22   2010039193

ISBN 978-0-521-84157-3 Hardback
ISBN 978-0-521-60243-3 Paperback

Additional resources for this publication at www.cambridge.org/neelin

To my parents, who gave me the Earth, and to my kids, who will inherit it.

Contents

Preface

Climate change and climate variability have become important topics in atmospheric, oceanic and environmental sciences. Recent developments in understanding, modeling and prediction of El Niño have brought seasonal-to-interannual climate predictions into everyday life. Projections of global warming as a consequence of human activity have been in the public consciousness for some time, even if the understanding of the scientific issues may not be as deep as would be desirable. There is a need to prepare science students for participation in environmental decision making by teaching the physics of the phenomena and the physical basis of computational climate models. This text aims to teach students current scientific understanding of global warming and of important natural climate variations such as El Niño, while laying out the essentials of how climate models are constructed.

Most of these students are not likely to become climate model builders. Some may become users of climate model output, others simply need to be aware of the strengths and limitations of climate modeling. Thus a course need not be so specialized that it aims only at future climate modelers, but should be at a level where some science background can be assumed. The treatment does not shy away from writing down the equations for a climate model, but they are explained in a way that students with calculus for biologists as a background have no trouble following.

This book arises from a course I have taught and continuously revised over the past dozen years at UCLA. It serves (i) as an initial core course for majors in Atmospheric, Oceanic and Environmental Sciences, and an option in the Environmental Sciences major; and (ii) as an introduction to this field for majors in other science fields, notably biology, with some students from social sciences and engineering. The second group is more numerous and reaching out to them has greatly increased the undergraduate population served by our upper division classes. The course grew from a handful of students initially until it routinely hit the enrollment cap with as many as 90 students. The mixture of students works well once a little extra background is provided for non-majors; typically the highest grade in the class goes to a non-major.

Climate science has grown too large to be fully treated in a single course, and this text reflects its origins as part of a larger curriculum. Following growth of the course on which this textbook is based, our department developed further courses for the upper division science audience, including courses on paleoclimate and biogeochemical cycles, atmospheric chemistry, and oceanography. As a result, certain topics related to these areas are treated briefly here. It seems likely that a similar sequence can reach across departmental boundaries at other universities. However, for an instructor planning an all-in-one course, other resources exist to extend the areas abbreviated here. If a shorter treatment of the physical climate system is desired, this book is written so that certain pieces can be condensed in a modular manner. For instance, Chapters 3 and 5 each have a summary section that assists abbreviation (sections 3.8 and 5.1, respectively). Chapter 5 can be treated succinctly with sections 5.1, 5.4 and 5.5, while still covering essentials of climate models, their evaluation and sources of error. Section 4.6 can be skipped or skimmed for a shorter treatment of El Niño that still captures the bottom line for forecasts and impacts.

Endnotes for each chapter are used to provide more rigorous underpinning and connection to the research literature. These are aimed largely at advanced students and instructors. In some cases the endnotes are used to provide definitions or elaborations that would weigh down the text.

The text sticks to the science of these issues and does not directly address policy questions, following the traditional approach that climate science should provide the best available information for policy decisions, but maintain reserve with respect to advocacy. Topics that follow the news cycle change too much from year to year to be suitable material for a textbook, but the background provided here can aid in discussing some of these as they arise. For instance, a recurring suggestion that possible warming on Pluto might be relevant to earthly climate change affords students an opportunity to assess this for themselves with the material in Chapter 2 and the information that Pluto has a 249-year, highly elliptical orbit. Substantial effort is made to provide students with a sense of where real uncertainties or limitations of climate models arise, including climatological simulations in Chapter 5, global climate sensitivity in Chapter 6 and 7, and regional sensitivity in Chapter 7. No one is more humble before the complexities of the climate system than the climate modeler trying to improve his or her model's simulation of rainfall in a particular region, or making real-time forecasts of climate variations. The students leave the course with a more concrete understanding of the capabilities and challenges of climate modeling.

Acknowledgments must begin with Joyce Meyerson, whom I first met when she was a student in the climate modeling course, and who has become a key member of my research group, assisting in innumerable ways. The extensive set of illustrations based on a combination of material from the scientific literature, material developed from scientific presentations, and schematics to illustrate key points attests to her skill at taking a scribbled sketch or description and turning it into a clear and aesthetic scientific illustration. This has offered the opportunity to redo even traditional figures, such as ocean current systems, with updates from more recent data. Grayscale versions of figures are included in the text because of a student preference for low cost but color versions and associated PowerPoint presentations are available for all figures online, as are examples of problem set and exam questions. Climate science changes rapidly, tending to leave textbooks behind, so updates to these online materials will be made periodically.

Thanks to former students from the course B. Tang, T. Rippeon, K. Roy, S. Chin, J. Park and others who have contributed corrections or pointed out areas that needed clarification, and to all of my former teaching assistants. Comments from D. Waliser, who has taught from a draft version, and from K. Hales, C. Chou and H. Su are appreciated. For discussion, I thank I. Held, G. Philander, all my UCLA colleagues, and many others. Of the many sources noted in the text I would particularly like to acknowledge a graduate-level volume edited by K. Trenberth, and the reports of the Intergovernmental Panel on Climate Change. To any colleagues whose work is not sufficiently referenced, my apologies – despite a substantial bibliography, some important work is bound to be left out and references are weighted towards works that summarize parts of the literature, are associated with figures, or are from areas where I have less direct expertise. Federal grants from the National Oceanographic and Atmospheric Administration, the National Aeronautics and Space Administration, and the National Science Foundation have supported my research over the years. Aspects of the preparation of the course and of material for this textbook have formed part of the contributions to undergraduate education and outreach of my National Science Foundation grant. A fellowship from the John Simon Guggenheim Memorial Foundation contributed to completion of this work.

1 Overview of climate variability and climate science

1.1 Climate dynamics, climate change and climate prediction

Climate is commonly thought of as the average condition of the atmosphere, ocean, land surfaces and the ecosystems that dwell in them. Every one knows what is meant by "Baja California has a desert climate" in terms of average temperature, average rainfall, average moisture in the air, and vegetation. Climate also includes the average wind direction and strength, average cloud cover, the temperature of the sea surface nearby, which affects the previous quantities, and the ocean currents that affect the sea surface temperature, and so on. While we might care most about the local climate in the land regions where we live, this interconnectedness of the climate system implies that we have to study it globally.

In contrast to climate, *weather* is the state of the atmosphere and ocean at a given moment in time. As the saying goes, "climate is what you expect, weather is what you get." However, climate includes not only average quantities, such as average precipitation, but also average measures of weather-related variability. These would include, for instance, the probability of a major rainfall event occurring in July in Baja, the range of variations of temperature that typically occur during January in Chicago, or the number of hurricanes that typically hit the US coast per year. Climate may thus be considered to include all quantities defined by averaging over the weather, i.e. over time scales of many weather events. Since the Earth has very strong changes with season, this means that an average must be taken, for instance, over January of many different years to obtain a climatological value for January, over many Februaries to obtain February climatology, and so on.

The importance of climate has increased with the realization that *climate change* is not restricted to past eons but is occurring on time scales that affect human activities. While climate is an average over weather events, the time period used in the average will affect the climate that one defines. For instance, the climate defined by an average from 1950–1970 will differ from the average from 1980–2000. We know that this average changes from one decade to another, and even more so between different centuries or millennia. These changes are referred to as *climate variability* – essentially all the variability that is not just weather. This includes ice ages and the long-term warm climate enjoyed by the dinosaurs, as well as events such as the drought that has plagued the Sahel region in Africa over the past decades, and *El Niño*, in which the tropical Pacific Ocean warms and cools every few years. Climate change has taken on a new dimension now that human activities can change the climate. This is referred to as *anthropogenic climate change* to distinguish it from natural climate variability such as El Niño. Examples of anthropogenic change include the ozone hole, acid rain and global warming.

Although climate has been of interest to humans since ancient times, the science that studies the processes that keep our climate in its current state, and cause climate changes, is new enough that there is not even agreement on what to call it. *Climate science* or *climate dynamics* are coming to be the preferred names. An older term, "climatology," is still also used but unfortunately has connotations of static, unchanging climate and old geographers poring over maps. The term *climatology* is now standardly used to refer to the average variables themselves, for instance "the January precipitation climatology." It can thus be confusing when it is used for the field of study. *Climate modeling* is a very important area of climate science, since much current work uses mathematical models. These *climate models* are mathematical representations of the climate system which typically consist of equations for temperature, winds, ocean currents and other climate variables, and which are almost always solved numerically on computers. Climate modeling necessarily interacts with the part of the field devoted to making and analyzing observations. Many climate scientists come from physics, mathematics, chemistry, engineering or biology, and bring the tools of their fields to bear on this rapidly developing area.

Climate system or *Earth system* are used to refer to the global, interlocking system of atmosphere, ocean, land surfaces, sea and land ice, and the parts of the biosphere and solid earth that are relevant for the problems of interest. The *biosphere* is the plant and animal component of the planet. Some jargon enthusiasts go so far as to call oceans, lakes, etc. the hydrosphere. Often the term *Earth system* is used to emphasize the simultaneous study of all parts of this system, including the important role of chemical reactions and biological contributions. The *physical climate system* is sometimes used to distinguish the parts of the system that can be studied while assuming that most of the chemistry and biology is unchanging. For instance, if one assumes that the composition of the atmosphere is roughly constant except for specified changes in carbon dioxide, one can examine the still very complex interplay of atmospheric circulation, heat balances, clouds, and oceanic circulation separately from the chemistry and biology of carbon dioxide uptake and release, and separately from other questions such as the ozone hole. A model that simulates a part, for instance, of the physical climate system is still termed a climate model, even if not all aspects are included. *Earth system model* is usually used for models that attempt to include physical, chemical and biological aspects at the same time.

Global warming is the predicted warming, and other associated changes in the climate system, that the vast majority of scientists in the field are convinced is beginning to occur in response to the increased amounts of *greenhouse gases* that are being emitted into the atmosphere by human activities. Greenhouse gases, such as carbon dioxide, methane and chlorofluorocarbons, are trace gases that absorb *infrared radiation* and thus affect the Earth's *energy budget* of incoming sunlight (solar radiation) and outgoing infrared radiation to space. This produces a warming tendency, known as the *greenhouse effect*. *Global change* more generally describes human-induced changes in the large-scale climate system, including the ozone hole. *Environmental change* is even more general, including air and water pollution, deforestation, soil erosion, and endangerment of individual species or ecosystems by loss or pollution of habitat. Some of these problems occur at *regional scales* such as the area of a few states, or even of a single city.

Climate prediction, on the other hand, includes the endeavor to predict not only human-induced changes in the global environment but also the natural variations of climate that affect us. El Niño is the most notable example of a phenomenon that was scarcely known two decades ago, and now is considered at least partly predictable because of advances over the past decade. Climate prediction relies heavily on physically based climate models, although for some purposes statistical models have also been used.

The current predictions of human-induced climate change are sufficiently grave that they demand decisions on response, mitigation strategies, government policy, international protocols and conventions. Predictions of natural climate variations raise questions of how the public interprets predictions, which climate variables (precipitation, temperature, etc.) are useful to which countries or interest groups, and what use will be made of the information. This is known as the *human dimension* of climate science.

1.2 The chemical and physical climate system

1.2.1 Chemical and physical aspects of the climate system

Changes in the chemical constituents of Earth's atmosphere and oceans are very important in environmental change from regional to global scales. This includes air pollution and the changes that human activities are creating in atmospheric concentrations of carbon dioxide, methane and other greenhouse gases that contribute to global warming.

The study of *environmental chemistry* includes the sources, reactions and pathways that contribute to setting the chemical composition of our atmosphere and ocean. It also includes the variations in chemical composition that have occurred during the history of the climate system.

Equally important are the many variations in the *dynamical* or *physical climate system*: the winds, the temperature, cloud amount, ice cover, ocean currents. Many of these variations do not depend on variations in the chemical composition of the atmosphere. There is no need, for instance, to model changes in ozone in a model aimed at predicting El Niño or in a weather prediction model. In studying a complex system, we need to make simplifications wherever this can be done without distorting the phenomenon of interest. So a conceptual separation is often made between these aspects of the Earth system. Examples of phenomena or topics of study associated with these subsystems include:

- Physical climate system: weather, El Niño, North Atlantic Oscillation, Asian monsoon variations, North American monsoon variations, droughts, floods, processes maintaining circulation of the atmosphere and oceans for current atmospheric composition, deep ocean circulation, ice ages…
- Environmental chemistry: the ozone hole, urban air pollution, aerosol formation, haze…

- Biosphere: evolution of the atmosphere, oxygen production, carbon cycle between biomass and carbon dioxide and other atmospheric and oceanic constituents, land surface processes, biodiversity…
- Linkages: the effects of the carbon cycle on carbon dioxide concentration and thus on the greenhouse effect, effects of dynamical processes on ozone hole formation (the stratospheric polar vortex, stratospheric ice clouds), vegetation effects on absorption of sunlight and evaporation from land surfaces…

Mathematical models of global climate can reproduce many aspects of the physical climate system without directly dealing with chemistry or biology. In this approach, the climatology of chemical and biological constituents is specified without going into the details of what maintains them. For instance, many models simulating the current state of ocean and atmospheric circulation take the current concentration of chemical constituents as given, such as oxygen, nitrogen, ozone and carbon dioxide. The total mass of the atmosphere and oceans is also taken as given, although early in Earth's history these were different.

Even in global warming, where changes in the concentration of carbon dioxide are crucial, it can be useful to *specify* an expected increase and study the response of the physical climate system. By specifying the carbon dioxide concentration as a function of time as an input to the physical climate system, one defers having to understand and correctly model the set of processes, involving chemistry and ecosystems feedbacks, that determine the concentration. While these processes are important, the burning question at the initial stage is what the physical climate system will do in response to the expected increases. Modeling and understanding this response can be extremely complex, as we shall see.

In studying such a complex system, what constitutes a good approximation depends on the question you are asking. It also depends on whether you are interested in understanding the overall behavior or if a highly accurate answer is required about particular details, and which set of interactions are key to the question being addressed. It is common to make one approximation to understand leading effects, and to improve on this in the next approximation. For instance, when first modeling global warming, initially approximations were used that amounted to specifying fixed ocean circulation. Current models applied to global warming typically include a full ocean model, but many specify carbon dioxide concentrations. Next-generation models now exist (and are being improved) that include interactive carbon cycles. At each stage the previous, simpler class of model remains valuable for understanding the results of the next class. Studying the components of the climate system separately is useful to make progress – as long as one never loses sight of the fact that there is one Earth system. Information from the chemists and biologists is essential to the dynamists and vice versa.

1.2.2 El Niño and global warming

El Niño is the largest *interannual* (year-to-year) climate variation. The source of the phenomenon involves an interaction between the tropical Pacific Ocean and the atmosphere above it. It can be reproduced in models in which the chemical composition is entirely fixed, i.e. it is a phenomenon of the physical climate system. In fact, the essential aspects of El Niño can be understood in models that include only the tropical Pacific region, although the impacts of what happens in this region are felt worldwide. As well as being a prime example of natural climate variability, El Niño was the first phenomenon for which the essential role of dynamical interaction between atmosphere and ocean was demonstrated. El Niño cannot be reproduced in an atmosphere model alone, nor in an ocean model alone – unless aspects of the observed El Niño evolution are specified in the other component. A coupled ocean–atmosphere model, on the other hand, can produce El Niño oscillations internally. Thus the study of El Niño has brought about an interdisciplinary interaction between atmospheric scientists and oceanographers.

In global warming, many of the complex effects created by an increase in greenhouse gases occur in the atmosphere. To a first approximation, these may be studied with relatively rudimentary effects of the ocean, although at the next level of understanding oceanic effects must be included. The number of subtle processes that must be modeled is daunting. These include, for example, the average on large scales of the effects of small-scale clouds, and how these change as the planet warms. These effects lie within the study of the physical climate system.

Scope of this text

The causes of El Niño, and many of the most important uncertainties affecting our assessment of global warming, lie in the physical climate system. This text aims to provide an understanding of what these processes are, of the strengths and weaknesses of climate models, and of the extent of our ability to predict the climate system, including limits to accuracy at regional scales. The focus is on global-scale aspects of environmental change and variability, and on the physical, rather than chemical, components of the climate system. Other books are available that deal with air pollution and atmospheric chemistry and related topics. The task of linking the physical and chemical climate components to the biological and human dimensions of climate would require several courses. This book presents the basic science on the physical climate side, to provide a solid background for students in a wide range of science disciplines who may go on to work in these other areas. A knowledge of the capabilities and limitations of climate models can be useful background as society debates how (and whether) to limit the eventual magnitude of global warming and prepares to adapt to its impacts.

1.3 Climate models: a brief overview

The motions, temperature and other properties of the atmosphere and ocean are governed by basic laws of physics. These can be written as equations, which one can then attempt to solve. The results are too complex for a general solution to be written, but approximations to these equations can be solved numerically on computers. One common type of approximation is to divide the atmosphere and ocean into discrete grid boxes, writing the balance of forces, energy inputs etc. for each box as an equation that permits one to obtain the acceleration of the fluid in the box, its rate of change of temperature and so on. From this one computes the new velocity, temperature, etc. one *time step* – say, 20 minutes for the atmosphere or an hour for the ocean – later. The equations for each box depend on the values in neighboring boxes, so the computation is done for a million or so grid boxes over the globe. This is then repeated for the next time step, and so on until the desired length of simulation is obtained. Since it is common to simulate decades or even centuries in climate runs, *computational cost* is obviously a factor to be considered.

The basic method of solving for the motions of a gas or liquid described above has much in common with what one encounters in many fluid dynamics applications in engineering, such as flow over an aircraft wing. The atmospheric component of a climate model also has a close relationship to weather forecasting models. Major differences arise from the complexity of the climate system, and from the range of phenomena at different time scales that must be addressed, as will be discussed in Chapter 2. The climate system is "messier" than typical fluid dynamics problems since the impacts of such things as clouds, aerosols and even vegetation are important. Compared with a weather forecast model, a climate model must pay much more attention to processes that affect the long term. For instance, an error in calculation of infrared radiation emitted from the atmosphere might have little effect in a weather forecast that begins from observed initial conditions and runs for a week, but in a climate model that must simulate the global energy balance correctly, this effect would be important.

The most complex climate models, described above, are known for historical reasons as *general circulation models* or GCMs (some authors reinterpret GCM as *global climate model*, except that one also uses the term "ocean GCM" to describe an ocean model of the same level of complexity). Even once a phenomena has been simulated in a GCM, it is not necessarily easy to understand the underlying physical mechanism, since a GCM includes so many effects. *Intermediate complexity* climate models are also used, in which the aim is to construct a model that is based on the same physical principles as a GCM and is also directly comparable to observations, but in which only the aspects of the system important to the target phenomenon are retained. Usually approximations are made that further simplify the solution of the equations. Intermediate complexity models are used for analyzing phenomena and also for exploring new phenomena. Intermediate complexity ocean–atmosphere models of the tropical Pacific region, for instance, were first used to simulate, understand and predict El Niño, while GCMs were still struggling with the difficulty of accurately simulating all aspects of the climate in the region. Part of our discussion of El Niño in Chapter 4 will be based on such a model, after the equations that govern the balances of the climate system are introduced in Chapter 3. Chapter 5 elaborates in more detail how climate models are constructed and the flavors of model used for different applications. It also addresses issues of model accuracy. In Chapter 6 we use a very simple climate model, a globally averaged energy balance model, to understand essential aspects of the greenhouse effect. We then return to simulations of global warming in the most complex climate models in Chapter 7.

1.4 Global change in recent history

1.4.1 Trace gas concentrations

Although we will concentrate on the physical climate system, we need to begin with some atmospheric chemistry. *Trace gases* form a tiny fraction of the atmosphere's mass, but that makes their concentration more susceptible to variation. In particular, human activities can significantly change the trace gas composition of the atmosphere. Table 1.1 gives recent typical concentrations of some of the trace gases that are susceptible to such variations. The units, parts per million (ppm) by volume, indicate what fraction of the molecules in air each gas constitutes. For instance, carbon dioxide, at about 370 ppm early in this decade, accounts for 0.037% of air molecules. The major components of air are nitrogen (N_2) at 78.08% and oxygen (O_2) at 20.95% (for dry air). Some trace gases, such as argon (0.93%), are essentially unchanging on our time scales, so are less important to global change. Water vapor, at typical concentrations of 1 to 20 parts per thousand, is an extremely important constituent in all aspects of Earth's climate. Because its concentration varies strongly in time and space, and because it changes phase to produce clouds, rain and snow, water is treated separately from other chemical constituents in climate models. All of the trace gases can have variations in time and space. Gases that have long residence time in the atmosphere tend to be mixed by atmospheric motions, so the variations are much smaller than the mean concentration. For instance, annual average carbon dioxide concentrations in northern and southern hemispheres differ by less than 1%, even though sources and sinks of CO_2 differ between the hemispheres. Gases with short residence times, such as ozone, are much more closely tied to the location of sources or sinks and can vary strongly in the vertical and horizontal.

Table 1.1 Typical concentrations (and chemical formulae) of some of the trace gases that are important to global change.

Trace gas name	Formula	Concentration
Carbon dioxide	CO_2	377 ppm
Methane	CH_4	1.8 ppm
Nitrous oxide	N_2O	0.32 ppm
Ozone	O_3	0.000 251 ppm (average; max 10 ppm in stratosphere)
CFC-11 (Freon)	CFC3	0.000 254 ppm

Note: Units are parts per million by volume. CFC denotes chlorofluorocarbon. Values are for 2004 from Clerbaux and Cunnold (2006); for magnitude of stratospheric maximum values see Randall *et al.* (2005).

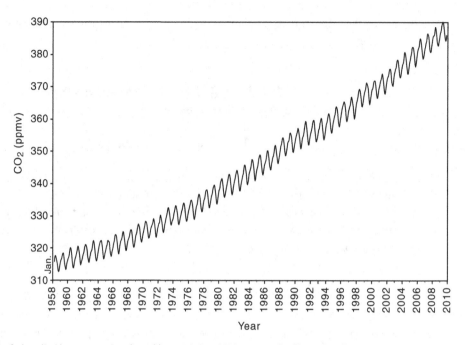

Fig. 1.1 Carbon dioxide concentrations (monthly mean) since 1958, measured at Mauna Loa, Hawaii. Units are parts per million by volume, and tick marks occur at January of the indicated year. From the National Oceanographic and Atmospheric Administration (NOAA) Climate Monitoring and Diagnostics Laboratory. Data prior to 1974 are from Keeling *et al.* (1976).

Figure 1.1 shows a measurement that has helped launch much of the current concern over global warming. The concentration of carbon dioxide has been consistently measured at a point far from continental effects, Mauna Loa, Hawaii, since 1958. It shows a dramatic, continued increase throughout the time series, evidence that the emissions of carbon dioxide from human use of fossil fuels is indeed having the expected effect of increasing atmospheric concentrations of carbon dioxide. Subsequent measurements have borne out that this holds on a global scale. The yearly variations in concentration are due to the seasonal cycle and biological effects. In summer, there is more sunlight available for photosynthesis, so there is more carbon dioxide fixed into plant biomass in the summer hemisphere. Since there is asymmetry between the amount of land and ocean in the northern and southern hemisphere, seasonal effects also occur in the global average, not just in the individual hemispheres. These small spatial and seasonal or interannual variations can give clues to the biological contributions to the *carbon cycle*. In Figure 1.1, smaller interannual variations may also be seen. These occur because of interannual climate variations, especially El Niño, affecting the biological systems.

Figure 1.2 shows how selected trace gases have varied over a longer time period, namely the last centuries, since industrialization made it possible for humans to release them in substantial quantities into the atmosphere. The concentrations may be taken as typical of global average values, but are estimated from various sources. These include direct atmospheric measurements at particular locations, for parts of the record, and measurements

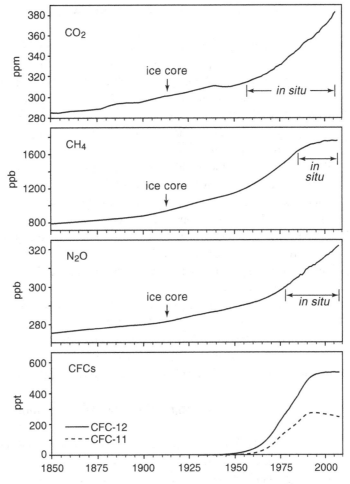

Fig. 1.2 Concentration of various trace gases, carbon dioxide, methane, nitrous oxide and two chlorofluorocarbons, respectively, estimated since 1850. The part of the record from direct atmospheric measurements is marked "*in situ.*" Data from Goddard Institute for Space Studies following Hansen *et al.* (1998).

of concentrations in air bubbles trapped in ice cores taken from the Greenland and Antarctic ice caps.[1] These permit the atmospheric concentrations to be estimated for times before measurements were being recorded for the atmosphere. Overall, they show increases in concentration, with a greater rate of increase in recent times, following rising population and industrialization. The CFCs are man-made compounds, so their concentration is zero before about 1950. All the gases shown contribute to the greenhouse effect, while the CFCs have an additional effect – stratospheric ozone loss. The human sources of nitrous oxide are imperfectly measured, but biomass burning and fertilizer use are believed to contribute. Methane is produced by cattle, sheep, rice paddies and waste disposal and as a by-product of fossil fuel use, which are all increasing with human population. Natural sources of methane include wetlands and termites.

1.4.2 A word on the ozone hole

The role of CFCs in ozone destruction was predicted by Sherwood Rowland and Mario Molina in 1974. In 1985, J. C. Farman and coworkers published observations of Antarctic ozone depletion in southern spring – that has since grown into what we all know as the *ozone hole*. The Montreal Protocol in 1987 set a timetable (since revised) for phase-out of CFC emissions.[2] CFC concentrations have indeed leveled out or begun to decrease slowly in recent years. Because of the reservoir effect of existing CFCs in the atmosphere, it may be 50 years before ozone levels recover. This is a relative success story compared with the discussions that the world is embroiled in regarding responses to the threat of global warming. It was aided by the ground work laid by the "spray-can ban" in the late 1970s, in which such countries as Canada, Sweden and the United States limited non-essential uses of CFCs, and by the development of alternative products. It is also worth noting that the prediction of ozone destruction involving CFCs was correct overall, but that nature still held a twist. The degree of ozone destruction producing the Antarctic hole turned out to be enhanced by the presence of polar stratospheric clouds whose ice crystals provide a surface on which the reactions occur more rapidly.

The ozone hole and global warming are separate environmental threats, in the sense that one can occur without the other, and they have different causes. Although there can be some small, hypothesized modifications of each by the other, we should not confuse the two. The ozone hole is essentially a chemical effect, and so is not treated further here. However, it provides a potent example of human impacts on our climate.

1.4.3 Some history of global warming studies

The threat of global warming by the greenhouse effect has been postulated since the beginning of the century. In contrast to the ozone hole, definitive identification of anthropogenic warming is not something that can occur in one step, but rather is a matter of slowly amassing evidence of a gradual warming. The quantification of future potential warming is a task that involves many climate scientists slowly pushing back the frontiers of what is known of the climate system. Table 1.2 gives a few of the events in the timeline of global warming.

Beginning in 1990, a concerted effort was made by climate scientists and others to summarize the current state of knowledge. The resulting Intergovernmental Panel on Climate Change (IPCC) Reports are consensus documents that capture the "center of gravity" of scientists' current understanding.[3] They are intended to make available to world leaders and decision makers the best estimates of how large global warming might become. At the same time, these reports, and other statements by climate scientists, must convey some sense of the uncertainty, often given as a range of possible outcomes, that remains. Chapters 6 and 7 examine in detail this consensus of what is known, what is uncertain, and why. Taking action to mitigate global warming entails enormous economic efforts and, therefore, political implications. The IPCC reports have therefore been the subject of considerable debate in both media and government.

Table 1.2 Some events in the history of global warming studies.[4]	
1850s	Beginning of the industrial revolution.
1861	John Tyndall notes that H_2O and CO_2 are especially important for infrared absorption and thus potentially for climate. The warming effect of the atmosphere and the analogy to a greenhouse had already been noted by J. B. Fourier in 1827.
1868	Jozef Stefan develops his law for blackbody radiation.
1896–1908	Svante Arrhenius postulates a relation between climate change and CO_2 and that global warming may occur as a result of coal burning.
1917	W. M. Dines estimates a heat balance of the atmosphere that is approximately correct.
1938	G. S. Callendar attempts to quantify warming by CO_2 release by burning of fossil fuels.
late 1950s	Popularization of global warming as a problem, notably by Roger Revelle.
1958	Start of C. D. Keeling's monitoring of CO_2 at Mauna Loa.
1975	First three-dimensional global climate model of CO_2-induced climate change by Suki Manabe.
1979	The "Charney Report" (US National Academy of Sciences).
late 1980s	Seven of eight warmest years of the century to that point.
1990 and 92	Intergovernmental Panel on Climate Change (IPCC) Report and Supplement.
1992	Rio de Janeiro United Nations Conference on the Environment Development; Framework Convention on Climate Change. *"The ultimate objective...is...stabilization of greenhouse gas concentrations in the atmosphere at a level that would prevent dangerous anthropogenic interference with the climate system."*
1995–96	Second Assessment Report of the IPCC: *"The balance of evidence suggests a discernible human influence on global. climate. [...] There are still many uncertainties."*
1995	Start of ongoing series of Conferences of the Parties to the Climate Convention: discussion of short term objectives in terms of greenhouse gas emissions by developed countries.
1997	Kyoto Protocol sets targets on greenhouse gas emissions at 5% below 1990 levels by 2008–2125.[5]
2001	Third Assessment Report of the IPCC.
2004	Nine of the ten warmest years since 1856 occurred in past ten years (1995–2004) (1996 was less warm than 1990).
2005	Kyoto Protocol enters into force.
2007	Fourth Assessment Report of the IPCC. Nobel Peace Prize awarded to the few thousand scientists of the IPCC process and one politician.

Some perspective can be provided by referring back to the 1979 "Charney report" by a committee convened by the US National Academy of Sciences. Although three-dimensional climate models were still in early stages of development, it is remarkable how generally consistent the report's conclusions appear with those of subsequent decades. Perhaps more remarkable is that the report came at a time when global temperatures had shown little increase for the prior three decades. The concern in the scientific community at that time was based on what was known of the climate system and the radiative impacts of increasing carbon dioxide, rather than on the warming trend that has been observed since then. Despite much re-examination of the issues, over the decades subsequent consensus reports have repeatedly come to similar conclusions, with incrementally increasing precision, and vastly more detail. There are also now accumulated data sets of temperature and other changes, whose significance can be evaluated.

1.4.4 Global temperatures

Figure 1.3 shows estimates of global average temperatures over the past century and a half. Whereas in recent times we have good measurements of surface temperature, with near global coverage, in earlier times the data are sparse. Substantial corrections have been applied to the data from earlier sources in an attempt to compensate for both coverage and methods of measurements. Sea surface temperature (SST) was measured by different types of vessels, using various devices, over the time period shown. For instance, in earlier times buckets were lowered over the side of the ship. The type of bucket used by different vessels will have affected the cooling rate as the bucket was drawn up, and estimates of how to account for these differences, relative to more modern vessels using exhaust intake temperature, have been carried out. Although these corrections are likely imperfect, they represent the best available estimate of how global temperatures have changed.[6]

Several points may be noted in the behavior of globally average temperatures in Figure 1.3:

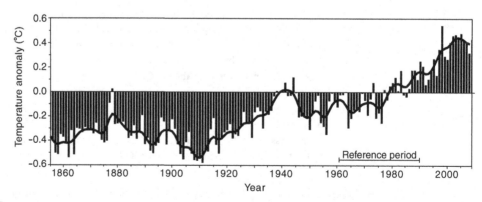

Fig. 1.3 Global mean surface temperatures estimated since preindustrial times shown as anomalies relative to the 1961–1990 mean. Bars give annual average values of combined near-surface air temperature over continents and sea surface temperature over ocean. The solid curve gives a smoothing similar to a decadal running average. From the Climatic Research Unit, School of Environmental Sciences, University of East Anglia (Brohan *et al*. 2006; Rayner *et al*. 2006; Jones *et al*. 1999).

- The amplitude of the variations (fractions of a degree) seems small compared with the temperature variations we experience locally, but since these are global averages they indicate temperature shifts over huge regions.
- Temperatures have been rising, although not uniformly.
- The overall warming from the 1860s to 1990s is less than 0.6 °C.
- There is considerable variability on time scales of years, decades and centuries that is natural in origin.

The presence of natural variability, especially on decadal and centennial time scales, makes detection of a trend due to human influence more difficult, since it is necessary to distinguish long-term trend from other variations. One way of detecting global warming with certainty would be to wait until it becomes much larger. By that point, however, the Earth would have been committed to experiencing even greater warming, as discussed in Chapter 7. Other methods include searching for spatial patterns consistent with natural variability or anthropogenic warming and comparing to the observed record. By these methods, it has been estimated that recent warming is attributable to human factors with reasonable levels of statistical significance as discussed in Chapter 7.[7] Increased confidence of having observed human-induced warming may be a matter of years or decades.

1.5 El Niño: an example of natural climate variability

One of the most important sources of climate variability on year-to-year time scales is El Niño, or rather, the El Niño/Southern Oscillation (*ENSO*) phenomenon. The term ENSO is often used within the field because El Niño is associated with the warm phase of a phenomenon that is largely cyclic (i.e. tends to repeat), and because originally El Niño was thought of as the oceanic part, while the Southern Oscillation referred to the atmospheric part. Since ENSO is the prime example of a phenomenon that depends fundamentally on ocean–atmosphere interaction, a term that includes both ocean and atmosphere seems apt. However, the term El Niño now is generally used for both atmospheric and oceanic aspects during the warm phase of the cycle. To emphasize the relationship to the warm phase, George Philander of Princeton University coined the term *La Niña* for the cold phase. Because El Niño has greater name recognition, this name is sometimes applied to the entire phenomenon, e.g. the "El Niño cycle."

The heart of the El Niño lies in the tropical Pacific Ocean along the equator. Changes in sea surface temperature, ocean subsurface temperatures down to a few hundred meters in depth, rainfall, and winds all contribute to produce the ENSO cycle. The variations in the Pacific basin within about 10–15 degrees latitude of the equator are the primary variables driving ENSO and are the most important factors in ENSO prediction. ENSO influence spreads much more broadly, including over North America and surrounding tropical continents. These remote effects are known as *teleconnections*. After presenting the history of ENSO studies and the essential observational characteristics of an El Niño event in this chapter, we will treat the dynamics of El Niño and teleconnections in Chapter 4.

In discussing ENSO, it is often useful to discuss a variable in terms of its departure from normal climatological conditions, or *anomaly*. An anomaly is calculated by taking the difference between the value of a variable at a given time, such as pressure or temperature for a particular month, and subtracting the climatology of that variable. The climatology includes the normal seasonal cycle, so for instance an anomaly of summer rainfall for June, July and August 1997 would take the average of the rainfall over that period and subtract the averages of all June, July and August values over a much longer period, such as 1950–1998. To be precise, the averaging time period for the anomaly and the averaging time period for the climatology should be specified. For instance, one might display a series of monthly averaged SST anomalies relative to a mean over a certain decade.

1.5.1 Some history of El Niño studies

Historically, both El Niño and the Southern Oscillation were known long before it was realized that there was any connection between the oceanic and atmospheric aspects. The coastal aspects of El Niño were experienced by Peruvian fishermen, because of its influence on fisheries and other phenomena in the coastal zone. The name El Niño refers to the Christ child and was originally applied to a warming of the coastal waters that begins around Christmas.[8]

On the atmospheric side, in the early 1900s, Sir Gilbert Walker was Director of Observatories in India and set his assistants to the task of searching for significant correlations among variables in all the available meteorological data. The aim was to find relationships that might help predict monsoon rainfall, since interannual variations in this have a large impact on the local economy and population. Among the things he found was an negative correlation between atmospheric surface pressure in the western and eastern Pacific Ocean. Since pressure in both regions was seen to vary irregularly from year to year, and since the equatorial Pacific seems southern to an Englishman, he termed it the Southern Oscillation. A more complete spatial picture of the pattern Walker discovered is seen in Figure 1.4, which is based on the work of Berlage (1957), decades later. Since data were scant, even at that time, Figure 1.4 contains much interpolation between data points, for instance between the limited number of islands in the eastern Pacific. However, the larger features of the pattern are borne out by later work.[9] The pressure gradient along the equator, such that the eastern Pacific has low pressure when the western Pacific has high pressure, is particularly important, since this tends to be associated with winds blowing along the equator.

In order to display time variations of this pressure difference, an index known as the Southern Oscillation Index (SOI) has been used historically. It makes use of normalized surface pressure anomalies at Tahiti minus those at Darwin, Australia (locations indicated in Figure 1.4 and Figure 1.5), since pressure data have been available as far back as the late 1800s. Tahiti is not ideally located, since it is somewhat south of the main ENSO region, but historically that is where observations were available. The time series of SOI in Figure 1.6 consists of monthly averaged anomalies relative to the mean over the whole period shown. In Figure 1.5, regions are indicated that have been used (more recently) as

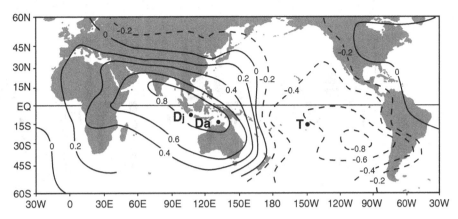

Fig. 1.4 The Southern Oscillation large-scale atmospheric pattern associated with El Niño as originally seen in surface pressure. Similar to work by Walker (1923), this figure from Berlage (1957) correlates pressure data at points everywhere on the map with pressure at one point (Djakarta, Indonesia, marked Dj). Maximum correlation of 1.0 occurs at that point necessarily, but the large negative correlations in the eastern tropical Pacific indicate a strong organized pattern of variability. Tahiti (T) and Darwin (Da) are also marked. Pressure data from these points are used to construct the Southern Oscillation Index (SOI).

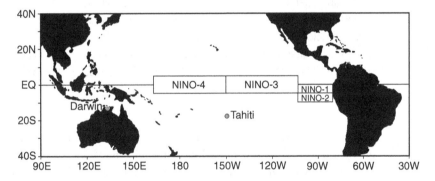

Fig. 1.5 Schematic indicating commonly used index regions for ENSO SST anomalies. Average SST anomalies over these regions are refered to as Niño-1 through Niño-4. Averages over the Niño-3 region are the most commonly used, since this area is where the largest anomalies occur during the typical El Niño or La Niña event.

indices of sea surface temperature (SST) variations associated with El Niño. SST data are averaged over the boxes shown, which are numbered westward from the South American coast. The small coastal boxes have longer data records, based on coastal stations, but the east-central Pacific box, Niño-3, is a better indicator of the main part of the El Niño signal. The time series of monthly Niño-3 anomalies since 1950 is shown along with the SOI in Figure 1.6. The close relation between atmospheric and oceanic aspects of ENSO may be seen in the negative correlation of variations in sea surface temperature and surface pressure gradient. When the SST in the Niño-3 region is warm during El Niño, the SOI

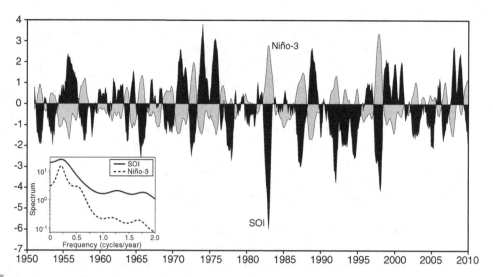

Covarying atmospheric and oceanic indices. The figure shows anomalies (departures from climatological mean) of sea surface temperature, and of the Southern Oscillation Index (SOI). The SST is averaged over a region in the eastern-central Pacific at the equator, and this index is known as Niño-3. The SOI consists of normalized surface pressure difference between Tahiti, in the mid-Pacific, and Darwin, Australia, near the equator. This provides a measure (available in relatively long records) of the pressure gradient across the Pacific, along the equator, which is in turn related to wind variations over the Pacific. During negative phases of the SOI, the anomalous winds blow from the west ("westerly") from high to low pressure along the equator. The SOI is normalized by the standard deviation, while SST is in degrees centigrade. Power spectra of these time series (inset) exhibit a broad but robust peak centered at approximately 4-year period (axis in cycles/year). A smaller (and less statistically robust) peak near 2-year period is sometimes noted but is not resolved here.[10]

is negative, i.e. pressure is low in the eastern Pacific relative to the west. This pressure gradient tends to produce anomalous winds blowing from west to east along the equator. The reverse holds during periods of cold equatorial Pacific SST (La Niña), when the pressure gradient reverses.

It may also be seen that the alternation between warm and cold phases is quite irregular, but that there is nonetheless a tendency toward a preferred time scale for recurrence, typically 3 to 5 years. This visual impression is confirmed by the power spectrum shown in the inset of Figure 1.6. A power spectrum treats a time series as if it were composed of a sum of sinusoidal oscillations at different frequencies and shows a measure of the squared amplitude at each frequency.[11] If there really is a dominant oscillation, a peak of power occurs at that frequency. If the series were composed instead of white noise, with each time uncorrelated with the last, then there would be equal amounts of power at all frequencies. In the spectrum for both SOI and Niño-3, there is a spectral peak at around 0.25 cycles per year, that is, around a 4-year period. It is a broad peak, with the power high from periods of about 3 years to 5 years. Furthermore, there is considerable power that is spread among all the frequencies, characteristic of a noisy (or chaotic) time series. In climate time series, obtaining even a broad spectral peak is unusual. Because it indicates an oscillation, it suggests that there must be some interesting mechanism at work causing the cycle to tend to repeat. And furthermore, it suggests the possibility of predictability, because if a phenomenon has a well-determined life cycle, it may be possible to predict its evolution for some time, subject to limitations set by the irregularity of the system. Both the mechanism for the oscillation and the basis and limitations of predictability will be treated in Chapter 4.

The Bjerknes hypothesis

A key step in the history of El Niño studies, summarized in Table 1.3, was the development that laid the basis for current understanding. In 1969, Jacob Bjerknes of the University of California, Los Angeles (UCLA) first postulated that ocean–atmosphere interaction was essential to the phenomenon. The data available to him were not nearly as extensive as those in Figure 1.6, but showed similar covariation of atmospheric and oceanic variables. His hypothesis, outlined in more detail in Chapter 4, was that the SST gradient across the Pacific affected the atmospheric pressure gradient and the winds, and other aspects of the atmospheric circulation. In turn, these tended to affect the oceanic circulation that created the SST anomalies in the first place. This feedback loop has proved essential to the modeling of El Niño in recent work with coupled ocean–atmosphere models.

Not long after Bjerknes' work, Klaus Wyrtki of the University of Hawaii, working with tide gauge data from island stations, added another piece to the puzzle. He noted that sea level tends to have a small (a few centimeters) but sustained rise in the western Pacific shortly before the onset of El Niño events. Since he was working with only a few events, this was a bold conjecture at the time, but has again been borne out in more detail by recent observations. Unfortunately, with limited modeling capability available at that time, it was difficult to reconcile the oceanic observations with Bjerknes' ocean–atmosphere interaction hypothesis. Wyrtki's conjectures about the mechanism, phrased in terms of ocean dynamics, are less consistent with present-day understanding than those of Bjerknes.[15] As with many climate phenomena, El Niño was too complex to understand simply by looking at observations. Self-consistent mathematical models of both ocean and atmosphere were required before the apparently opposing views could be reconciled, and such models began to be developed in the late 1970s. Oceanographers and atmospheric scientists gradually became aware that tropical regions have interesting dynamics that is very different from the typical midlatitude behavior that had been studied in traditional oceanography and meteorology. Interest in discovering the mechanisms of El Niño was already on the rise when the 1982–83 event began to develop, even as a meeting of experts was concluding that nothing major was happening that year.

Partly as a result of this, an international program was launched under the umbrella of the World Meteorological Organization, although funded by scientific programs in each member country. In the US, the National Atmospheric and Oceanographic Administration and the National Science Foundation played leading roles. Individual scientists had already begun research in the area, so the program quickly met with success. In 1985 and 1987, Mark Cane and Stephen Zebiak of the Lamont Doherty Earth Observatory of Columbia University produced a simulation of El Niño in a coupled ocean–atmosphere

Table 1.3 Some events in the development of El Niño studies, as an example of how a climate science area moved from early forays, to understanding, to routine forecasts.	
late 1800s	Peruvian sailors refer to a coastal current that appears after Christmas in certain years as the current of "El Niño," the Child Jesus.
1923	Sir Gilbert Walker, working in India on monsoon predictors, publishes negative correlation of pressure in western and eastern Pacific Ocean. He later shows that this irregular oscillation is associated with changes in rainfall and winds. He names it the *Southern Oscillation.*
1957	H. P. Berlage follows up on Walker's work but receives scant notice.
1969	Jacob Bjerknes (UCLA) looks at both atmospheric variables and ocean surface variables and hypothesizes that ocean–atmosphere coupling is essential to the development of El Niño (the *Bjerknes hypothesis*).
1975	One step forward: Klaus Wyrtki (University of Hawaii) notices that an increase in sea level height in the western Pacific tends to precede El Niño warm phases and notes the potential role of oceanic dynamics in communicating this to the eastern basin. But one step back: he blames the ocean entirely.
late 1970s to early 1980s	Developments in tropical oceanography and modeling[12]
1982–83	The biggest El Niño of the century catches experts unawares.
1985	The Tropical Ocean–Global Atmosphere program is launched.
1985–87	Mark Cane and Stephen Zebiak (Columbia University) develop first coupled ocean–atmosphere model with realistic El Niño (CZ model).
1986	First El Niño forecast with a physically based coupled model forecast (CZ). At the time, there was controversy over whether to trust it since the phenomenon was still not understood.
late 1980s to early 1990s	Developments in ENSO theory, including reconciling the role of subsurface ocean memory with the Bjerknes hypothesis.[13] Development of more complex ocean–atmosphere models including the first successful coupled general circulation model simulation of El Niño by George Philander and coworkers.[14]
1997–98	El Niño becomes a household word. Forecasts by national weather services and the newly established International Research Institute for Seasonal-to-Interannual Climate Prediction.

model that was realistic in its main features. They also produced the first El Niño forecast with a coupled ocean–atmosphere model, discussed below. In climate models it is
often the case that simulating a phenomenon does not necessarily imply that one immediately understands it. Even in the model world the dynamics can be very complex, and
theoretical examination of the simulated phenomenon often takes time. Most of the theoretical understanding of the mechanism for El Niño, discussed in Chapter 4, was developed
after the first simulation. There were also great advances in observations that occurred

during the Tropical Ocean–Global Atmosphere (TOGA) program, both in oceanographic measurements and satellite observations. In addition to the relatively less complex model used by Cane and Zebiak, more complete climate models began to simulate ENSO, and prediction schemes were developed for these. By the end of the TOGA program in 1995, prediction schemes had passed from the research community to national centers. An International Research Institute for Seasonal-to-Interannual Climate Prediction was formed, and in the US the National Meteorological Center renamed itself the National Centers for Environmental Prediction (NCEP), in recognition of the fact that meteorological prediction had become only a subset of a more inclusive mission, predicting environmental impacts. Climate prediction of seasonal-to-interannual time scale variations had become a reality.

1.5.2 Observations of El Niño: the 1997–98 event

Here we focus on the essential aspects of the ENSO phenomenon: the anomalous conditions within the tropical Pacific. Excellent observations are available for the 1997–98 event, and its spatial structure is quite typical of El Niño events. Although the magnitude is the largest of the past century, it thus provides an example of the form and evolution of the warm phase of the ENSO cycle. Here we focus on the fully developed stage of the event; we will return to the evolution in Chapter 4, once we have the modeling tools to understand it.

Sea surface temperature anomalies, shown in Figure 1.7 for December 1997, exhibit a large warming in the whole eastern and central part of the equatorial Pacific, by up to 5 °C. This is the essential signature of El Niño. Warming up and down the west coast of North and South America is a common by-product. Anomalies of SST seen in other ocean regions

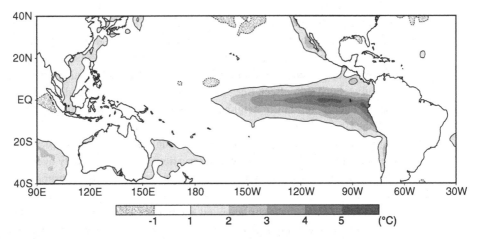

Fig. 1.7 December 1997 anomalies of sea surface temperature during the fully developed warm phase of ENSO are up to 5 °C warmer than normal along the equator over the eastern Pacific. In terms of total temperature (i.e. climatology plus anomaly), this implies that the cold waters that usually occur in this region are almost as warm as the western Pacific.[16]

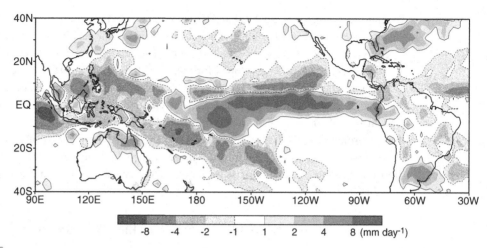

Fig. 1.8 December 1997 anomalies of precipitation during the fully developed warm phase of ENSO show a large increase over the anomalously warm waters.[17]

are less closely related. Smaller variations of SST commonly occur in all oceans driven by random weather fluctuations but are not as coherent and long-lasting as El Niño. The warm SST anomalies in Figure 1.7 occur in a region along the equator that is normally relatively cold, called the *equatorial cold tongue*. During some El Niño events the warm SST anomalies might have a maximum in the central Pacific rather than in the eastern part of the basin, but they always occur along the equator. The area of the warm anomaly is very large – the region of 1 °C or larger warming is roughly the area of the continental United States.

The change in the distribution of warm SST creates a shift in the regions of strong convection during El Niño, as seen in the precipitation map in Figure 1.8. The region with increased rainfall tends to occur over the region with anomalously warm SST and is associated with rising motion and convergence of the surface winds. The reduced precipitation over much of the western Pacific and Indonesian region and parts of equatorial South America is a strongly related side-effect. Because the convection has extended over a larger region in the eastern Pacific, rainfall in neighboring regions tends to be reduced. Precipitation impacts over the United States are not visible in Figure 1.8, partly because the scale is set for the large tropical anomalies and partly because the impacts outside the tropics (in "midlatitudes") are statistical in nature. Since these observations are averaged only over one month, there is also considerable variability that is weather-related, and would disappear in a longer-term average.

Figure 1.9 shows near-surface wind anomalies typical of an El Niño. The winds tend to converge into the rising region with increased rainfall. Because the wind to either side of the equator is affected by the rotation of the Earth, the largest wind anomalies tend to occur along the equator, with westerly wind anomalies blowing into the convergent region.

A three-month average is shown for the wind anomaly because in an individual month features associated with weather variations would appear, in addition to those associated

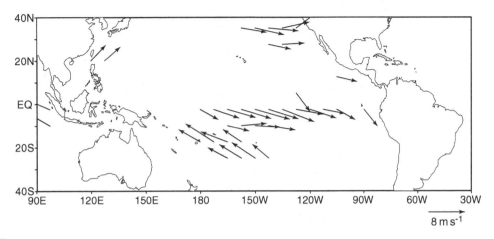

Fig. 1.9 Low-level wind anomalies (averaged December 1997 to February 1998) during the warm phase of ENSO. Anomalies smaller than 4 m s^{-1} are omitted.[18]

with El Niño SST anomalies. The wind anomaly off California in Figure 1.9 may be asso-ciated with an El Niño teleconnection, but would not be as reproducible in different El Niños as the winds at the equator. In a smaller El Niño event, both wind and precipitation anomalies would tend to be confined to the region near the *International Date Line* (180° longitude), and would not extend as far eastward. The wind anomalies in turn set in motion a complex adjustment process in the ocean. East of the westerly wind anomalies, the warm water in the upper ocean flows eastward along the equator, affecting subsurface temperature in the ocean.

Besides the SST, an important aspect of the oceanic side of ENSO involves changes in temperature that are occurring below the ocean surface in a layer about 100–200 m down known as the *thermocline*, which separates the deep ocean from the upper ocean. The waters in the upper layer of the ocean above the thermocline are much warmer than those below. Currents flowing in the upper layer can change the depth of the thermocline. Direct measurements of subsurface temperature structure are available, but recently it has become possible to obtain more detailed horizontal maps by measuring *sea surface height* from satellite. Small changes in sea surface height correspond to large changes in thermocline depth.

Figure 1.10 shows a surface height map for the mature El Niño. The region of increased sea surface height in the eastern Pacific corresponds to a deeper than normal thermocline. This is a leading factor in producing the warm SST anomalies in that region. The region of decreased surface height in the western Pacific (see Figure 1.10) does not impact SST in that region, but plays a role in subsequent evolution into a cold phase of ENSO. The thermocline shallows in the west because currents corresponding to the wind anomalies in Figure 1.9 are transferring warm water to the eastern side of the basin along the equator. Due to effects of the Earth's rotation, this shallowing occurs preferentially off the equator. Thus the horseshoe pattern of decreased sea surface height extending around the west Pacific with maxima just off the equator is characteristic of tropical ocean dynamics undergoing a

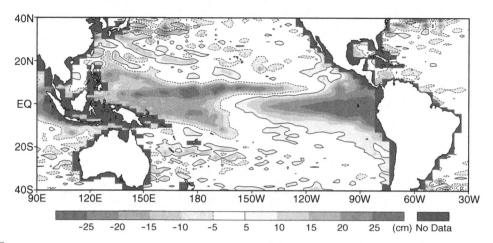

Fig. 1.10 December 1997 anomalies of sea level height (centimeters) during the fully developed warm phase of ENSO. In the eastern Pacific the thermocline is deeper (sea level is higher) than normal owing to the affects of westerly wind anomalies near the Date Line. West of the wind anomalies, the thermocline shallows (sea level drops) as warm water is transferred to the east. Data from NOAA Laboratory for Satellite Altimetry following Cheney *et al.* (1994).[19]

slow adjustment process to El Niño wind anomalies. It is this slow adjustment that causes the coupled system to oscillate between warm and cold phases, as we will see in Chapter 4. The regions of low sea level tend to propagate westward and make their way slowly back to the equator. It is this effect that brought about the termination of the 1997–98 El Niño warm phase. The decreased surface height in the west during El Niño, and thus prior to La Niña, is the counterpart in modern data of Wyrtki's observations of sea level at island tide gauge stations increasing prior to El Niño.

During the cold La Niña phase of the ENSO cycle, anomaly patterns would be similar, but with reversed signs, for each of the variables in Figures 1.7 to 1.10.

1.5.3 The first El Niño forecast with a coupled ocean–atmosphere model

The Cane and Zebiak coupled model of ENSO was used for an experimental forecast of El Niño conditions as early as 1986.[20] The researchers had atmospheric conditions in the form of wind measurements,[21] but, at the time, few measurements of subsurface temperature or sea surface height were available to set the initial conditions of the ocean component of their model. They found that they could get around this problem by running the ocean model first, with atmospheric conditions specified. The past history of the surface winds caused the ocean to undergo variations in thermocline depth that, while not perfect, were accurate enough to set the initial conditions at the time of the forecast. They then ran the coupled model forward in time to see if anomalous warm or cold conditions would develop from these initial conditions. To compensate for uncertainty in the initial forecast conditions they ran several forecasts, beginning from slightly different initial conditions, and then averaged the forecasts. They also took 3-month averages of the

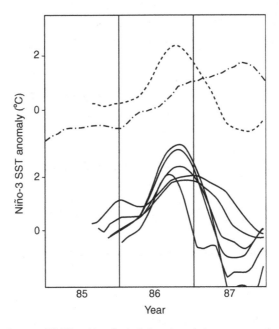

Fig. 1.11 First published real-time forecast of El Niño with a physically based coupled ocean–atmosphere model, published June 1986. The forecast was made using data up to January 1986. Forecasts are for 3-month average of the SST anomaly averaged over the Niño-3 region in the central equatorial Pacific. An average of several forecasts, beginning at slightly different times, is used to reduce scatter in the forecast. The lower set of curves shows individual forecasts, while the upper dashed curve shows the ensemble average of these. The upper dash-dotted curve shows observations added for comparison. After Cane *et al.* (1986).

resulting equatorial Pacific SST anomaly. The resulting forecast (Figure 1.11) was thus for a climate variable, not a weather variable. They tested the success of this system by "forecasting" past SST anomalies from 1970 to 1985. This testing procedure, sometimes known as hindcasting, suggested the system could produce forecasts with enough skill to issue a real-time forecast.

Their first forecast predicted that an El Niño would develop late in 1986. As may be seen from the observed Niño-3 index that has been added to Figure 1.11 for comparison, 1986 indeed marked the transition from cold toward warm conditions, but the El Niño continued through 1987, peaking near the end of 1987. Note that the SST forecast had substantial error even at the initial time because SST was not used in initializing the model. The model simulated initial SST and subsurface anomalies in response to the past history of wind stress. Current forecast systems incorporate observations of surface and subsurface temperatures with data assimilation methods that fit the model initial conditions to available data.

Although the forecast could not be considered a success by the standards of today, indications of potential skill in forecasting a climate variable at lead times of months can be taken as a milestone, in retrospect. Chapter 4 will discuss more quantitative evaluation of the success or failure of forecast systems. Furthermore, even if the main aspects of ENSO can be forecast in the Pacific, it can still be a substantial step to understanding and predicting the ENSO impacts on other parts of the globe, as we shall see.

1.6 Paleoclimate variability

A fascinating area of climate dynamics is *paleoclimate* (or paleoclimatology), the study of past climates of Earth. In this text, for the sake of brevity, only a brief summary of this area is presented. This at least illustrates the most important lesson from paleoclimate: that climate can change quite dramatically on a global basis and on all time scales. Figure 1.12 summarizes the major events in Earth's history both from a geological perspective and in terms of the evolution of life forms and ecosystems. It also provides the terminology used to describe past geological periods, as developed by geologists and paleontologists based on the dominant life forms of the era. Much of what we know of paleoclimate is inferred from species that lived at the time but still exist today or have some counterpart in the more recent record. If, for example, the temperature tolerances of a species of foraminifera (one-celled sea animals with calcareous shells) are known and can be assumed the same in the past, then the zone where they lived can be assigned an approximate temperature range. By using a combination of such species, it is sometimes possible to roughly map out past temperature zones based on the fossil record. The further back in time, the more difficult this becomes. Additional evidence can include such geological indicators as ice rafting, in which tiny amounts of sedimentary material typical of continents are carried out into the deep ocean trapped in drifting ice, which melts and releases them. Detecting such materials in cores of sedimentary rock drilled from the ocean floor can suggest that sea ice existed in a past epoch. Another set of techniques used to study paleoclimate is based on isotopic composition. For instance, oxygen exists primarily as the isotope ^{16}O but there is a small fraction of oxygen atoms of the isotope ^{18}O that have two extra neutrons and are thus heavier, with an atomic weight of 18. When water evaporates, the molecules with heavier oxygen atoms are slightly less likely to evaporate, affecting the ratio of ^{18}O isotope to ^{16}O, a process known as fractionation. This process depends on temperature. Similarly, the ratio of deuterium, ^2H (hydrogen with one extra neutron), to normal hydrogen, ^1H, depends on temperature. Such isotope measurements can thus, when calibrated properly, be used to infer information about global ice mass from ocean sedimentary rock cores in some applications, or atmospheric temperature above an ice cap in others. Many techniques involving isotope fractions of various elements have been developed to study paleoclimate.

Some features of past climate that are of interest from the point of view of studying modern climate change include the following (refer to Figure 1.12 for time scale).[22]

- Climate has undergone very considerable variations in past ages.
- While climate further in the past is more difficult to infer, it appears that Earth may have been ice-free until roughly 2500 Myr ago when there is evidence of a first glaciation

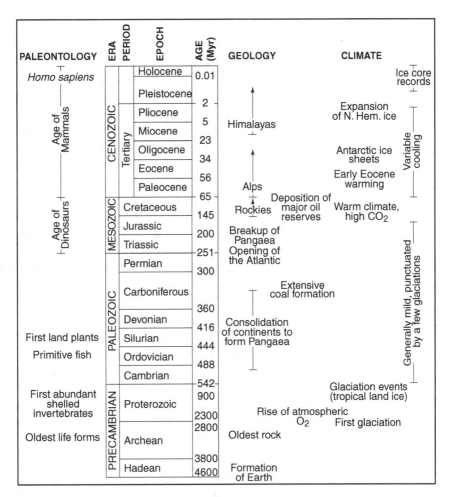

Fig. 1.12 Geological time scale, names and events in Earth's history, with selected paleoclimate events added in the right hand column. Adapted from Crowley (1983) with added information based on Crowley and North (1991), Zachos *et al.* (2001), Gradstein *et al.* (2004) and Royer (2007). Note that the time scale is expanded towards present, with Myr = millions of years. Major periods of fossil fuel deposition are noted between geology and climate columns, since these provide the source for current anthropogenic CO_2 input to the atmosphere.

event. Between roughly 900 and 600 Myr ago, at least three major glaciation events occured. Evidence of ice has been found on land masses that magnetic data indicate were in the tropics at the time. The term "snowball Earth" is sometimes applied to these events, although it is not clear that tropical oceans were ice-covered.

- During the Paleozoic era, it is believed that climate was generally warmer, although glaciations occurred, for instance, in the late Ordovician and parts of the Carboniferous and Permian periods.
- The climate of the Mesozoic era is sometimes referred to as the Mesozoic optimum: it was much warmer and there was little ice (geological evidence from effects of glaciers,

ice rafting, or lack thereof; plant types related to present tropical forms found in high midlatitudes, oceanic foraminifera, etc.).

- Abundant life existed in the Mesozoic climate, including what became sources of significant stores of fossil fuels (oil, natural gas, coal).
- Approximately 60% of all known oil reserves are from the Cretaceous period. Coal formed at various times, notably the Carboniferous.
- In the Mesozoic the continents were in a rather different configuration. For instance, the Atlantic Ocean was just opening as continental drift moved the Americas apart from Europe and Africa. The Atlantic remained small into the Eocene epoch.
- Carbon dioxide concentration in the Mesozoic was considerably higher than in the preindustrial human era. The concentrations are inferred by several methods (from isotopes in ancient soils or phytoplankton, from boron isotopes, and from concentrations of stomata) each of which has substantial error bars, but which place concentrations in the range 500 to over 2000 ppmv.
- The Paleocene and early Eocene epochs were also warmer and with higher CO_2 than present. The early Eocene optimum, roughly 50–54 million years ago, had ocean bottom temperatures (inferred from oxygen isotopes) more than 10 °C warmer than present, arguably reflecting similar warming at high latitudes (which are bottom water source regions). A shorter (roughly 100 000 years), relatively sudden warming around 55 million years ago is known as the Paleocene–Eocene Thermal Maximum. Carbon isotope measurements indicate a massive release of carbon into the climate system at that time, possibly from sediments or volcanic activity.
- Gradual cooling through the Eocene led to glaciation and Antarctic ice sheets in the Oligocene. After smaller warmings in late Oligocene and mid-Miocene, cooling toward our recent climate with ice age cycles occurred.

One might ask the obvious question: if we burn all the fossil fuels laid down over past eras, such as the Mesozoic, could we be headed for similar levels of CO_2 and a warmer climate? This argument, although qualitatively reasonable, by itself is too simplistic because there are many differences in the deep past from the present conditions. To name just one significant difference, the continents in the Mesozoic were substantially different in shape and position than they are today and the Atlantic Ocean scarcely recognizable. Paleoclimate can, however, give a feeling for the range of variability of climate, and thus of the sensitivity of the climate system. Furthermore, the the paleoclimate perspective does remind us that the carbon dioxide that is currently being released into the atmosphere over the period of a few human lifetimes has been sequestered over hundreds of millions of years in the form of fossil fuels. Thus it should not be surprising that we find rapid increases in atmospheric CO_2.

Figure 1.13 gives a detailed view of much more recent times, merely the past 650 000 years (compared with 65 million since the end of the Cretaceous). Ice laid down in Antarctica over this time period has been retrieved as ice core samples. The chemical and isotopic composition of gases trapped in these ice cores has been carefully reconstructed and calibrated to produce estimates of CO_2 concentration and isotopic

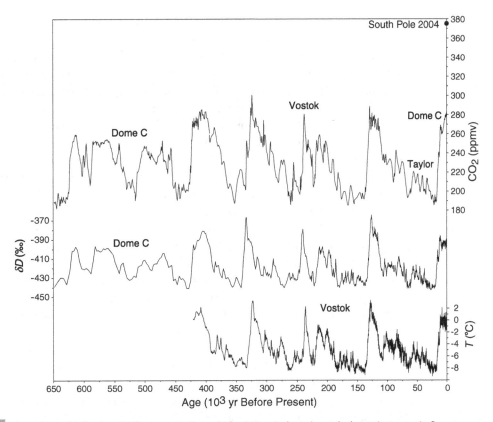

Fig. 1.13 Ice core records of carbon dioxide concentration and of variations in deuterium to hydrogen isotope ratio. Recent concentrations of CO_2 have been added for perspective. Data following Siegenthaler *et al.* (2005) are replotted from the National Climate Data Center archive. The carbon dioxide record is a composite from three Antarctic core sites: Dome C, Taylor Dome and Vostok. Deuterium ratio (δD, relative to standard, in per mil), a proxy for temperature above the ice, is from Dome C. The bottom curve gives an estimate of Antarctic air temperature difference relative to present, inferred from deuterium ratios at the Vostok core site from Petit *et al.* (1999).[23]

proxies for temperature. The temperature reconstruction is based on isotopic measurements of deuterium to hydrogen ratios and ^{18}O ratios and so should be viewed only as a rough indicator of temperature above Antarctica. Several features may be noted in this record:

- There is a great deal of variability even on long time scales. The largest changes are between glacial periods (ice ages), seen as low deuterium ratios, and interglacials, when there were no major ice sheets outside Antarctica and Greenland, such as the interglacial we are currently in.
- The most recent ice age ended around 10 000 years before present. It reached its maximum about 21 000 years before present, with its onset around 116 000 years ago. The previous interglacial started about 129 000 years ago.

- The variations are irregular, but there are some preferred timescales. For instance, the last five glacial cycles occur at roughly 100 000-year intervals.
- There is an association between high temperature and CO_2 concentration.
- While the changes from glacial to interglacial appear abrupt on this time scale (with tick marks at 25 000-year intervals), they are far slower than the recent anthropogenic CO_2 increases, and the predicted resulting temperature increases.
- Carbon dioxide concentrations near the end of the record are consistent with estimates of preindustrial CO_2 levels from other records.
- Present-day CO_2 concentrations, added for comparison, are considerably higher than anything produced by natural variations in the past 650 000 years. Glacial values are roughly 180 ppmv; the high values during interglacial periods are about 280 to occasionally 300 ppmv, compared with present (2004) Antarctic values of 375 ppmv.

The main driver of the ice age cycle is slow variation in Earth's orbital parameters. The tilt of Earth's axis and the eccentricity of Earth's orbit around the Sun have variations with important periodicities at 19, 23, 41, 100 and 400 kyr (1 kyr = 1000 years). These affect the seasonal and latitudinal distribution of insolation, i.e. the incoming flux of solar radiation, which sets in motion feedbacks in the climate system. For instance when northern hemisphere summer receives less sunlight, conditions are favorable for ice sheets to grow slowly since less summer melting occurs. The orbital parameters have little effect at time scales less than a thousand years, and orbital effects suitable for creating the next ice age are not due for roughly 30 000 years.[24] Sea level was about 120 m below present level at the last glacial maximum, and 4 to 6 m above present during the previous interglacial. The sea level rise at the end of the last ice age was not smooth, but included several large meltwater pulses.

The cause of the low CO_2 concentrations in glacial periods is not yet established, and the roles of various feedbacks affecting CO_2 and temperature are not yet clear. On the one hand, CO_2 can affect temperature through the greenhouse effect. This likely contributed substantially to the radiative forcing needed to transition between glacial and interglacial states. On the other hand, temperature changes can be correlated with changes in ocean circulation (the Southern Ocean is a prime suspect), or effects on the rate of functioning of the biosphere, each of which can affect CO_2. Other gases, such as methane, also tend to covary with CO_2 and temperature in these cores. The past greenhouse gas variations are not a primary source of evidence of the greenhouse effect, for which there is more direct physical evidence. These variations do raise the concern that feedbacks between the biogeochemical and physical climate systems might amplify anthropogenic CO_2 release, a subject of recent research that is as yet too uncertain to be included in the recent consensus of climate projections.

The key point from the ice core CO_2 records is that the recent increase to the high present-day concentrations (not to mention anticipated future concentrations) is unprecedented in the past 650 000 years. This implies that we are undertaking an experiment with the climate system unprecedented since the very distant past. The magnitude of the naturally occurring temperature variations in paleoclimate records shows that climate can indeed change substantially, and may do so in the future.

Notes

1 The ice core records of greenhouse gases from Antarctica and Greenland match quite well to the curves for *in situ* atmospheric measurements (Lorius *et al.* 1990; Etheridge *et al.* 1998). At the Vostok and Concordia (Dome C) stations in Antarctica, cores more than 3 km deep have been taken (Petit *et al.* 1999; Siegenthaler *et al.* 2005).

2 The amended Montreal Protocol is described in UNEP (2000). For further detail on the ozone hole, CFCs and the Montreal Protocol at the undergraduate level see Turco (1997) and Masters (1998).

3 The IPCC process is conducted under the auspices of the World Meteorological Organization and the United Nations Environmental Programme. The IPCC has three working groups addressing different aspects of the global warming problem: Working Group I on the science of climate change, Working Group II on impacts, adaptation and vulnerability, and Working Group III on mitigation of climate change. This text focuses on the climate science, rather than mitigation or adaptation strategies, and thus, unless otherwise specified, references to IPCC reports in the text are to Working Group I reports. In these reports, a group of lead authors is designated for each chapter, and a broad collection of scientists in each research area who provide input are listed as contributors. A separate group of known scientists in each area are selected as reviewers, who critically assess the material and return general and specific comments. The resulting documents have some of the shortcomings of a report written by committee, but nevertheless represent a very earnest attempt to state both the consensus of what is known and the areas of scientific uncertainty. One consequence of the completeness of the IPCC process is that, because a large portion of the relevant scientific community is involved, most experts have some connection to it. The author of this text, for instance, was a contributor in IPCC (1996) and an external reviewer of IPCC (2001 and 2007). Lead authors of each chapter bear the brunt of the workload, taking time off from their normal activities to carry out this service.

4 The Conferences of the Parties of the United Nations Framework Convention on Climate Change (UNFCCC) occur every year and usually receive some attention in the press. They center on diplomatic and societal considerations rather than scientific input, the latter being organized under the IPCC. They have been respectively held in: Berlin, March 1995; Geneva, July 1996; Kyoto, December 1997; Buenos Aires, November 1998; Bonn, October 1999; The Hague, November 2000; Marrakesh, November 2001; New Delhi, October 2002; Milan, December 2003; Buenos Aires, December 2004; Montreal, December 2005; Nairobi, November 2006; Bali, Indonesia, December 2007; Poznań, Poland, December 2008; Copenhagen, November 2009.

    The Kyoto Protocol to the United Nations Framework Convention on Climate Change was one of the more significant agreements to emerge from these conferences. Some of the conditions were fairly general, for instance that each party would "elaborate policies and measures in accordance with its national circumstances, such as: (i) Enhancement of energy efficiency in relevant sectors of the national economy; (ii) Protection and enhancement of sinks and reservoirs of greenhouse gases [...]." Article 3 contained more specific targets for developed nations: "The Parties included in Annex I shall, individually or jointly, ensure that their aggregate anthropogenic carbon dioxide equivalent emissions of the greenhouse gases listed in Annex A do not exceed their assigned amounts, calculated pursuant to their quantified emission limitation and reduction commitments inscribed in Annex B and in accordance with the provisions of this Article, with a view to reducing their overall emissions of such gases by at least 5 percent below 1990 levels in the commitment period 2008 to 2012." Targets for the key industrial powers were similar – 8% below 1990 emissions levels for the European Union, 7% for the United States and 6% for Japan. The United States caused some ruffles in the international community when it rejected the Kyoto Protocol in March 2001. The protocol entered into force (for those countries that have ratified it) on February 16, 2005. The first commitment period of the Kyoto Protocol ends in 2012, by which time a new international framework would have to have be negotiated and ratified if continuity is to be maintained. Since 2003, a number of US states have introduced laws, Executive Orders or strategic plans aiming to regulate greenhouse gases. Many of these include provisions along the lines of: reduce greenhouse gas emissions to 2000 levels by 2010, to 1990 levels by 2020, and to 80% below 1990 levels by 2050 (this example based on California Executive Order S-3-05 and Bill AB-32).

5 Arrhenius (1896) first attempted to quantify the warming effects of CO_2, then called carbonic acid. His estimate of warming in response to doubled CO_2 (in absence of climate feedbacks) was larger than present estimates and in this paper he was primarily interested in possible sources for ice ages, although anthropogenic effects on CO_2 are mentioned. Nonetheless, the paper reads astonishingly well a century later and the model he used is closely related to that described in Chapter 6. While appreciation of this pioneering work was left to later generations, he did receive a Nobel prize for other work in chemistry.

Revelle and Suess (1957) was notable among works reviving interest in CO_2 impact in climate and anthropogenic CO_2 increase. Revelle, as director of Scripps Institution of Oceanography, had considerable clout in the larger community, for instance influencing Keeling's routine monitoring of CO_2.

Early three-dimensional models of Earth's climate included Manabe *et al.* (1965), Kasahara and Washington (1967) and Arakawa *et al.* (1969). These models, now known as General Circulation Models (GCMs), were first applied to impacts of CO_2 increase by Manabe and Wetherald (1975). Hansen *et al.* (1988) first used a GCM for time dependent response experiments with scenarios for future greenhouse gas changes. Hansen *et al.* (1985) had noted that owing to slow ocean response, warming due to current CO_2 increase would be delayed. Useful sources for history include Handel and Risbey (1992) and Chapter 7 of IPCC (2007).

The Nobel Peace Prize for 2007 was shared between the Intergovernmental Panel on Climate Change and Albert Arnold (Al) Gore Jr., for their efforts to build up and disseminate greater knowledge about man-made climate change. The citation reads, in part, "Through the scientific reports it has issued over the past two decades, the IPCC has created an ever-broader informed consensus about the connection between human activities and global warming. Thousands of scientists and officials from over one hundred countries have collaborated to achieve greater certainty as to the scale of the warming." To see the names of the scientists involved, hunt down the contributor lists of all the IPCC reports.

6 Sea surface temperature data from ships have reasonably good coverage globally, except for some regions near Antarctica, since the 1950s (Worley *et al.* 2005) and require more care in interpolation and bias correction as one moves back in time (Folland and Parker 1995; Hansen *et al.* 1999; Jones *et al.* 2001). From 1982 on, satellite measurements with high-resolution global coverage are combined with *in situ* data (Reynolds and Smith 1994). Global temperature data sets are revised periodically as additional historical data are located, and as methods are refined. The impact of these revisions is typically not more than 0.1 degree in particular decades, and smaller in longer-term trends. See for example Jones and Moberg (2003) compared with the newer version (known as HadCRUT3; Brohan *et al.* 2006) used in Figure 1.3. The decadal averaging is done with a 21-point binomial filter (Folland *et al.* 2001). As noted in Rayner *et al.* (2006), the availability of data back to the 1850s owes much to the Brussels Maritime Conference of 1853, when several nations agreed to standard observations from ships.

7 Evaluations of the statistical significance of global warming rely on what are known as "fingerprint" methods (Hegerl *et al.* 1996). For instance, spatial patterns of model predicted warming and observed warming are compared to known spatial patterns of natural variability to estimate the probability that the observed pattern could have arisen by chance. See Chapter 7 for further discussion.

8 The introduction of Philander (1990) provides a marvellous discussion of some historical descriptions of El Niño. This includes letters describing the effects of the El Niño of 1891 on Peru, and the documenting, by the President of the Lima Geographical Society, of the term El Niño as used by sailors of Paita, Peru.

9   See Trenberth and Shea (1987) for more recent pressure correlation maps. They also discuss historical records of the Southern Oscillation Index. The precise definition of the SOI is:

$$SOI = \left(\frac{p'_T}{sdev(p'_T)} - \frac{p'_D}{sdev(p'_D)} \right) \left(sdev \left[\frac{p'_T}{sdev(p'_T)} - \frac{p'_D}{sdev(p'_D)} \right] \right)^{-1}$$

where p'_T and p'_D denote surface pressure anomalies at Tahiti and Darwin respectively, and $sdev(x)$ denotes the standard deviation of x over the time series. It is an odd, unwieldy index since the entire series must be recalculated when changing the base period used to define anomalies and standard deviation.

10 Niño-3 data are from the Reynolds data set, SOI data are from the NOAA Climate Data Center. Anomalies are relative to the 1971–2000 mean. Power spectra are calculated using the maximum entropy method with software described in Dettinger et al. (1995).

11 The power spectrum of a time series of a variable is most simply calculated by taking its Fourier transform, i.e. decomposing it into a sum of sinusoidal oscillations of a large range of frequencies. The averaged squared amplitude in a particular frequency band is the power spectral density and the power spectrum displays this for all frequencies. For a short time series, it is important to estimate the statistical significance of a peak in the time series, i.e. whether it is an artifact of the particular method and time segment used, or is a true feature of the observed system. For more detail see, for instance, Press et al. (1992).

12 Wyrtki (1975) went so far as to state, "In total El Niño is the result of the response of the equatorial Pacific Ocean to atmospheric forcing by the trade winds." Wyrtki correctly identified the role that ocean dynamics plays in connecting wind anomalies in the western Pacific with SST anomalies in the eastern Pacific, and the role of the thermocline (see section 4.6).

13 Good reviews of the ocean model contributions to equatorial oceanography are found in Cane and Sarachik (1983), McCreary (1985) and Stockdale et al. (1998). On the atmospheric side, a relatively simple model developed by Gill (1980), and related subsequent models, helped explain and simulate ENSO wind anomalies despite some drastic approximations. Observational work by Rasmusson and Carpenter (1982) provided a picture of a "standard" El Niño event by compositing surface observations for several events.

14 In 1988–89 Schopf and Suarez (1988), Suarez and Schopf (1988) and Battisti and Hirst (1989) developed the "delayed-oscillator model" for ENSO. This is a simplified model that showed for the first time how oceanic adjustment processes, with subsurface ocean temperature anomalies evolving slowly, could provide the transition between warm and cold phases of ENSO. Jin and Neelin (1993) showed how the spatial structure of ENSO is strongly determined by the ocean–atmosphere feedbacks of the Bjerknes hypothesis, with ocean memory adding the oscillatory tendency. The reasons for the irregular evolution of ENSO were worked out in the 1990s, as described in Chapter 4.

15 Philander et al. (1992) described the first GCM simulation of ENSO that can be said to be qualitatively correct. Many of the early coupled GCM simulations suffered from errors in simulating the mean climate, known as climate drift. The simulations yielded a variety of internal variability that was ENSO-like in some respects and not in others. A sequence of model intercomparisons (Neelin et al. 1992, Mechoso et al. 1995, Latif et al. 2001) shows the progress in these models from early coupled simulations to the present generation.

16 SST data are from the Reynolds data set which includes satellite data for detailed spatial coverage and in situ data (which help keep the satellite-based values accurate) following Reynolds (1988) and Reynolds and Smith (1999). The anomalies in Figure 1.7 are relative to the November 1981 to March 1998 climatology.

17 Data from the NOAA Climate Prediction Center following Xie and Arkin (1996). The data in Figure 1.8 are a combination of satellite retrievals (based on the relation of precipitation to cloud top temperature) and ground observations of precipitation. The anomalies shown are relative to the July 1987–1995 climatology.

18 The winds in Figure 1.9 are at the 925 mb level and are from the National Centers for Environmental Prediction (NCEP) analysis data set (Kalnay *et al.* 1996). This means that a numerical weather prediction model is constrained to fit available data as well as possible. The result is a combination of model and true data. This is useful because it gives full coverage with no gaps, but must also be considered with caution since model error may enter. In the case shown, the overall pattern is well established from many sources. The anomalies shown are relative to the 1958–98 climatology.

19 Data for sea level height are from the TOPEX/POSEIDON altimeter from the NOAA Laboratory for Satellite Altimetry. Anomalies shown are relative to the 1993–95 mean. The measurement of variations of sea level to within a few centimeters error from a satellite is a considerable technical accomplishment (Cheney *et al.* 1994).

20 The Cane and Zebiak coupled model (Cane and Zebiak 1985; Zebiak and Cane 1987) is an intermediate complexity model that simulates climate anomalies in the tropical Pacific basin. Because the mean climate of this model is largely specified, it was easier to focus on the ENSO simulation than for GCMs during early work. Other intermediate coupled models of that time included Anderson and McCreary (1985), Battisti (1988) and Xie *et al.* (1989).

21 The Florida State University (FSU) surface winds (Legler and O'Brien 1984), provided by James O'Brien for the Cane–Zebiak model and other early forecast systems, were hand-interpolated from ship and other measurements. An important feature was that they were available on a near real-time basis, i.e. in time to make monthly forecasts.

22 For further reading on paleoclimate, see the undergraduate textbook by Ruddiman (2001). Hartmann (1994) and Graedel and Crutzen (1993) each provide a succinct chapter on paleoclimate at the undergraduate level. Discussion here follows Crowley and North (1991), Sarmiento and Gruber (2006) for biogeochemical cycles, papers listed in the next two footnotes, and Chapter 6 of IPCC (2007), which provides a detailed overview of recent paleoclimate work. Although not the subject of this course, it is worth noting that periods of extensive ecosystem change occur, and in some cases these coincide with periods of climatic change. In addition to the well-known extinction of roughly 75% of species at the end of the Cretaceous, other examples of multi-species extinctions include benthic (deep ocean) extinctions at the end of the Paleocene, the decline of Archaic mammals at the end of the Eocene and coral extinctions at the end of the Oligocene. These approximately coincided with a warming and two glaciation events, respectively.

23 The orbital forcing of glacial cycles is often referred to as Milankovitch theory (Crowley and North 1991). Orbital effects on insolation are given in Berger and Loutre (1991). Some of the strongest orbital effects on insolation involve changes in the seasonal cycle and latitude distribution of insolation, rather than changes in the annual average, and the larger land mass of the northern hemisphere creates an different reaction of the climate system to changes in northern versus southern hemisphere insolation. For discussion of glacial sea level and meltwater pulses, see, for example, Fairbanks (1989) and Tarasov and Peltier (2005). Cores from the Greenland ice cap (Johnsen *et al.* 1992; GRIP Project Members 1993) have complemented findings from Antarctic cores in a number of respects, though over a shorter period. Among other things, they reveal fascinating millennial-scale variations, but for reasons of brevity are not treated here.

24 In Figure 1.13, the carbon dioxide record is a composite from three Antarctic core sites. The periods from each site, in kyr BP (10^3 years before present), are: Dome C: 0.4–20 kyr BP and 396–650 kyr BP; Taylor Dome 20–41 kyr BP; and Vostok 41–396 kyr BP. Dome C ice core data are from the European Project for Ice Coring in Antarctica (EPICA), described in Monnin *et al.* (2001), Flückiger *et al.* (2002), Jouzel (2004) and EPICA community members (2004); Taylor Dome data are described in Indermühle *et al.* (1999), with the combined record in Siegenthaler *et al.* (2005). The estimation of the time scale is a major part of the ice core analysis, described in these and references therein. The method for the time scale in the temperature estimate by Petit *et al.* (1999) from the Vostok core differs slightly from that of the other two curves shown in Figure 1.13. The temperature estimate is based on deuterium to hydrogen ratios in the ice and oxygen isotope ratios in marine core records (Petit *et al.* 1999; Jouzel *et al.* 2003). The present-day

(2004) CO_2 concentrations (375 ppmv) are 2004 annual average South Pole *in-situ* data from the NOAA Earth System Research Laboratory Global Monitoring Division, following Gillette *et al.* (1987). The South Pole values are used simply to make the comparison to Antarctic ice cores clear. Current CO_2 concentrations in the northern hemisphere are about 2 ppmv higher because most fossil fuel emissions are in the northern hemisphere; there is a slight lag in southern hemisphere values due to the time for atmospheric transports to mix the CO_2 (Dargaville *et al.* 2003).

Basics of global climate

2.1 Components and phenomena in the climate system

In this section, we introduce some examples of phenomena within the major components of the climate system and some of the ways they interact. In attempting to understand this complex system it is useful to introduce the notion of characteristic time and space scales of a phenomenon. One of the fundamental difficulties faced by climate models is that a huge range of scales turns out to be important. The examples here set the stage for later discussion of how this range of scales affects climate models.

Figure 2.1 summarizes some of main features of the climate system. A common way of listing the components of the climate system is: the atmosphere, the ocean, land surfaces, the cryosphere, the biosphere and the lithosphere. The *cryosphere* consists of land ice (including ice shelves and glaciers), snow and sea ice. The *biosphere*, the sum total of all living things on the planet, is obviously spread throughout the oceans and land surfaces. Indeed when considering land surfaces as a climate component, a leading effect is the vegetation type. The *lithosphere*, i.e. the "solid" Earth, creates the distribution of ocean basins, mountain ranges etc., not to mention the occasional volcanic eruption. Chemical composition can be viewed as a component of the climate system as well. One could view the chemical composition to be an additional set of variables associated with each of the other climate system components, but chemical interactions relevant for climate often involve interactions across these traditional boundaries. Thus the term *biogeochemistry* is used for the complex interactions of biology with the chemistry of the climate system. For instance, carbon dioxide and other carbon compounds are exchanged at the ocean and land surface in a manner that depends on ecosystems as well as temperature and other physical climate variables.

Within these components, there are a great number of climate processes, some of the most important of which are schematized in Figure 2.1. A fundamental effect in setting Earth's climate is that solar radiation tends to get through the atmosphere and be absorbed at the land surface and in the upper 10 meters or so of the ocean. Because the ocean is heated from above, it tends to be quite stable to vertical motions. The solar heating acts to create a warm surface layer, which in most regions is less dense than the colder deep waters below, and thus tends to remain near the surface. Mixing near the ocean's surface by turbulent motions created by wind or near-surface currents keeps the ocean temperature relatively constant through an upper mixed layer on the order of 50 meters depth. This mixing carries surface warming down as far as the *thermocline*, the layer of rapid transition of temperature to the colder abyssal waters below.

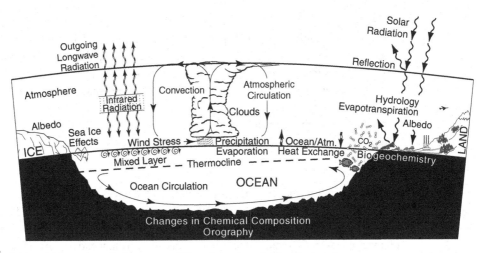

Fig. 2.1 Schematic of some important processes within each component of the climate system.

On the other hand, the absorption of solar heating at the surface results in the atmosphere being heated from below by heat fluxes from the surface. This creates a great deal of thermally driven circulation in the atmosphere. Like a pot boiling, heating from below gives rise to *convection* – overturning motions at small horizontal scales that carry the heat to higher levels in the atmosphere from which infrared radiation can reach space. Deeper convective motions form clouds by condensation of moisture; such *moist convection* dominates vertical transfer of heat in large parts of the atmosphere. Contrast between warmer and colder latitudes creates thermally driven circulations at larger scales. Infrared radiation, also known as *longwave radiation*, participates in vertical transfer of energy in the atmosphere. At the top of the atmosphere, emission of infrared radiation back to space is the sole means by which the Earth system as a whole can balance energy input from the Sun. This upward infrared radiation at the top of the atmosphere is termed *outgoing longwave radiation (OLR)*, and is important not only for its role in the energy budget but because it can be directly measured by satellite, and is thus used as an observed variable to examine many phenomena.

Exchanges between the atmosphere and the upper ocean include exchanges of several forms of heat energy, largely cooling the ocean and warming the atmosphere to balance the solar input. Another exchange is transfer of momentum via *wind stress*, by which the more rapid motion of the atmosphere tends to accelerate ocean currents. Various aspects of Figure 2.1 will be elaborated on later in this chapter.

2.1.1 Time and space scales

The notion of characteristic time scales is not exact, so the easiest way to introduce it is by example. The term "on the order of" 100 years means closer to 100 than 1000 or 10.

- *Period:* For phenomena that oscillate in a periodic manner. Sometimes this is externally determined, e.g. the seasonal cycle (period one year, highly periodic). Sometimes this period is internally determined, e.g. El Niño (period approximately 3–4 years, but quite irregular).

- *Response time:* When the sun comes out, land surfaces heat up in hours, but the ocean surface has a much longer response time. In presence of a sustained anomalous heat source, the ocean surface layer warms up to the new equilibrium temperature on a time scale of months. Chemical systems also have response times: if we stop emitting CFCs, the characteristic time for the chlorine to be transferred out of the stratosphere is roughly half to one century.
- *Lifetime:* For phenomena that have an identifiable beginning and end. This is usually a more loosely defined time scale than the above two but still useful. For instance, the lifetime of a convective cloud is on the order of hours.

Later we will introduce other time scales, such as measures of the time scale of loss of predictability. The time scale for an initial error to grow in a weather prediction would be one example. Another simple measure would be the time scale over which the correlation of a variable to earlier values tends to decrease. For instance, daily sea surface temperature in a given location tends to correlate fairly well with its value a few weeks before, whereas daily precipitation in a midlatitude location would correlate poorly with its value even a few days before.

Table 2.1 lists a sampling of time scales for phenomena found in each component of the climate system. These scales must be regarded as very rough. Many have quite a range of variation. For instance, the response time of the upper ocean to heating depends on a number of factors that influence heat exchange, on the depth of the layer through which the heating is mixed, and on the horizontal spatial scale of the region, since the scale affects feedbacks with the atmosphere. Furthermore, even a phenomenon with a well-defined time scale in one sense may contain a large number of time scales in another sense. For instance, a sequence of events with short lifetimes that are randomly distributed in time can produce an effect that, when viewed from long time scales, appears like white noise, and contains all frequencies. Thus a typical weather system may have a lifetime of days, but a random sequence of weather events can still affect climate at longer time scales (which will be relevant to ENSO prediction in Chapter 4).

If time scales are well separated for the phenomenon of interest, this can sometimes be used to simplify study. For instance, on time scales of anthropogenic climate change, continents and mountains can be taken as fixed because their time scales of variation are so much slower. This would be an example of *scale separation* where the slower component can be treated as constant.[1] Similarly, for many purposes, land surface vegetation type can be taken as fixed, or changes of vegetation can be prescribed from one decade to the next, even though the separation of time scales is not as good as in the case of fixed continents. At the other extreme, the very smallest scales can be averaged over. For instance, we do not have to calculate the motion of every molecule in a gas to know the pressure, temperature and density. The molecular motions are fast enough and small enough that we can treat just their aggregate effects, averaging over a very large number of molecules. This is an example of treating fast motions by averages that depend on the slower motions, as elaborated below and in Chapter 5.

Typical spatial scales can often be defined for a phenomenon as well, some of which are listed in Table 2.1. For instance, a typical frontal system for midlatitude weather variations

Table 2.1 Typical time scales for various phenomena in the climate system (space scales in brackets where useful).

Atmosphere	
Overall response time to heating	months
Typical spin-down time of wind if nothing is forcing it	days
Frontal system lifetime (1000s of km)	days
Convective cloud lifetime (100 m to km horizontal; up to 10 km vertical)	hours
Time scale for typical upper-level wind ($20\,\mathrm{m\,s^{-1}}$) to cross continent (a few 1000 km)	days
Ocean	
Response time of upper ocean (above thermocline) to heating	months to years
Response time of deep ocean to atmospheric changes	decades to millennia
Ocean eddy lifetime (10s to 100 km)	months
Ocean mixing in the surface layer	hours to days
Time for typical ocean current ($\mathrm{cm\,s^{-1}}$) to cross ocean (1000s of km)	decades
Cryosphere	
Snow cover	months
Sea ice (extent and thickness variations)	months to years
Glaciers	decades to centuries
Ice caps	centuries to millennia
Land surface	
Response time to heating	hours
Response time of vegetation to oppose excess evaporation	hours
Soil moisture response time	days to months
Biosphere	
Ocean plankton response to nutrient changes	weeks
Recovery time from deforestation	years to decades
Lithosphere	
Isostatic rebound of continents (after being depressed by weight of glacier)	10 000s of years
Weathering, mountain building	1 000 000s of years

develops over thousands of kilometers, while a convective cloud would be on the order of a kilometer or less. There is a tendency for smaller space scales to be associated with smaller time scales, although this does not always hold.

2.1.2 Interactions among scales and the parameterization problem

Separation of time and space scales does not always guarantee that one can treat only the scale that is of interest by itself, and the scales are not always well separated. This is the case for weather systems and convective clouds interacting with the larger scales of the global

Fig. 2.2 Instantaneous satellite image of Earth showing clouds (white) associated with weather and convective systems, overlaid with a latitude–longitude grid to illustrate grid cells used in climate models. For clarity, a coarse resolution grid is shown (5 by 5 degrees), but even with a finer grid, the cloud systems would have many features smaller than the grid size. The satellite image is computer-enhanced, especially over continents, but is mainly from visible light. Satellite image from GOES-8, courtesy NASA Remote Sensing Division.

circulation. Figure 2.2 shows an instantaneous satellite image in which many complex weather phenomena may be seen. The image is for visible light that is reflected from clouds and surfaces on the Earth and detected at the satellite (although the image is computer-enhanced and a particular band of visible light has been used to create a vegetation map for the land surfaces). Later in the chapter, we will see that the climatology of precipitation (Figure 2.13) consists of much broader-scale, smoother features than the myriad of complex convective systems that appear as tiny dots in Figure 2.2. The regions of large climatological convection in the tropics can actually be seen on the particular day in Figure 2.2, but they are made up of many smaller-scale features. At midlatitudes, the main regions of climatological precipitation are known as storm tracks, occurring about 30–50° N in both Pacific and Atlantic oceans. On a particular day, what one sees is individual storms, and only by averaging over many of these does the shape of the storm tracks emerge.

A major question is: if we wish to model the effects of clouds on climate change over a period of decades or centuries, is it necessary to model many individual weather systems and then take the average? Or to model millions of small clouds just to get the mean effects? In the case of weather systems, the answer is, unfortunately, yes (or fortunately, if you are a supercomputer salesman). In the case of convective clouds, the answer seems to be that we can approximate the average effects directly, although it is challenging. In fact, we have to, because we cannot yet afford to represent the smallest scales in current global models.

The grid in Figure 2.2 illustrates how climate models represent the atmosphere, dividing up the continuous atmosphere into a series of discrete boxes. Rates of change of the average values of temperature, moisture, wind, etc. within each grid box are computed, including the effects of all the other boxes. A new value of each variable is computed a short time later, and the operation is repeated until a simulated year, decade or century has been reached. Now consider the area inside one grid box of Figure 2.2. In the computer representation, only an average across the grid box is included. In the observations, many fine variations occur inside. These include phenomena such as squall lines, mesoscale convective complexes, tower-anvil cumulonimbus clouds, etc. And yet the average of these small-scale effects has important impacts on large-scale climate. As an example, consider that clouds primarily occur at small scales, and yet the average amount of sunlight reflected by clouds, and the average latent heat released in them, will affect the average heating of a whole grid box. These average effects of the small scales on the grid scale must be included in the climate model. Furthermore, these averages must change with the parameters of large-scale fields that affect the clouds, such as moisture and temperature. The method of representing average effects of clouds over a grid box interactively with the other variables is an example of what is known as *parameterization*. The successes and difficulties of such parameterization will be important factors in evaluating the trustworthiness of climate models.

The grid used in Figure 2.2 to illustrate a climate model is not as fine as in some current models. It is roughly comparable to the coarsest resolution model used in the IPCC (2007) Fourth Assessment Report, while some global models in this report ran with roughly 1 degree grid cells. Weather forecast models operate with substantially smaller grid boxes. A finer grid implies either greater computational costs or shorter simulation times, so there are strong barriers to making the grid extremely fine. As computers become faster, modelers can afford finer grids, and every few years climate models decrease in their grid size. However, close examination of the small scales in the figure leads one to realize that climate models will never escape the need to deal with the effects of phenomena that are smaller than the grid size. There is always some smaller scale, and there is a tendency for the scales to interact. This scale interaction is one of the main effects that makes climate modeling so challenging. Figure 2.3 shows an example of a deep convective system, which would be just one of the many small-scale convective elements contributing to the patterns in Figure 2.2. The horizontal scale of the system would be on the order of 10 km, so the area covered would fit on the order of 100 times into the area of a single GCM grid box of, say, 100 km × 100 km. The complexity of this convective tower and anvil combination suggests some of the difficulties faced in cloud parameterization, especially when one considers that the area, thickness and height of different parts of the cloud affect how it will reflect solar radiation and absorb or emit infrared radiation, two effects that will be important for global

Fig. 2.3 A cumulonimbus cloud in the tropics. Note the tower which contains a strong updraft where much of the condensational heating is taking place, the rain falling from below the tower, and the anvil of cirrostratus cloud being sheared out at a height of 10–12 km by upper level winds. Courtesy T. A. Toney.

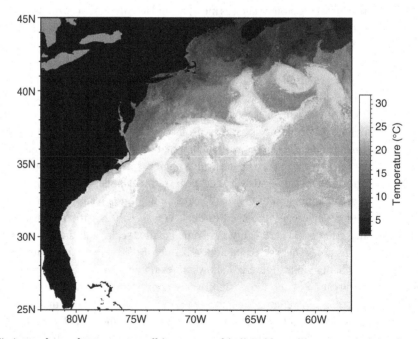

Fig. 2.4 Satellite image of sea surface temperature off the east coast of the United States. Warm waters in the southern portion of the region are carried northward in the Gulf stream, which flows northeast from Cape Hatteras in a series of meanders. Note the sudden transition to colder waters to the north. Courtesy of Ocean Remote Sensing Group at the Johns Hopkins University Applied Physics Laboratory.

warming. Yet other effects of clouds are comparatively simple, such as transporting heat from the surface through a deep layer in the atmosphere.

These challenges of modeling or parameterizing many spatial scales arise in the oceans as well, as seen in Figure 2.4. The sharp line between warm waters leaving Cape Hatteras and cold waters to the north is the Gulf stream. Ocean eddies may be seen as swirls and variations along this boundary. The eddies have spatial scales on the order of 50 km and time scales of weeks to months. To capture such small-scale features in an ocean model is computationally very costly, and yet they have substantial effects on the average climate. In this case, it may be seen that the eddies are tending to move colder waters southward and warmer waters northward across the line that tends to separate them on the larger spatial scales. If the motion simply oscillated forward and back, there would be no average effect, but the eddies tend to twist up into very small-scale features that then become mixed with surrounding waters, such as the tongue of cold water that is getting mixed into the warm water in the center of the picture. Overall this process transports heat poleward and tends to reduce the temperature gradient from equator to pole. These *"eddy transports"* of heat can be as important as the transport by large-scale, long-lived features, such as the Gulf stream.

2.2 Basics of radiative forcing

The essence of radiation in the Earth's atmosphere is simple: solar radiation comes in, mostly reaching the surface. Infrared radiation (IR) is the only way this heat input can be balanced by heat loss to space. Since IR emissions depend on the Earth's temperature, the planet tends to adjust until it reaches a temperature where this balance is achieved. This occurs when the upward flux of long wavelength infrared radiation, integrated over the Earth, balances the flux of incoming short wavelength solar radiation. Throughout this discussion, we will measure the intensity of the radiation (light) as a flux in units of watts per square meter ($W\,m^{-2}$).

We thus need to say something about how light behaves in the atmosphere as a function of wavelength and how its emission depends on temperature. Recall that different types of electromagnetic radiation differ by the wavelength of the light and that light also has characteristics of a particle, coming in units called photons. Shorter wavelengths have more energy per photon. The *electromagnetic spectrum* includes, in order of decreasing wavelength: radio waves, microwaves, IR, visible light, ultraviolet, X-rays, and gamma rays. How different wavelengths of light tend to be absorbed or emitted depends in part on the particular types of molecules in the air. Before turning to this, let us discuss how radiation depends on temperature in an approximation that does not depend on the particular substance doing the emitting. This is known as *blackbody radiation*.

2.2.1 Blackbody radiation

If a body is black in the sense of perfectly absorbing light at every wavelength, then it will emit in a way that only depends on temperature. For any substance not at absolute zero (0 K), molecules have kinetic energy and bounce against each other, which excites electrons, which in turn emits radiation. The hotter the substance is, the more radiation is emitted overall (greater flux) and more radiation is emitted as energetic photons (shorter wavelengths). The wavelength at which the peak emission occurs is given by Wien's law:

$$\lambda_{peak} = \frac{2897}{T} \qquad (2.1)$$

Here λ is the wavelength in micrometers, also known as microns (1 μm $= 10^{-6}$ m), and T is the temperature in kelvin (add 273.15 to degrees Celsius to get kelvin). The Sun, with surface temperature of about 6000 K, will thus have a peak emission around 0.5 microns (500 nanometers), while the Earth will tend to emit at much longer wavelengths. The blackbody emission per wavelength has a dependence on temperature which is shown in Figure 2.5a for temperatures characteristic of the Sun and the Earth.[2] About 45% of solar radiation is in the visible (wavelengths from 400 to 750 nm); about 10% of solar is in the ultraviolet. Even in the infrared, the solar emissions are mainly at much shorter wavelengths than the terrestrial emissions.

The atmosphere is quite transparent in the visible but much less so in the infrared. Figure 2.5b and c quantify this by showing the fraction of radiation absorbed at each wavelength as it passes through all or part of the atmosphere. The fraction absorbed in the visible wavelengths is small, although parts of the ultraviolet (at shorter than visible wavelengths) are completely absorbed. In much of the infrared almost all the radiation is absorbed going through the depth of the atmosphere (this holds true whether the radiation starts at the top and travels downward or starts at the bottom going upward). The molecules that have strong absorption at particular wavelengths are noted along the axis. There are a few bands of wavelengths where the atmospheric absorption is less than 100%, sometimes referred to as "atmospheric windows." Molecules that have absorption bands in these windows will be disproportionately important in adding to the atmospheric greenhouse effect. Note that the absorption between 11 km and the top of the atmosphere is much smaller, so the Earth can emit effectively in the IR from such upper levels. Many of the satellite estimates of quantities in the atmosphere make use of the particular preferred bands of emission, for instance by water vapor, to make estimates of quantities in the atmosphere. We will pay most attention to the energy flux integrated across all wavelengths, since the overall energy gain or loss is what matters most to the Earth's energy budget.

If we consider the total energy flux integrated across all wavelengths of light for a blackbody emitter, we get the *Stefan–Boltzmann law*

$$R = \sigma T^4 \qquad (2.2)$$

where $\sigma = 5.67 \times 10^{-8}$ W m^{-2} K^{-4} is a constant of physics and R denotes IR flux. Since actual surfaces or gases do not absorb or emit as fully as a perfect blackbody one can define an *emissivity* ϵ for each substance and use

$$R = \epsilon \sigma T^4 \qquad (2.3)$$

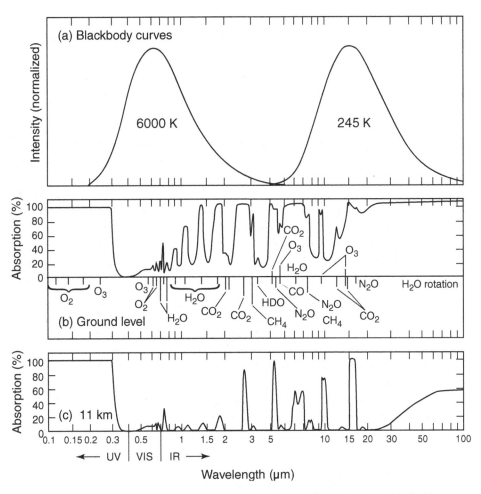

Fig. 2.5 (a) Blackbody radiation curves for temperatures characteristic of the Sun's surface and upper levels in Earth's atmosphere from which infrared radiation escapes to space. (b) Absorption of radiation at each wavelength if the beam passes through the entire atmosphere from top to ground level or vice versa. (c) As in b, but for radiation passing from the top of the atmosphere to 11 km or vice versa. After Goody and Yung (1989).

The *absorptivity* of a substance is equal to its emissivity. We will later use Eq. (2.3) with bulk emissivity ϵ_a for an atmospheric layer to approximate atmospheric emission. This is conceptually equivalent to averaging the percentage absorption in Figure 2.5b across all infrared wavelengths. We will take the emissivity of ground surfaces as approximately equal to 1 for simplicity. While we will use Eq. (2.2) and Eq. (2.3) often in a simple model of the greenhouse effect, it is important to note that this is a simplification to understand the general behavior. Full climate models do a much more detailed computation that takes into account the variations of absorption as a function of wavelength seen in Figure 2.5. They do such calculations for every level in the atmosphere, for the local temperature, pressure and concentrations of various gases (a very significant part of the computational cost).[3]

2.2.2 Solar energy input

The total solar energy flux (integrated across all wavelengths) coming in from the Sun, at the distance of Earth's orbit (averaged over the year), is approximately

$$S_0 \approx 1366 \, \text{W} \, \text{m}^{-2}$$

In the past it was referred to as the "solar constant," although it is now known to change by a few $\text{W} \, \text{m}^{-2}$ over the course of the 11-year solar cycle, and to vary even on weekly time scales. These short-term variations have little climate impact because of ocean heat storage and large natural climate variability. Over the lifetime of the Sun, it may have varied considerably, and there has been some concern that it might have variations on time scales of centuries that could affect climate. Effects of solar variation are estimated to be modest in recent climate variability and change (see Chapter 7), though they may be important in paleoclimate. Such variations can easily be taken into account in climate models over the period for which they have been measured. Satellite observations of the solar energy flux are available from the late 1970s.[4]

A useful number to bear in mind is the value of insolation averaged over one day, and over all latitudes. This amounts to averaging over the whole surface of the Earth, since the Earth rotates once per day. A quick trick to do this calculation is to realize that this amounts to taking the energy arriving at a disk the size of the Earth, such as the one shown in Figure 2.6, and then spreading it over the spherical surface of the Earth. If the Earth's radius is a, the area of the disk is πa^2, whereas the area of the sphere is four times as large. Thus the global average solar flux is

$$S_0/4 = 341.5 \, \text{W} \, \text{m}^{-2}$$

The solar flux S_0 would apply to a surface above the atmosphere perpendicular to the Sun's rays, such as a portion of the disk shown in Figure 2.6. The value arriving at the

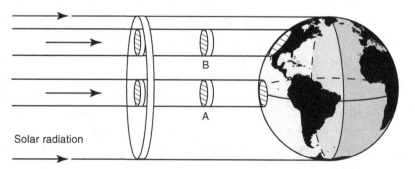

Fig. 2.6 Schematic of Sun's rays arriving at disk and spherical Earth. The intensity of the solar flux depends on the angle at which the rays hit the Earth. For the equinox configuration shown, a unit area of rays A arriving at the equator has the original intensity, but the unit area of rays B arriving at a higher latitude has lower intensity because the same energy is spread across a larger area.

spherical surface of the Earth (or rather a spherical surface just at the top of the atmosphere, since we are not yet considering complications such as reflection by clouds) depends on the angle at which it strikes the surface. Consider that a flux is a measure of energy per second per unit area, hence the units $W\,m^{-2}$. If a bundle of rays crossing a disk of unit area perpendicular to the disk is spread over a larger area, the same energy divided by a larger area implies a smaller flux. In Figure 2.6, shown for equinox (approximately April 21 or September 21), when the equator is perpendicular to the Sun's rays, the intensity depends on latitude because of this effect. In other seasons, the tilt of the Earth's axis of rotation relative to the plane of the Earth's orbit around the Sun comes into play. A reminder of how this tilt produces the seasonal cycle is schematized in Figure 2.7. Briefly, when the axis tilts so that the North Pole is away from the Sun, the line where the Sun's rays are perpendicular to the Earth's surface (the thermal equator) lies in the southern hemisphere. Two factors then reduce solar heating of high latitudes in the winter hemisphere. The intensity of insolation at noon drops off with increasing angle between the Earth's surface and the rays as in Figure 2.6. Furthermore, a smaller portion of a given latitude circle lies in the sunlight so days are shorter. The region near the North Pole is in the dark the full day. At the northern winter solstice (December 21), this region extends to the Arctic circle at 66.5° N, since the tilt of the Earth's axis is about 23.5°.

There are small variations in the incoming flux as Earth goes around its orbit since the orbit is elliptical. Currently the closest approach to the Sun (perihelion) occurs in January, at which time the solar flux at the top of the atmosphere is about 7% larger than when the Earth is farthest from the Sun (aphelion). Such effects are, of course, included in climate models. On time scales of centuries or less, these parameters change little, but their variations are important to paleoclimate, notably to the ice age cycle.[5] The eccentricity varies on 100 000 and 400 000-year time scales; the axial tilt varies (between 22.1° and 24.5°) on 41 000 year scales; and axial precession – that is, the position relative to perihelion on the Earth's orbit at which a given season occurs – varies on 23 000-year time scales.

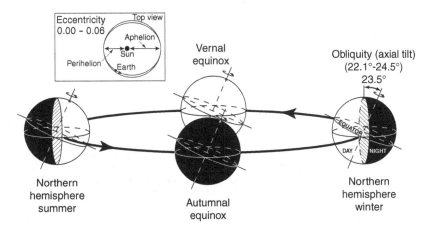

Fig. 2.7 Schematic of how seasonal dependence of insolation arises as a function of latitude. Orbital parameters that vary on paleoclimate time scales are also shown.

2.3 Globally averaged energy budget: first glance

The globally and annually averaged energy budget might seem simple from the point of view of the solar forcing. However, the pathways by which this input of energy is transferred from the Earth's surface through the atmosphere and eventually back to space include many of the complications that pose problems for assessment of global warming. Figure 2.8 shows the energy transfers, beginning from the approximately 342 W m^{-2} average input of solar radiation. Not all of this energy is absorbed by the Earth system, since part is immediately reflected. Clouds and *aerosols* (suspended particles) reflect part from the atmosphere, and part is reflected from the surface. *Albedo* is defined as the fraction of incident solar radiation that is reflected. The albedo for the global average reflection of solar radiation (the planetary albedo) according to Figure 2.8 is thus 0.31 (107 W m^{-2} reflected out of 342 W m^{-2} input). Different parts of the surface have different albedos. For instance, deep clouds have albedos of roughly 0.9. The ocean has an albedo of 0.08, which is much lower than that of snow, ice, desert or clouds. Variations in the extent of snow or cloud cover can thus potentially affect the average albedo.

Some sunlight is absorbed in the atmosphere and goes directly to atmospheric heating. Part of this absorption occurs in the ozone layer (ozone is created by absorption of ultraviolet radiation), and part occurs lower in the atmosphere, largely by clouds, water vapor and suspended particles. The exact amount of absorption is still the subject of observational programs such as the Atmospheric Radiation Measurement (ARM) program, so the 67 W m^{-2} noted in the figure may be revised in coming years. Most of the incoming sunlight that is

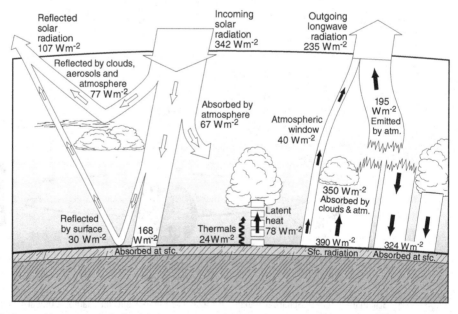

Fig. 2.8 Pathways of energy transfer in the global average energy budget. After Kiehl and Trenberth (1997).

not reflected passes through the atmosphere and is absorbed at the ground. This heating must be balanced by heat transfer from the ground up to the atmosphere.

Three forms of heat transfer occur from the ground upward: sensible heat, latent heat and infrared emission. *Sensible heat* transfer is due to contact between molecules of air with the hot ground and subsequent upward transfer by parcels of hotter air. One of the main factors transferring this form of heat effectively through about the lowest kilometer of atmosphere are hot plumes known as *thermals*. This is a form of convection without a phase change of moisture (sometimes referred to as "*dry convection*"). The depth through which these thermals penetrate defines the *atmospheric boundary layer*, which will be familiar to those living in cities like Los Angeles or Denver as the top of the brown haze layer on smoggy days. Although sensible heat is the first form of heat transfer one tends to think of, it is only a modest contribution to the surface heat budget.

Wherever water is available, *evaporation* is a more effective means of cooling the surface over most of the Earth. The phase change of liquid water to water vapor stores energy in the form of *latent heat*. This latent heat is subsequently released when the water vapor condenses to form clouds, so wherever there is precipitation the latent heat remains in the atmosphere (if a cloud condenses and then re-evaporates there is no net heating). Cloud formation is most often associated with overturning motions, and this process, known as *moist convection*, transfers heat through a deep layer in the atmosphere. Since the overturning motions also transfer heat between layers, in addition to condensation adding latent heat, the overall effect is referred to as *convective heating*. Over land, vegetation plays such an important role in evaporation that the process is called *evapotranspiration*. This involves plants accessing ground water through their root depth and actively regulating water loss through their leaves.

Infrared radiation plays a double role in the surface energy budget, as seen on the right half of Figure 2.8. The warm surface loses energy by emitting IR upward. But the atmosphere also emits IR downward, which is absorbed at the surface. The net IR loss from the surface is only 66 W m^{-2}, much smaller than the 390 W m^{-2} that is emitted upward. This is our first view of the *greenhouse effect*, as it appears in Earth's current climate. It will be discussed in detail in Chapter 6, but can be briefly summarized as follows:

- The upward IR from the surface is mostly trapped in the atmosphere, rather than escaping directly to space, so it tends to heat the atmosphere.
- The atmosphere warms to a temperature where it emits sufficient radiation to balance the heat budget, but it emits both upward and downward, so part of the energy is returned back down to the surface where it is absorbed.
- This results in additional warming of the surface, compared with a case with no atmospheric absorption of IR.
- Both gases and clouds contribute to absorption of IR and thus to the greenhouse effect.

At the top of the atmosphere, in the global average and for a steady climate, the IR emitted at the top of the atmosphere must balance the incoming solar. Global warming involves a slight imbalance in which a change in the greenhouse effect results in slightly less IR emitted from the top of the atmosphere. The imbalance required to give a slow warming is so small that it cannot currently be directly measured. Even smaller imbalances occur associated with natural climate variations, causing slight variations in ocean heat storage.

It is worth summarizing the three roles that we have noted above for clouds and convection:

(i) heating of the atmosphere (through a deep layer);
(ii) reflection of solar radiation (contributing to albedo);
(iii) trapping of infrared radiation (contributing to the greenhouse effect).

As indicated in Figure 2.8, clouds produce the largest contribution to the global albedo. Of the total reflection of solar radiation, 72% (77 W m^{-2}) is due to clouds and aerosols. The contribution of clouds alone, as measured by clear sky versus cloudy sky reflected solar radiation, is about 50 W m^{-2}. Figure 2.8 does not specify the breakdown of IR absorption by clouds versus gases because it is more complicated to separate the effects. Nonetheless, the importance of clouds in infrared trapping is comparable to their importance in solar reflection.[6] Clouds and convection will thus figure substantially in all our discussions of the maintenance of Earth's climate and of global warming.

2.4 Gradients of radiative forcing and energy transports

The differences in input of solar energy between different latitudes tend to create temperature gradients. In fact, if it were not for heat transport and storage in the ocean and atmosphere, these gradients would be huge. Consider the surface temperatures that would occur at the North Pole in winter time if there were no atmosphere or ocean. There is no incoming sunlight, so there is no direct solar radiative heating. If there were no ocean, there would be little heat storage since land surfaces have small heat capacity. If there were no atmospheric or oceanic motions transporting heat, the infrared radiation required to balance tiny inputs from starlight would be very small. The temperature would be extremely cold, as it is on parts of the Moon where, when in the dark, it reaches only a few degrees above absolute zero.

Figure 2.9 gives the solar radiative input to the climate system for December climatological average as measured from satellite during the Earth Radiation Budget Experiment, a program designed to accurately map the top-of-atmosphere energy budgets. The net input is shown, i.e. the part of the solar energy that is reflected back to space has been subtracted from the incoming solar. The energy input varies from essentially zero north of the Arctic circle to about 385 W m^{-2} near 30° S.[7] Without the ocean or atmosphere, the temperature difference would be several hundred degrees Celsius if the temperature at each latitude had to warm or cool to the point where outgoing longwave radiation would be equal to the solar input. The observed outgoing longwave radiation varies much less as a function of latitude. This is because (i) the atmospheric and oceanic transports are very effective at reducing temperature gradients and (ii) the ocean stores some heat from the previous summer and returns it to the Arctic atmosphere in winter as temperatures start to cool. Both effects act to reduce extremes.

The annual average case corresponding to Figure 2.9 is shown in Figure 2.10. In the annual average climatology, the rate of heat storage by the ocean is small. The difference between the solar input and the compensating infrared radiation is due to north–south heat transport

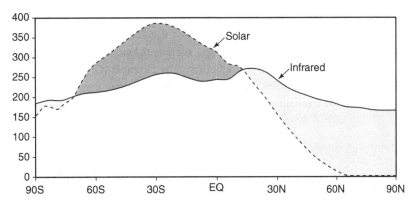

Fig. 2.9 Net solar energy input (dashed) and outgoing longwave radiation (solid) for December climatology as a function of latitude. Units are W m^{-2}. The net heat input into the climate system at each latitude is shown by shading. Shading above the solid line shows positive net input, shading below solid line shows net heat loss. Data from the Earth Radiation Budget Experiment (averaged January 1985 to December 1989).

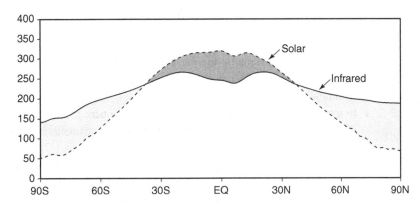

Fig. 2.10 Annual average net solar energy input (dashed) and outgoing longwave radiation (solid) as a function of latitude. Units are W m^{-2}. Shading above solid line shows positive net heat input by radiation, shading below line shows net heat loss. Data from the Earth Radiation Budget Experiment (averaged January 1985 to December 1989).

into or out of each region by the atmosphere and ocean (shaded areas in Figure 2.10). The tropics have about 60 W m^{-2} more solar input than infrared loss to space, and this excess heating is balanced by heat export to the poles. The poles have small solar energy input but, because atmospheric and oceanic transports keep them from becoming very cold, they remain warm enough to emit substantial amounts of infrared radiation. As an indication of the importance of transports, note that in polar regions the annual average heat input by transport is about three times as large as the solar input (since the latter is small). Because the Earth is approximately in energetic balance when averaged over a year, the net energy loss at the poles approximately balances the net energy gain in the tropics.

The oceanic heat transports are accomplished by moving warm water into cooler regions and cooler water toward warmer regions. This can occur through horizontal motions, or through warm surface waters moving into cold polar regions, sinking and returning southward. It can be due to climatological currents, such as the Gulf stream, which would carry warm water northward even if it did not vary in time, but heat transport can also be due to time varying currents, such as the ocean eddies seen in Figure 2.4. The atmospheric circulation is strongly driven by the radiative heating differential between tropical and higher latitudes, so the mechanisms of energy transport are described simultaneously with the circulation in the following section.

2.5 Atmospheric circulation

2.5.1 Vertical structure

Before beginning our discussion of circulation, it is worth considering the vertical structure of the atmosphere. Figure 2.11 gives a one-dimensional view of the atmospheric structure in terms of temperature. The following points are worth noting:

- Pressure can be used as a vertical coordinate (commonly used units are millibars (mb)).
- Pressure is proportional to the mass (per unit area) above each level (see section 3.1.5. Because density also decreases with pressure, pressure drops off roughly exponentially with height.
- In terms of mass, the upper layers of the atmosphere are small. For instance, only 10% of atmospheric mass is found above 15 km. The coordinates in Figure 2.11 are stretched to make the upper layers visible.
- The troposphere is very important for climate variations; the stratosphere counts somewhat because of ultraviolet radiation absorption by ozone. The thermosphere and mesosphere have very little mass. Interaction with charged particles from the Sun and the Aurora Borealis (Northern Lights) occurs at these levels, but these regions are not important to our discussion of climate variations.
- Heating by absorption of ultraviolet radiation by stratospheric ozone is the reason for the temperature increase with height in the stratosphere.
- In the lowest kilometer, the atmospheric boundary layer, temperature drops at 10°C per km.
- Through the rest of the troposphere, the temperature decrease per increase in height (the *lapse rate*) is about 6°C per km. This lapse rate is smaller than that in the atmospheric boundary layer partly owing to the release of latent heat in convective clouds.
- The deep convective cumulonimbus clouds can reach the tropopause (top of the troposphere), about 10 km at midlatitudes, 12 km in the tropics. This depth of the layer through which moist convection can occur is important in setting the properties of the troposphere and determining the level of the tropopause.
- Moist convective cumulus clouds span a range of heights. Some other significant cloud types include cirrus ice clouds (typically just below the tropopause) and stratus clouds just above the boundary layer.

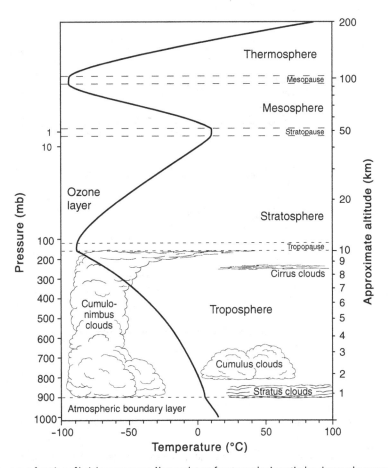

Fig. 2.11 Temperature as a function of height or pressure. Nomenclature for atmospheric vertical regions and common cloud types is also given. Temperature data are from the Standard Atmosphere except in the troposphere where it has been modified to a profile typical of deep convective regions. The vertical scale is linear in pressure from the surface to 100 mb, but stretched in the upper half.

2.5.2 Latitude structure of the circulation

Since latitudinal gradients in solar energy input are a dominant driver of atmospheric circulation, the primary features depend on latitude. Figure 2.12 schematizes the main features of the circulation, both in a latitude–height plane and in the horizontal, without yet considering longitudinal variations due to continents, oceans etc. This represents the circulation in an average over all longitudes, known as a *zonal average*. The figure is also idealized to appear symmetric about the equator, whereas in reality there are asymmetries between hemispheres. The main features are as follows.

- The Hadley cell is a thermally driven, overturning circulation that tends to rise in the tropics and sink at slightly higher latitudes. Warming from the surface near the equator is transferred upward through a deep layer by convection (schematized as a single deep

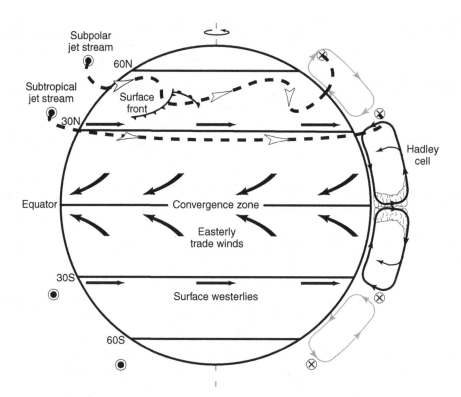

Fig. 2.12 Schematic of major features of the atmospheric circulation. The average circulation is idealized as being independent
of longitude and symmetric about the equator. The subpolar jet (dashed curves for upper-level features) is shown as a
snapshot for a particular day that includes weather variations: an upper-level wave and an associated surface front. In
a climate average, these features would not be present, yet they are important for maintaining climate. Along the
right side, the circulation in the vertical and in latitude is shown as a sequence of overturning cells. The upper-level
subtropical jet occurs near the poleward, descending branch of the Hadley cell. Circled crosses/dots represent flow
into/out of the plane of the paper.

cloud; in reality, of course, the circulation is the average of a multitude of small-scale
convective motions). The rising air spreads poleward and cools slowly, returning to the
surface and then toward the equator. Overall, heat is transported poleward.

- The descending motion occurs over a broader region than the ascending motion. The
descending region generally occurs in the subtropics (and in some ways defines what are
the subtropics). These descending regions have little rainfall; roughly speaking, they are
on the losing end of a competition with the deep convective regions that involves moisture
and heat transport feedbacks. Moisture is transported from the descending region into
the ascending branch, greatly enhancing rainfall there. Poleward heat transport tends to
warm the upper levels in the descending branch, making it difficult for convective plumes
to rise since they would be cooler than surrounding air.

- Poleward of the Hadley cell there is a weak zonal average overturning circulation (light
curve) that appears to run backwards from the point of view of heat transfer by mean

motions. It is really just a residual circulation resulting from averaging many weather disturbances. The disturbances transport so much heat poleward that what this average cell does is almost irrelevant, but traditionalists call it the Ferrel cell. A zonal average polar cell (not shown) is also traditionally noted but is not very important.

• Roughly speaking, the Hadley circulation transports heat to about 30 degrees latitude, and the average effect of the transient weather disturbances transports the heat further poleward.

• The trade winds in the tropics blow westward (so are known as easterlies, since they come from the east) as they converge slowly toward the equator.

• The region they converge into is known as the Intertropical Convergence Zone (sometimes abbreviated ITCZ) or the tropical convection zone. This region is the base of the Hadley cell rising branch, closely associated with convection.

• At midlatitudes, the surface winds are westerly (from the west). At upper levels, there is actually more westerly wind than easterly, i.e. overall the circulation goes around the planet slightly faster than the rotation of the planet.

How does the Hadley cell transport heat poleward? Since air at upper levels is colder than air near the surface, it might seem that cold air is moving poleward. Because air is compressible, temperature is not a good measure of energy transport if the pressure changes. One way of explaining this is to note that an air parcel has internal energy proportional to temperature and potential energy due to gravity. Because the potential energy increases with the height above ground, the air at upper levels actually has greater energy associated with it even though it is cooler. One can define a quantity called potential temperature (see Chapter 5) that takes this into account. This is simply the temperature the air would have if it were brought back down to ground level while conserving energy. Air heated in the tropics rises and moves poleward with high potential temperature. As it cools to a lower potential temperature (by emitting infrared radiation), it sinks and subsequently returns to the tropics. This has the overall effect of transferring heat out of the tropics and keeping the temperature relatively constant at each level in the atmosphere between the tropics and the subtropics.

Transient weather disturbances tend to transport cold air equatorward and warm air poleward in different regions in the horizontal, accomplishing poleward heat transport in the average. These are referred to as *eddy transports* of heat (in both atmosphere and ocean). Latent heat is transported in the form of water vapor by these circulations as well. In the Hadley cell, water vapor, which is largest at lower levels, is transported toward the convection zones. Transient eddy transports of moisture tend to carry moisture from the tropics to higher latitudes.

Momentum transport also occurs in these circulations and plays a significant role in determining wind patterns. As the air flows poleward in the Hadley cell, the sphericity of the planet causes it to become slightly closer to the axis of rotation. By conservation of angular momentum, it must spin faster relative to the Earth's surface. That is, the subtropical jet attempts to move at a speed more like that of the equator (farther from the center of the Earth) than that of the ground at 30 degrees latitude. When the air returns towards the equator, conversely, it must reduce its westerly momentum and become more easterly relative to its subtropical value.

An equivalent way of discussing this turning of the winds is in terms of the *Coriolis force*, which will be discussed in more detail in Chapter 3. This is an apparent force associated with objects tending to conserve momentum in a rotating reference frame, from the point of view of an observer fixed in the rotating frame. This occurs simply because the Earth is rotating while the object attempts to move in a straight line. Away from the surface, friction in the atmosphere is fairly small, so the effects of the Coriolis force are noticeable. In the northern hemisphere, the Coriolis force tends to turn moving objects to the right for an observer standing with his feet on the ground; in the southern hemisphere, to the left. Near the equator, the horizontal component of the force is zero for horizontal motions, since the vertical is perpendicular to the axis of rotation. The winds in the poleward branch of the Hadley cell turn to the right in the northern hemisphere, so they accelerate eastward (westerly). They turn to the left in the southern hemisphere, so they also accelerate eastward, likewise producing the upper-level westerly jet. In the lower branch, returning toward the equator, the winds in both branches, turning right and left respectively, are accelerated westward producing winds that are relative to those in the subtropics.

This picture sounds simple, but in fact, it glosses over one important question. The argument above says that the tropical winds should be easterly relative to the westerlies in the subtropics, but why do they actually become easterly? This has to do with eddy transports at midlatitudes and frictional effects at the surface.

Consider the wave pictured in Figure 2.12, consisting of transient weather disturbances, which is shown for the northern hemisphere case. It has faster westerly flow in the northward moving segments than the southward moving segments. It thus transports westerly momentum poleward. Corresponding weather disturbances in the southern hemisphere likewise create poleward momentum transport. This is the principal type of disturbance connecting the tropics and midlatitudes. Since westerly momentum is transported out of the tropics, tropical winds must be easterly relative to midlatitudes. At the surface, there is a roughly equal area of westerly and easterly winds because frictional effects from wind drag at the surface must balance (the winds cannot speed up or slow down the planet, on average). Thus even the steady, dependable surface trade winds in the tropics actually depend on the climate effects of midlatitude storms.

The trade winds in the tropical climate turn out to be crucial to El Niño. The jet streams are important for the effects of El Niño at midlatitudes. For global warming, the heat transports are the leading contribution of these circulations for large-scale effects, but additional features of atmospheric circulation become important when considering possible regional impacts.

2.5.3 Latitude–longitude dependence of atmospheric climate features

The idealized picture of the climate depending on latitude is a reasonable first approximation, but land–ocean contrasts and other variations in longitude are also important. Among other things, El Niño depends on east–west contrasts in the Pacific. Figure 2.13 gives a view of the latitude–longitude dependence of the precipitation climatology for January and July, corresponding to southern and northern summer. We note the following features.

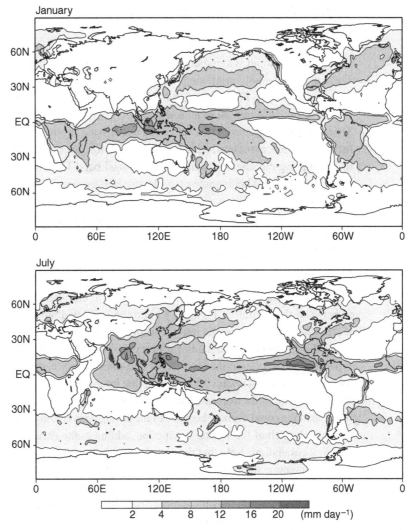

Fig. 2.13 Precipitation climatology (July 1987 to July 1995) in mm per day. Estimated from a combination of satellite and ground rain-gauge data (data from the Xie and Arkin (1996) data set).

- The intertropical convergence zones (ITCZs): these are the elongated precipitation features in a belt deep in the tropics, though not always near the equator. The term convergence refers to the low-level winds that converge into these regions. An equivalent but perhaps better term is *tropical convection zones*.[8]
- The tropical convection zones move northward in northern summer, and southward in southern summer, especially over continents. These are the monsoon circulations. From the global scale view, there is the Asian-Austral monsoon, the Pan-American monsoon and the African monsoon. Local regions also have specific names, e.g. Mexican monsoon, Indian monsoon, although these are just part of a larger system. Traditionally monsoon was defined in terms of local reversals of wind, but this has been generalized

to include all the circulation changes associated with seasonal movements of the tropical convection zones.

- The climate is not perfectly symmetric about the equator. This holds even in spring and fall when solar input is symmetric, largely owing to the asymmetry in continents.
- There are substantial variations in longitude (technically referred to as departures from "*zonal symmetry*"). For instance, the eastern Pacific has little rain, whereas the western Pacific has intense rainfall in all seasons.
- There is thus more convection and associated rising motion in the western Pacific and little convection, associated with descending motion, over the eastern Pacific. This results in overturning circulations along the equatorial band known as the *Walker circulation* (see Figure 2.14). This is somewhat like the Hadley circulation but in longitude instead of latitude.
- Around 30°–40° N, midlatitude rainfall is organized in *storm tracks* which are most intense over the oceans.
- Midlatitude continents look relatively dry, simply because the tropics get such intense rain.

Deep convection has a strong influence on tropical circulation, so a relationship may be seen between the rainfall in Figure 2.13 and the low-level winds shown in Figure 2.15b. Surface winds tend to converge into the tropical convection zones, although not always in the most direct manner, since the Earth's rotation is also a strong influence on their motion. The part of the circulation that converges and rises is only a small part of the total wind, although it is important because it is the part that interacts most directly with the convection. Overall, the winds in the tropics are primarily easterly. The easterlies in the Pacific are much stronger than the zonal average because of the strong convection in the western Pacific. This is the lower branch of the Walker circulation, adding to the easterly tendency produced by the Hadley circulation. Note that the actual winds are considerably more complex than the schematic diagrams for either Hadley (Figure 2.12) or Walker (Figure 2.14) circulations.

At midlatitudes the low-level winds are primarily westerly. In upper-level winds (Figure 2.15a), the dominant features are the subtropical jets centered at about 30°–40° N (the winter hemisphere) and 40°–50° S (the summer hemisphere) for the December–February (DJF) season shown. An annual average map would show generally

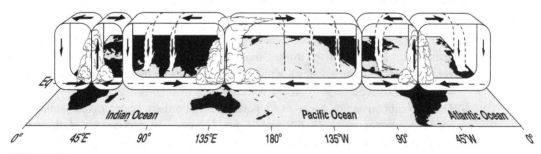

Fig. 2.14   Schematic diagram of the Walker circulation. Overturning cells along the equator in the east–west direction, with rising branches occurring in the convergence zones, associated with large amounts of convection. Note that the wind shown has the longitudinal average around the equator removed. Total wind tends to be predominantly easterly. Adapted from Madden and Julian (1972), and Webster (1983).

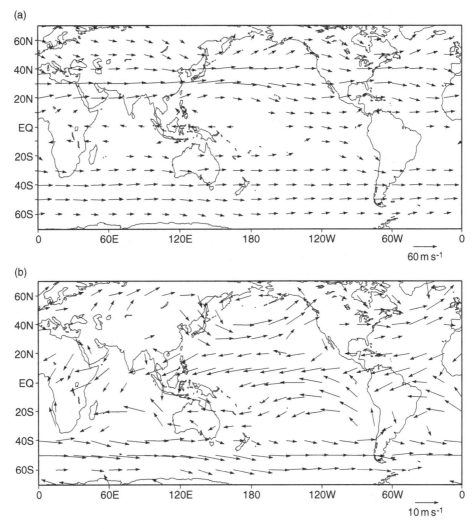

Climatology for December–February of (a) upper-level wind (200 mb) and (b) low-level wind (925 mb). Data are from the National Center for Environmental Prediction (NCEP) analysis averaged from 1958 to 1998. The scale for wind vectors is shown in each panel (note the scale for the upper levels is six times larger).

the same features, but DJF is chosen to illustrate seasonal variations, and because this is the season in which the strongest ENSO teleconnections to North America occur. The jet is stronger in the winter hemisphere because the high latitude regions are colder and thus the temperature contrast to the tropics is strongest. It shifts slightly equatorward in winter since the region of strongest temperature gradients shifts toward the tropics.

Unlike the tropics, where winds associated with deep convection tend to be roughly opposite at upper and lower levels, the winds at midlatitudes tend to be the same sign in upper and lower troposphere. In the northern hemisphere, the strongest regions of the jet stream are found over the Pacific and Atlantic oceans. This coincides with the storm track regions noted in precipitation in Figure 2.13. There is a two-way interaction affecting this: the storms are favored by regions of strong temperature gradients and thus of strong upper-level wind, and the jet is strengthened by momentum transports by the storms.

2.6 Ocean circulation

2.6.1 Latitude–longitude dependence of oceanic climate features

The most important oceanic variable for interaction with the atmosphere is the sea surface temperature (SST), seen in Figure 2.16, since this is what affects the heat fluxes exchanged between the atmosphere and the ocean surface. In the January and July climatology of SST, we note:

- SST is warmest in tropics, as expected since the solar input is highest. Temperatures approach freezing in higher latitudes (sea water freezes at about $-2\,°C$). The strongest SST gradient occurs at midlatitudes.
- SST is not perfectly symmetric about the equator. This holds even in spring and fall when solar input is symmetric.
- There are substantial variations in longitude. For instance, the eastern Pacific is relatively cold, whereas the western Pacific is warm in all seasons.
- The pattern of rainfall over oceans has a close, though not perfect, relationship to this SST pattern (compare to Figure 2.13). Roughly speaking, in the tropics it tends to rain over SST that is warmer than SST in neighboring regions. For instance, there is an Atlantic convergence zone, even though the SST there is cooler than the western Pacific, because it is warmer than nearby SSTs in the Atlantic.
- Relatively cold waters extend along the equator in the Pacific, in what is known as the *equatorial cold tongue*. This is important in the dynamics of El Niño. It is maintained by *upwelling* of cold water from below, i.e. by ocean heat transports. The upwelled water diverges from the equator, transferring heat poleward, until it is warmed by solar radiation at the surface.

The climatology of ocean currents at the surface is shown in Figure 2.17. The surface currents are primarily driven by the surface wind which is similar to the low-level wind of Figure 2.15b. Some features of the current systems relevant to our discussion include:

- Along the equator, the currents are in the direction of the wind, i.e. easterly winds drive westward currents. (Note that oceanographers and meteorologists adopted different terminology for direction. Imagine getting them to cooperate, as needed for climate studies.)
- Off the equator, the currents need not be in the direction of the wind. For instance, just north of the equator there are the equatorial countercurrents which go eastward in the opposite direction of the easterly winds. In the interior of the ocean, currents are strongly influenced by the rate of change of the *zonal* (east–west) wind with latitude. This occurs because pressure gradients across the basin tend to balance the direct effects of the wind stress at each latitude. The currents are then set by the rate of change of the wind and Coriolis force with latitude.

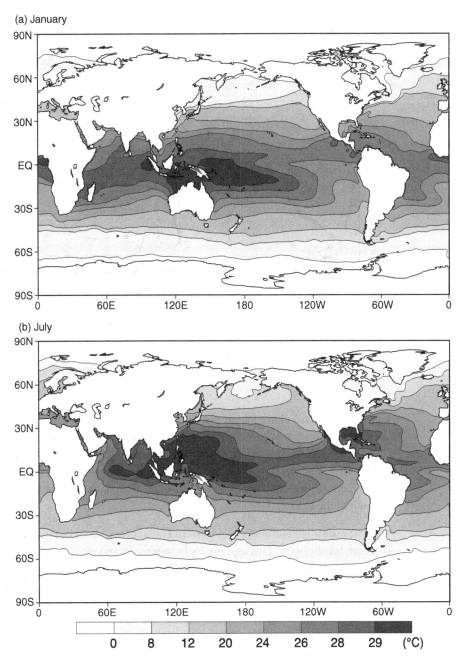

(a) January

(b) July

Sea surface temperature climatology for January and July (averaged 1982–2000). Note that the contour interval is smaller for warmer temperatures to show tropical features. Data are from the Reynolds *et al*. (2002) data set.

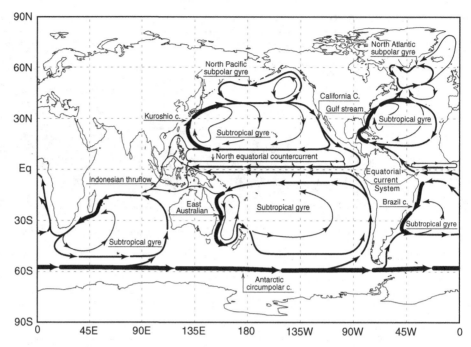

Fig. 2.17 Major systems of ocean surface currents. Thick arrows denote stronger currents. Thinner arrows within gyre systems indicate weaker, widespread flow.[10]

- Just slightly off the equator, the component of the current that moves poleward is relatively small compared with the westward component. Because it diverges, it is disproportionately important, since it is associated with the important upwelling motions. The wind is accelerating the water westward, but the water is also influenced by the Coriolis force, which turns water to the right of its motion north of the equator, and to the left south of the equator.

There are strong currents along western boundaries of both oceans, in both northern and southern hemispheres. This asymmetry between east and west is, perhaps surprisingly, due to the rate of change of Coriolis force with latitude.[9] Similar east–west asymmetries will prove important for El Niño dynamics, as discussed in Chapter 4. The currents away from the equator are organized in circulation systems known as *gyres*. In the subtropical gyres, currents flow slowly equatorward in most of the basin. Compensating return flow toward the poles occurs in the narrow, fast western boundary currents in both hemispheres. These include the Gulf stream, the Kuroshio current and the Brazil current. In the Antarctic Ocean, there is no continental boundary to create zonal pressure gradients, so the currents tend to flow in the direction of the wind (north–south pressure gradients balance the Coriolis force).

2.6.2 The ocean vertical structure

Curiously, one cannot understand the vertical structure of the ocean without understanding something of the three-dimensional circulation, including the source of deep waters.

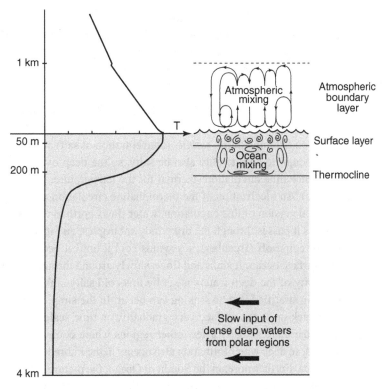

Fig. 2.18   Schematic of vertical temperature structure.

Because the ocean surface is warmed from above, this tends to produce lighter water over denser water. This is referred to as stable stratification, where *stratification* is the rate of change of density in the vertical. This may be seen in the typical vertical profile of temperature shown in Figure 2.18. Temperature is nearly constant in a layer of a few tens of meters near the surface. At a typical depth of a few hundred meters (100–200 m in the tropics, deeper at higher latitudes) it drops rapidly, and this transition defines the thermocline. Deep ocean temperatures are as low as a few degrees Celsius, since they originate in polar regions.

This structure is maintained as follows. Incoming solar radiation penetrates only on the order of 10 m into the water column and so warms the very surface. Mixing near the surface driven largely by wind-generated turbulence and instabilities of surface currents transports some of this warming downward. When the mixing reaches the thermocline, any mixing of the denser fluid below into lighter fluid above requires that work be done, expending the kinetic energy of the turbulent motions. This strongly limits the mixing from extending deeper. The deep waters tend to remain cold, and thus this becomes a self-reinforcing situation, maintaining the sharp temperature gradient of the thermocline. On long time scales, however, even a very small amount of mixing would tend to warm the deep waters if nothing were maintaining the cold temperatures. This is accomplished by import of cold waters from a few sinking regions near the poles. Thus the three-dimensional deep ocean circulation comes into the picture.

Another feature that may be noted in Figure 2.18 is the relation of oceanic and atmospheric temperature structure. Because the ocean surface, which is directly warmed by solar radiation, must lose heat to the atmosphere in most regions, air temperature a few meters above the surface tends to be slightly colder than the surface temperature.

2.6.3 The ocean thermohaline circulation

In addition to temperature, *salinity* (concentration of salt) affects the ocean density. Waters dense enough to sink tend to also be salty, so the deep overturning circulation is termed the thermohaline circulation (thermal for the temperature, haline from the greek word for salt, hals). An idealization of the thermohaline circulation is shown in Figure 2.19. In this simplified version of this circulation, water flows north in the Atlantic. Strong evaporation occurs as it passes through the dry windy subtropics, and it cools as it moves poleward. In a small region off Greenland, icy winds cool it until it becomes the densest water in the entire surface ocean. It sinks and flows slowly around through the entire deep ocean. Since the density of the deep waters is set by this cold salty sinking region, most other regions are stably stratified and no sinking can occur. In the simplest picture, the deep waters are mixed back up to the surface very gradually, on time scales of centuries to millennia. Of course, there are actually a few other regions where comparable surface densities can be achieved, so deep water formation also occurs in the Labrador Sea and some regions around Antarctica. Furthermore, in the Southern Ocean, the picture is complicated by effects of the wind on the density surfaces. Water can both rise and sink again in the region near Antarctica. The deepest flow schematized in Figure 2.19 is known as Antarctic bottom water, sinking in the Antarctic region and flowing northward in the Atlantic. This water is sufficiently dense that it actually flows below the North Atlantic deep water coming southward in the main branch of the thermohaline circulation. Overall, the principle remains that a few very small regions control the temperature of the deep ocean. The thermohaline circulation is thus a candidate for sensitive response in climate change, a source of concern for surprises in global warming. It has likely been a player in past climate variations.

☐ Warm shallow current   ■ Cold & salty deep current

Fig. 2.19 Schematic of the thermohaline circulation. Strong sinking in the North Atlantic and overturning in the Southern Ocean are important features in connecting the deep ocean to the surface.[11]

2.7 Land surface processes

Although the land surface covers only 30% of the Earth's surface, it is disproportionately important in human activities. When discussing land surface processes in climate, it is useful to distinguish among processes that feed back significantly onto the physical climate system, or onto chemical aspects of the climate system, and land surface processes that are impacted by the climate but do not have strong feedbacks on it. The latter may be of great interest for their impact on ecosystems or economics, but may be studied "offline," i.e. using output from climate models but not included at the time the climate simulation was made.

In terms of their impact on physical climate, the main effects are as follows.

(i) Land does not transport or store a significant amount of heat. This may seem simple but it has large impacts on the circulation, producing land–ocean contrast. The lack of heat storage produces contrast during the seasonal cycle, while the lack of transport produces contrast with certain ocean regions in the annual average. This property also makes certain aspects of the land surface very easy to model to a first approximation.

(ii) Albedo. The high albedo of certain land regions can affect regional circulation and reduces the average energy input into the climate system.

(iii) *Evapotranspiration* and surface hydrology. The land surface stores moisture as *soil moisture* in subsurface layers of soil, and in other reservoirs, such as lakes and rivers. The transfer of this moisture to the atmosphere occurs substantially via vegetation, which actively regulates moisture loss. This process thus behaves quite differently from evaporation from a body of water or a moist surface, and is termed evapotranspiration.

The annual average albedo of the Earth's surface is given in Figure 2.20. The polar regions stand out with high albedo, especially where ice caps are found. The annual average albedo is also high in high northern latitude land surfaces owing to winter snow cover. In these regions the albedo changes greatly with season, since during the summer the vegetated land surfaces are much less reflective. The Sahara desert also has high albedo (much higher than most other major desert regions).

Both the albedo and evapotranspiration depend on vegetation, and thus climate models must include some representation of the vegetation at each grid point. In traditional climate models, characteristics of observed vegetation are specified from observations. More recently, there is a move towards including vegetation models so that interactions with changes in vegetation are included in the simulation. Biological activity on land surfaces, as well as in the ocean, can have significant effects on trace gases. For instance, the uptake of carbon dioxide by vegetation has an important effect on the fraction of human fossil fuel emissions that remains in the atmosphere. Thus, models of terrestrial ecosystems now consider carbon storage. Such effects have initially been studied separately from the physical climate system, using the approximation that greenhouse feedbacks on the physical climate system tend to be more important on longer time scales. Inclusion of interactive carbon cycle models is an area of much current research.

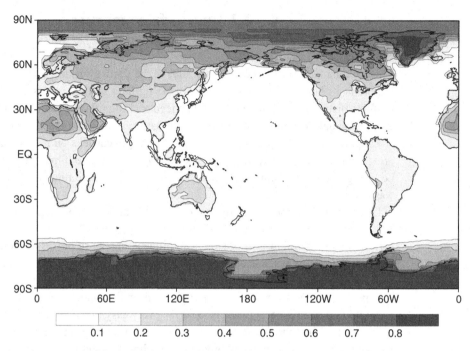

Fig. 2.20 Annual average surface albedo. Data from Darnell *et al*. (1992). Note that high values of albedo indicate very reflective surfaces.

Land surface hydrology has some impact on climate, since the reduction of moisture over subtropical desert regions can contribute to limiting the northward advance of summer monsoons, and hence to maintaining the desert. The details of hydrology, such as the flow patterns of rivers, are secondary in terms of their climate feedbacks for many purposes. Climate models have recently included this level of detail. Hydrologists interested in climate impacts in particular watershed regions can also take the output from climate models and use them as inputs into hydrological models for such quantities as river flow.

2.8 The carbon cycle

The carbon cycle is a key ingredient of the global climate system. It involves not only transport and storage in the physical components of the climate system but interactions with the chemistry and biology of the Earth system. Variations in the carbon cycle are important on paleoclimatic time scales – and are of immediate concern for the changes involved in global warming. Figure 2.21a shows the carbon cycle approximately as it would occur without human perturbation, for instance in preindustrial times (although the estimates here are made based on modern data, separating out the known human effects). The cycle involves transfers among reservoirs in the atmosphere, in the ocean and in land biomass. This cycle is shown as approximately in balance, although slight imbalances can cause

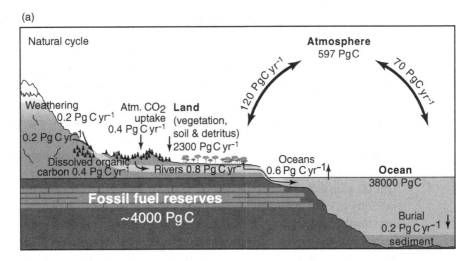

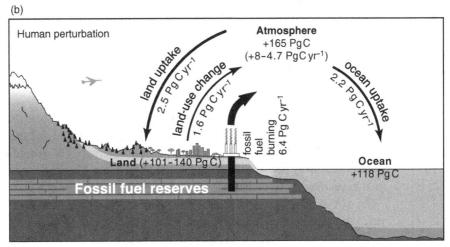

Fig. 2.21 The global carbon cycle. (a) The natural carbon cycle, typical of preindustrial times. The size of the main reservoirs of carbon is given in petagrams (PgC). Thick arrows give gross fluxes of carbon in PgC per year, thinner arrows give net fluxes. (b) The anthropogenic perturbation, using data for the 1990s. Increases ($+$ sign) or decreases ($-$ sign) of carbon from 1750 to 1994 are given in PgC for each reservoir; ($+8-4.7$ PgC yr^{-1}) denotes 8 PgC yr^{-1} total gain from fossil fuel and land-use change combined and 4.7 PgC yr^{-1} total loss by uptake in ocean and land biosphere. Arrows give anthropogenically perturbed fluxes of carbon in PgC per year. Values are based on Denman *et al.* (2007); format simplified from Sarmiento and Gruber (2002).

changes on long geological time scales. The size of each reservoir is given in petagrams of carbon. One petagram is equal to 1 trillion kilograms (1 000 000 000 000 kg) or one gigaton (1 000 000 000 metric tons). Petagrams of carbon (PgC) is the official unit, but gigatons of carbon (GtC) is commonly used, perhaps because it is easier to digest the enormous scale of 1 million tons. The fluxes are given in petagrams of carbon per year (PgC yr^{-1}). These budgets keep track of the mass of carbon, which is equivalent to keeping track of the number of carbon atoms, as they move from one form to another, even though they appear in different chemical compounds in the different reservoirs.

The ocean contains large amounts of carbon in various forms: dissolved inorganic carbon (i.e. the compounds into which CO_2 is converted when it is dissolved in sea water, largely carbonate and bicarbonate ions), dissolved compounds containing organic carbon, and particulate matter containing carbon. Of the total oceanic reservoir of 38 000 PgC, 900 PgC is held in the upper ocean, with a much larger portion in the intermediate and deep waters. Roughly 3 gigatons of the ocean carbon is in the form of marine biota, which has a strong effect in transferring carbon from the surface ocean to the deep ocean in the form of organic detritus. The land biomass reservoir, including vegetation, soils and vegetation detritus such as leaf litter on the forest floor, is substantially smaller than the ocean total, at 2300 PgC. The atmospheric preindustrial content weighs in at only 597 PgC. Finally, an important reservoir from our current perspective is the geological reserves of fossil organic carbon laid down in past geological eras. Although the size of these reserves is imperfectly known, it is large compared with the atmospheric reservoir. Estimates for coal, oil and natural gas exceed 4 000 PgC, and the estimated reservoir is much larger when non-conventional sources such as tar sands are included. Without human intervention, the interaction with this reservoir occurs very slowly.

Strong exchanges of carbon occur between the atmospheric reservoir and the land and ocean reservoirs. For instance, roughly 120 PgC per year are taken up from the atmosphere into plant material while respiration by plants, soil bacteria and animals returns approximately the same amount to the atmosphere each year. Similarly, about 70 PgC per year are taken up in the ocean as CO_2 from the atmosphere is dissolved and used by ocean biota, and another 70 are returned to the atmosphere on average. The inward and return fluxes do not necessarily occur in the same locations and the net flux can vary considerably from year to year, as discussed below.

As part of this overall balanced cycle, there are small net fluxes between some of the reservoirs. For instance, about 0.4 PgC per year is taken up from the atmosphere into land vegetation, compensated by 0.4 PgC per year of dissolved organic matter from land biomass carried by rivers to the oceans. Weathering processes, as minerals interact chemically with air and water, and as mountains and hills are eroded by rainfall, contribute another 0.2 PgC per year from each of the land surface and atmosphere, respectively. This is also carried to the oceans via river transport. About 0.6 PgC is released from the ocean to the atmosphere to complete this cycle, while 0.2 PgC per year are buried in ocean sediments, representing a slow loss process to the geologoical reservoir.

We turn now to the anthropogenic perturbation to the carbon cycle shown schematically in Figure 2.21b. The dominant effect is from fossil fuel burning. Fossil organic carbon that had long been sequestered from the climate system is being returned to the atmosphere at a rate of about 6.4 PgC per year. Roughly 40% of this comes from coal, another 40% from burning of oil and oil derivatives such as gasoline, with the remaining 20% from natural gas. The contribution from coal, which used to be slightly less than that of oil, has been climbing over the past decade, and now slightly exceeds that of oil. This anthropogenic flux includes a small contribution from cement production, roughly 0.1 petagrams carbon per year. Technically, this contribution is from rock carbonates, rather than fossil organic carbon, but since it is a small additional contribution from human industrial processes, it is convenient to lump it in with the fossil fuel perturbation to the carbon cycle. Another important source of carbon dioxide in the atmosphere is land-use change. Deforestation tends to put a considerable amount of carbon into the atmosphere, although the net effect depends on what type of land usage replaces the forest. For instance, burning to clear the forest for agriculture releases CO_2, partially counteracted by carbon uptake when the area regrows. However, when agricultural land replaces forest, the amount of carbon stored in the vegetation and soils tends to be smaller than when the region was forested. Overall, this results in a net emission of carbon dioxide into the atmosphere. The net effect of land-use change is estimated at 1.6 PgC per year.

Adding the contributions of fossil fuel burning and land-use change gives potential increase of about 8 gigatons of atmospheric carbon per year – if all of the human emissions remained in the atmosphere. Fortunately, the increase in atmospheric carbon dioxide leads to an increase in the flux into the ocean, as the surface ocean carbon concentrations increase toward equilibrium with the new atmospheric values. Further transfers occur into the deep ocean on longer time scales. The ocean uptake is approximately 2.2 PgC per year. Another approximately 2.5 PgC carbon per year is taken up in land vegetation. The details of where exactly in the land surface that carbon is taken up is as yet uncertain. The overall number is known as a residual from the estimates of how much is taken up by the ocean and the reasonably well-known flux into the atmosphere due to human activities. Overall about 4.7 PgC per year are taken up by the land surface and the ocean, i.e. over half of the carbon that humans emit into the atmosphere does not remain there. That is to say, the oceans and land surface have been buffering us from part of the potential impacts of our fossil fuel burning.

The fluxes in Figure 2.21 are given in terms of the mass of the carbon atoms, but two alternative measures of this flux are sometimes used. In the atmosphere, the carbon is principally in the form of carbon dioxide. Thus, human emissions are sometimes given in terms of their equivalent weight of carbon dioxide. In those units, the same emissions differ by a factor of 3.66 because this also counts the weight of the two oxygen atoms. To see this, note that the molecular weight of CO_2 is 44.01, while the atomic weight of carbon is 12.01; the ratio is 3.66. Thus if an amount of fuel containing 12 gigatons of carbon is burned, it produces 44 gigatons of CO_2 in the atmosphere. A second equivalence is perhaps more important. Atmospheric concentrations of CO_2 rise by about 1 ppm for each 2.1 gigatons of carbon that remains in the atmosphere. This is obtained by computing the number of molecules per gigaton of carbon compared to the number of molecules in the atmosphere. This conversion factor, 1 ppm CO_2 increase per 2.1 PgC, implies that the 3.3 PgC per year of human emissions that remain in the atmosphere in Figure 2.21b increase the CO_2 concentration by about 1.6 ppm per year.

These fluxes change as function of time owing to both human and natural effects. The values in Figure 2.21b are based on 1990s data. The anthropogenic fossil fuel emissions have been climbing, from around 2.5 PgC yr^{-1} in 1960 to 8.7 PgC yr^{-1} in 2008. The top curve in Figure 2.22 shows this increase, given as the carbon dioxide concentration change per year that this would imply if all the emissions remained in the atmosphere, using the conversion from PgC to ppm given in the previous paragraph. The actual accumulation rate is about 55% of this on average, with the remainder taken up by the oceans and terrestrial

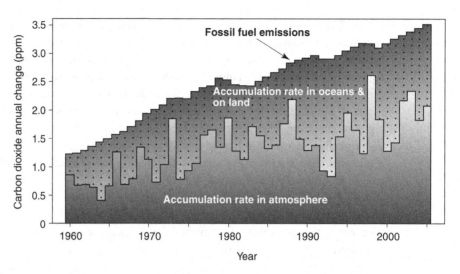

Fig. 2.22 Fossil fuel emissions and increases in atmospheric CO_2 concentrations. Top curve: rate of anthropogenic CO_2 emissions from fossil fuels given as the yearly increase in atmospheric CO_2 that would occur if these all remained in the atmosphere. Lower curve: rate of accumulation of CO_2 in the atmosphere, given as the observed increase for each year. The area between these curves (stippled) represents the accumulation in the ocean and in land carbon reservoirs. Values are from Denman *et al*. (2007); format follows Sarmiento and Gruber (2002).

biosphere. This fraction has remained fairly consistent from decade to decade so far, but exhibits considerable year-to-year variability. These interannual variations appear to be substantially due to variations in the functioning of the land biosphere. For example, a substantial contribution to the year-to-year variation is associated with El Niño. As El Niño teleconnections affect surrounding land areas with warmer temperatures and droughts, there tends to be a net release of carbon dioxide into the atmosphere on a temporary basis. This is compensated during cold phases of the ENSO cycle. These and other impacts of climate variations on the carbon dioxide uptake remind us that it is by no means guaranteed that the fraction of anthropogenic CO_2 taken up in land and ocean reservoirs will remain the same in future. There is current concern that this fraction may have decreased slightly over the past 50 years, implying that a larger portion of our emissions remain in the atmosphere. Thus, reducing uncertainty in the carbon cycle is the subject of much current research.[12]

It is worth underlining that the curves in Figure 2.22 represent rates of increase – constant values would represent a constant upward trend in atmospheric CO_2 concentrations. The increases over the decades imply that atmospheric CO_2 has been increasing at an ever faster rate because of emissions increases. To put some personal perspective on the global human emissions, 1 US gallon of gasoline yields about 9 kg of carbon dioxide, and it is not unusual in North American society to use, say, 1000 gallons in a year of driving. This would imply 9 metric tons of CO_2 emitted per year by one activity of one individual. Although this is higher than average, it makes it easy to see how the per capita emission of CO_2 is about 20 tons per person in the US, and roughly 10 tons per person in the United Kingdom, Germany or Japan. The world average per capita emission stood at almost 4 tons per person in 2002. This has been climbing in both the developed world and in developing nations.

Notes

1 The concept of separation of time scales is introduced informally here but there exists a set of powerful mathematical techniques based on multiple scales. Bender and Orszag (1978) provides a good introduction at the undergraduate level. These approximation techniques are sometimes used to create intermediate complexity models by systematic approximation to the equations used by GCMs.

2 The expression for the blackbody dependence on temperature T for each wavelength λ is $B = a\lambda - 5(e^{b/\lambda T} - 1)^{-1}$ (the Planck function), where a and b are constants. This is the basis of the two curves in Figure 2.5a which show B normalized so the maximum is the same.

3 Emissivity is defined as the ratio of the emission of a substance to that of a perfect blackbody. In Eq. (2.3) it is integrated across all wavelengths, but it is also defined for each wavelength. For a gas, emissivity is defined for small increments of distance of travel through the gas and an associated absorption coefficient is used to find absorption integrated along the path the radiation travels. Absorptivity is equal to emissivity (Kirchhoff's law). Hartmann (1994) provides a chapter on radiative transfer at the upper division undergraduate level.

4 There has been debate about speculated variations in solar forcing and their possible impact on global temperatures. The number of sunspots is known to have varied (for instance, there were very few sunspots in the late 1600s), so it is plausible that variation in solar intensity could have occurred. Although such forcings are extremely simple to incorporate in climate models when they are known, there is no accurate, directly measured data prior to the past few decades. See Hansen *et al.* (1998) or IPCC (2001) for an overview. Note that various sources give values of S_0 that differ by a few $\mathrm{W\,m^{-2}}$ because the total value is more difficult to measure than the relative variations. For solar irradiance data matched between different satellite missions, see Fröhlich and Lean (1998) or Fröhlich (2003).

5 Orbital effects have substantial impact on paleoclimate. For instance, the season at which perihelion occurs precesses slowly with a period of 23 000 years. Around 11 000 years ago, perihelion occurred in northern summer, resulting in about 10% more insolation for portions of the northern hemisphere in summer. Although the annual average insolation remains the same, the asymmetry in land masses between hemispheres can produce effects, for instance, on continental ice sheets. For discussion of ice age mechanisms and other paleoclimate variability at the undergraduate level see for example, Ruddiman (2001).

6 The contribution of clouds to solar reflection mentioned in the text is known as shortwave cloud forcing (SWCF) even though "forcing" is a misnomer for these highly interactive cloud effects. A longwave cloud forcing (LWCF) can likewise be defined from upward longwave radiation from clear or cloudy skies. Its value on global average is about $30\,\mathrm{W\,m^{-2}}$ (warming tendency). The clouds have an additional, indirect effect, shielding absorption and emission by greenhouse gases. This is of similar magnitude to the LWCF. See Kiehl and Trenberth (1997) for additional details.

7 The December net solar input in Figure 2.9 is maximum near 30° S because there is a great deal of solar radiation reflected in southern high latitudes. The incoming solar radiation has a broad maximum of over $500\,\mathrm{W\,m^{-2}}$ from 30° S to the pole, with over $550\,\mathrm{W\,m^{-2}}$ at the South Pole in December.

8 The term ITCZ is often applied to specific regions, such as the convection zone north of the equator in the eastern Pacific or in the Atlantic. Other regions of strong convection and surface convergence are known by other specific names such as the South Pacific convergence zone (SPCZ) or the South Atlantic convergence zone (SACZ). One solution for a general term would be to call all such regions convergence zones. The other solution, favored in this text, is tropical convection zones, or simply convection zones, since the presence of deep convection is the leading factor in formation of these regions.

9 Derivation of the westward intensification of ocean currents from the velocity equations in Chapter 3 is standard material in oceanography courses but is omitted here for brevity. See Pond and Pickard (1997).

10 The surface current map has been updated from classic maps such as in Sverdrup *et al.* (1942) by incorporating recent data as compiled by McPhaden *et al.* (1998) and Schmitz (1996a,b) and references therein.

11 The thermohaline circulation is highly simplified in the schematic in Figure 2.19, as in widely cited schematics by Broecker (Broecker 1987, 1990). The three-dimensional view is motivated by a version in Trenberth (1992), but the current systems are based on the careful compilation of Schmitz (1995; 1996a,b) which provides more detailed features and history for the interested reader. Schematics of the thermohaline circulation typically show only the sinking in the North Atlantic (which produces North Atlantic deep water, seen in Figure 2.19 as southward flow in the Atlantic) because it is relatively large – about 14 Sverdrup, where one Sverdrup $= 10^6$ cubic metres per second. The sinking in the Southern Ocean that produces Antarctic bottom water (schematized below the North Atlantic deep water) is roughly 4 Sverdrup but the overturning in the Southern Ocean is important for global change because it contributes to the role of the ocean in slowing global warming in this region.

12 Carbon cycle notes. For discussion of trends in CO_2 sinks see Le Quéré *et al.* (2009). Sarmiento and Gruber (2002) provides a succinct overview of the carbon cycle and carbon sinks; the textbook by Sarmiento and Gruber (2006) covers these at a graduate level accessible to advanced undergraduates. Denman *et al.* (2007) gives an authoritative chapter-length discussion, with careful treatment of uncertainties. For additional discussion and fossil fuel reserve estimates, see Falkowski *et al.* (2000); Sabine *et al.* (2004) list non-conventional fossil fuels such as tar sands as roughly an additional 6000 PgC. Contributions from land-use change are not included in the anthropogenic emissions in Figure 2.22 owing to uncertainties in the year-to-year fluxes, and because net effects of land-use change may not be apparent in a given year. For instance, forest clearing by burning in a given year may be compensated by regrowth in subsequent years. Atmospheric CO_2 data in Figure 2.22 are from observations by members of the Scripps Institution of Oceanography at the South Pole and Mauna Loa, Hawaii, updated from Keeling and Whorf (2005). One gallon of gasoline yields 8.87 kg of carbon dioxide (source: California Air Resources Board).

3 Physical processes in the climate system

Chapters 1 and 2 introduced the major players in the climate system, both for phenomena and processes. This chapter introduces more quantitative statements of the processes that shape climate and that are included in climate models. This will (i) provide tools that can be used to understand climate phenomena such as sea level rise, or El Niño motions, and (ii) provide the full set of equations used in a climate model. For students majoring in oceanic or atmospheric sciences, this constitutes an introduction to these equations, which will be used again in other courses. For students with other specialties, it can be very helpful to have a general feel for what these equations represent when it comes to understanding how a climate model is constructed. While traditional presentations emphasize understanding the equations per se, here the physical processes essential for climate applications are emphasized. An application is included for each process. These applications use approximations to the full equations to focus on the main balance needed to understand the process.

The equations are basically budgets of conserved quantities. The equations for a climate model presented here are: Newton's law (conservation of momentum), the thermodynamic equation (conservation of energy), conservation of mass (applied to air, salt and moisture), plus an equation expressing the effects on the density of temperature, pressure, etc. in air or water.

3.1 Conservation of momentum

The winds and currents are simply governed by Newton's law

$$ma = F \tag{3.1}$$

where a is acceleration and F is the total force acting on a body of mass m. In the atmosphere or ocean, the body will be a *parcel* of air or water. Since the law has to apply to every part of the fluid, we can divide it up into smaller parcels with imaginary boundaries. It is convenient to deal with density rather than the mass of these arbitrary parcels, so we divide both sides of Eq. (3.1) by mass. From here on forces will be in units of force per unit mass. The acceleration is the rate of change of velocity of the parcel with time, so

$$\frac{d}{dt}\text{velocity} = \text{Coriolis} + \text{PGF} + \text{gravity} + F_{drag} \tag{3.2}$$

where PGF is the pressure gradient force and F_{drag} denotes friction-like forces due to turbulent drag (mixing of momentum between neighboring parcels) or to surface drag on

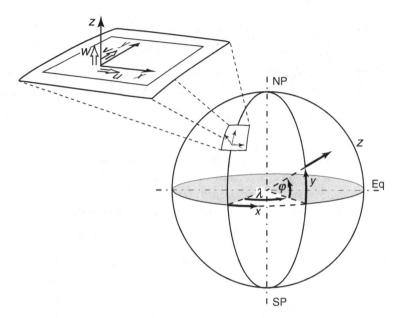

Fig. 3.1 Coordinate system for directions and velocities. The blown-up region shows how a local Cartesian coordinate system is defined for each region of the sphere. It also shows the velocity components u, v, w.

air parcels blowing along the surface. The Coriolis force is due to the rotation of the Earth and appears as a force because we choose our frame of reference to be fixed to the (rotating) surface of the Earth. Velocity is a three-dimensional vector, so this equation amounts to three equations for the components of velocity in each direction: eastward, northward and upward.

Figure 3.1 shows the coordinate system in which components of the velocity and forces will be expressed. The coordinate system is chosen with atmospheric and oceanic applications in mind, so it rotates with the Earth, and the vertical direction (z) is defined as upward, i.e. opposite the local direction of gravity on a surface at mean sea level. Latitude (ϕ) is defined as the angle relative to the plane of the equator, and longitude (λ) an angle relative to the longitude circle (or meridian) that passes through Greenwich, England. Although these are spherical coordinates, by defining horizontal coordinates x and y as distance northward from the equator (or other reference latitude) and eastward from a reference longitude along the Earth's surface, a local Cartesian coordinate system is created. Eastward, northward and upward components of velocity are denoted u, v and w, respectively. The equations will be expressed in this local Cartesian coordinate system for simplicity, occasionally neglecting some of the smaller terms associated with spherical geometry for clarity, although these are of course included in climate models. Distance or motion in the north–south direction is often termed *meridional*, while the term *zonal* is used for the east–west direction.

3.1.1 Coriolis force

The Coriolis force is an apparent force that acts on moving masses in a rotating reference frame. Since the Earth rotates once per day, the Coriolis force is negligible for motions (such

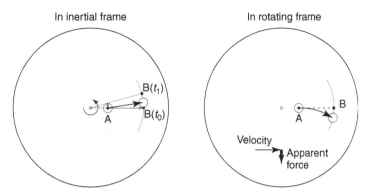

In inertial frame   In rotating frame

Fig. 3.2   Schematic of the Coriolis force. The left panel shows the motion of a body from the point of view of an observer in a non-rotating frame above the North Pole. Points A and B are fixed on the Earth's surface. At time t_0, a body is at point A with a velocity relative to the Earth that would carry it to point B by time t_1. It also has a velocity due to the Earth's rotation; combining the two velocities gives the heavy arrow. Viewed from the inertial frame, the body moves in a straight line, but by the time it arrives, the Earth's surface has rotated carrying point B with it. The right panel shows this from a point of view rotating with the Earth. The body appears to be deflected to the right by an apparent force acting perpendicular to the velocity.

as walking or driving) in which other forces, such as frictional effects, operate on much shorter time scales. For large-scale atmospheric or oceanic motions, however, frictional forces are much smaller, and the Coriolis force becomes a leading effect.

To an observer in the rotating frame, such as the surface of the Earth, the motion of a body appears to curve as if a force were acting. Figure 3.2 shows how this occurs from two points of view looking down on the Earth from space above the North Pole, one rotating with the planet, the other not. Although a body moves in a straight line viewed in the non-rotating frame, points on the surface rotate during the time of its motion. As a result, from a point of view in the rotating frame, the body appears to be deflected to the right of its motion. Here "to the right" means when facing along the direction of motion, with a head-upward convention relative to the surface of the Earth. If we consider what this implies in the southern hemisphere, imagine flipping Figure 3.2 over so one has the point of view looking downward from the South Pole. The direction of rotation will appear opposite since the viewer has changed orientation, and the same process will appear as an apparent force deflecting the body to the left. The Coriolis force applies only to bodies moving relative to the rotating frame.[1] It is proportional to this relative velocity times the rotation rate of the frame.

For most Earth system studies, it is only the horizontal component of the Coriolis force that matters. The vertical component is tiny compared with gravitational force and may for many purposes be neglected. The schematic in Figure 3.2 still applies if one considers the circle to represent the whole Earth, rather than simply a region near the pole. A body rolling on the surface of the Earth (neglecting friction) will experience the same type of apparent force shown, although it will be constrained to move on the spherical surface, held by gravity. If the motion is at the equator, the apparent force will be upward and will not affect the horizontal motion. The main properties of the Coriolis force may be summarized as:

- It turns a body or air/water parcel to the right (facing along direction of motion, head upward) in the northern hemisphere; to the left in the southern hemisphere.
- Exactly on the equator, the horizontal component of the Coriolis force (associated with horizontal motions) is zero.
- It acts only for bodies moving relative to the surface of the Earth and is proportional to velocity.
- The constant of proportionality is $f = (4\pi/1 \text{ day})\sin(\text{latitude})$, known as the *Coriolis parameter*; f is positive in the northern hemisphere, negative in the southern hemisphere, and zero at the equator.
- The northward force is $-fu$ (to the right of the eastward wind component).
- The eastward force is fv (to the right of the northward wind component).

The Coriolis force is so important in atmospheric and oceanic motions that even the rate of change of the Coriolis parameter with latitude has a symbol:

$$\beta = \frac{df}{dy} \tag{3.3}$$

The effect of this on motions is called the "*beta effect*." Since β is proportional to cos(latitude), it is always positive and maximum at the equator. Thus even though the Coriolis force is zero exactly at the equator, Coriolis effects can be very important even in equatorial dynamics.

3.1.2 Pressure gradient force

The pressure gradient force tends to accelerate air from higher toward lower pressures. Pressure is force per unit area. Consider two regions a small distance δx apart. If the pressure on both is the same, the pressure force will be in balance between the two regions and no net force will result. If there is a pressure difference δp in the x direction, as illustrated in Figure 3.3, the pressure gradient $\delta p/\delta x$ gives the force per volume in that direction. To get force per mass, divide by the density, ρ (mass per volume). Thus the pressure gradient force per mass in the x direction is $-\frac{1}{\rho}\frac{\delta p}{\delta x}$. Taking the limit of small δx gives derivatives. Considering a rate of change in x only is simply a partial derivative in x, $\partial p/\partial x$.

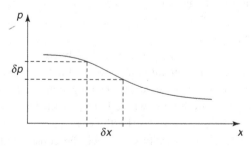

Fig. 3.3   Schematic of the pressure gradient. For pressure decreasing with x, the pressure gradient in the x direction is negative, and the pressure gradient force is positive, from high to low pressure.

The same applies in the y direction. If a curve similar to Figure 3.3 were plotted in the y direction, the changes in that direction alone would be the partial derivative in y, $\partial p/\partial y$. For a given pressure map, moving a unit increment in the x direction or in the y direction will each be associated with a different pressure change, so $\partial p/\partial x$ and $\partial p/\partial y$ provide two independent components of the pressure gradient. The magnitude of the gradient is defined along the direction of maximum rate of change.

3.1.3 Velocity equations

Using the above, the *horizontal velocity equations* are

$$\frac{du}{dt} = fv - \frac{1}{\rho}\frac{\partial p}{\partial x} + F^x_{drag} \tag{3.4}$$

$$\frac{dv}{dt} = -fu - \frac{1}{\rho}\frac{\partial p}{\partial y} + F^y_{drag} \tag{3.5}$$

Since Newton's law expresses conservation of momentum, these are also called the *horizontal momentum equations*. The time derivatives give the acceleration for a parcel of air. The implications of following the parcel are discussed in section 3.3.5. The terms F^x_{drag} and F^y_{drag} denote turbulent drag on the flow (in the x and y directions, respectively) due to mixing of momentum by small-scale motions. Because these arise at smaller scales than can be resolved in climate models (or even regional weather models), they must be parameterized. Typically they depend on the gradients of the wind, on its velocity and on atmospheric stratification (vertical density gradient). These terms also give the surface stress effects of the wind on the ocean.[2] The surface drag tends to slow the winds and to accelerate the ocean currents since the winds are much faster.

3.1.4 Application: geostrophic wind

At large scales at midlatitudes, and even approaching the tropics, the Coriolis force and the pressure gradient force are the dominant forces. The balance between these, neglecting the beta effect, friction, and acceleration, gives a steady balance where the flow goes around lows (or highs) just fast enough to balance the PGF. This is termed the *geostrophic balance*, and the velocity that obeys this is termed the *geostrophic flow* (or geostrophic wind or current). This is why on weather maps wind blows counterclockwise around low pressure regions in the northern hemisphere (clockwise in the southern). If the air is initially at rest, the PGF will accelerate it toward the low. As it picks up speed, the Coriolis force will turn it to the right, until eventually it reaches a speed where the Coriolis force is strong enough to balance the PGF. It will then blow approximately parallel to lines of constant pressure, rather than across them. The components u_g and v_g of the geostrophic flow are given by

$$fu_g = -\frac{1}{\rho}\frac{\partial p}{\partial y} \tag{3.6}$$

$$fv_g = \frac{1}{\rho}\frac{\partial p}{\partial x} \tag{3.7}$$

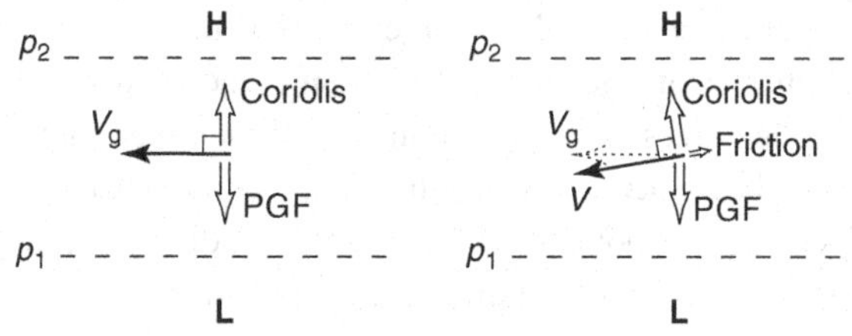

Fig. 3.4 Schematic of geostrophic wind v_g and "true" wind v as a result of balances between the pressure gradient force, the Coriolis force and frictional effects of turbulent drag. H and L denote regions of high and low pressure at a given height.

The eastward component of the geostrophic wind is particularly important in both atmosphere and ocean. The wind cannot be perfectly geostrophic or air would not flow into low pressure regions at all. For instance, the trade winds occur between the subtropical region of high surface pressure and the lower pressure near the equator. If the geostrophic balance held perfectly and pressure changed only with latitude, the trade winds would blow purely westward, and not partly toward the equator at low levels as illustrated in Figure 3.4, which would apply for the area between the equator and the subtropical high to the north. Surface drag tends to slow the wind slightly, so the Coriolis force does not perfectly balance the PGF. This causes some turning toward the low pressure region. Below we will connect the pressure gradient to the temperature gradient, and thus to the basic thermal driving of the system.

3.1.5 Pressure–height relation: hydrostatic balance

In the vertical direction, the velocity equation includes the gravitational force per unit mass, given by the gravitational acceleration $g = 9.81\,\mathrm{m\,s^{-2}}$ (the atmosphere is so thin that variations of this in the vertical are negligible). For motions with large horizontal scales, the vertical acceleration term is much smaller than this. For models of squall lines, thunderstorms and other intense, small-scale features, the vertical acceleration term would be retained, but for global models it can be dropped. Thus the vertical velocity equation just becomes a balance between vertical pressure gradient and gravity

$$\frac{\partial p}{\partial z} = -\rho g \tag{3.8}$$

This implies that the pressure at each level in atmosphere or ocean is just given by the amount of mass above it.[3]

3.1.6 Application: pressure coordinates

Since pressure and height are in a one-to-one relation, pressure can be used as a vertical coordinate instead of height in the atmosphere. Because pressure levels are related to mass, this has several advantages for atmospheric motions. The pressure gradient along surfaces of constant height is obviously related to the gradient of height along surfaces of constant pressure. Using Eq. (3.8), the pressure gradient force term $\rho^{-1}(\partial p/\partial x)$ in the velocity equations is just replaced by $g(\partial z/\partial x)$, and similarly for y. Noticing that the height is multiplied by g, and recalling that gz is the gravitational potential energy (per unit mass) of a body above the Earth's surface, let us define this as the *geopotential*

$$\Phi = gz \tag{3.9}$$

If we choose to measure winds along pressure surfaces, then the velocity equations, Eq. (3.4) and Eq. (3.5), look the same except that now gradients of geopotential appear.[4]

3.2 Equation of state

An equation of state relates the density to temperature, pressure and other factors for a particular substance. The atmospheric equation of state is used frequently in atmospheric applications because the density changes with pressure are large. Fortunately it is simply the ideal gas law.

3.2.1 Equation of state for the atmosphere: ideal gas law

For the atmosphere, the ideal gas law is used in the form

$$\rho = \frac{p}{RT} \tag{3.10}$$

where $R = 287\,\mathrm{J\,kg^{-1}\,K^{-1}}$ is the ideal gas constant for air, and the temperature T is in kelvin.[5]

3.2.2 Equation of state for the ocean

For the ocean, the density depends on temperature in a more complicated way. For most temperatures, density decreases with temperature, and more so for warmer water. Near the freezing point, water actually gets slightly less dense as it gets colder (consistent with ice being lighter than water). Density increases with salinity (for example, people float well in the Dead Sea). Pressure also affects density, although not nearly as much as in air since water is much less compressible. Computer models use look-up tables or polynomial fits to measurements of this relation. For the sake of keeping track of the equations required in a climate model, call this function $\wp(T, S, P)$ where S is salinity.

$$\rho = \wp(T, S, P) \tag{3.11}$$

One can define a *coefficient of thermal expansion* ϵ_T for sea water. This is just the percent decrease in density per degree of temperature increase. A typical value for the upper ocean is $\epsilon_T = 2.7 \times 10^{-4}\,{}^\circ\mathrm{C}^{-1}$ (which is appropriate for $T_0 = 22\,{}^\circ\mathrm{C}$ and upper ocean pressures). This implies that for a $1\,{}^\circ\mathrm{C}$ temperature increase water becomes 0.027% less dense.

For the deep ocean a typical value is $\epsilon_T = 1.5 \times 10^{-4}\,°C^{-1}$ (appropriate for $T_0 = 2\,°C$ and pressure of 300 bar).

For small changes in temperature and density relative to reference values T_0 and ρ_0 and for a case where salinity does not change significantly, one can use a linear approximation to Eq. (3.11):

$$\rho = \rho_0[1 - \epsilon_T(T - T_0)] \tag{3.12}$$

A typical upper ocean value of density ρ_0 is $1.03 \times 10^3\,kg\,m^{-3}$.

3.2.3 Application: atmospheric height–pressure–temperature relation

Since density depends on pressure as well as temperature in the atmosphere, the ideal gas law makes the hydrostatic equation much more useful. Combining the two by substituting Eq. (3.10) into Eq. (3.8) gives

$$\frac{\partial \ln p}{\partial z} = -\frac{g}{RT} \tag{3.13}$$

where we have used $\frac{\partial \ln p}{\partial p} = p^{-1}$. Recalling that temperature is in kelvin, to a rough approximation T is constant (i.e. to the extent that 300 K and 250 K can be considered similar). Then Eq. (3.13) just gives

$$p = p_0 e^{-z/H}, \quad H = \frac{RT}{g} \tag{3.14}$$

where H is known as the scale height. This is why pressure and density drop off approximately exponentially with height, as seen on the axes of Figure 2.11. To picture this physically, consider it as a feedback in the hydrostatic Eq. (3.8). Starting at a surface pressure and moving up in height, pressure will decrease, since there is less mass above, so density will decrease. In the next increment of height, pressure will therefore decrease a bit less, so density will decrease a bit further, and so on. Because the density decreases gradually at high altitudes, the "top" of the atmosphere is less clearly defined than the surface of the ocean.

The relation of height and pressure becomes more interesting when we consider temperature variations, since that is what drives the atmospheric circulation. Multiply Eq. (3.13) by RT and integrate in z, and recall that gz is the geopotential Φ:

$$\Phi - \Phi_0 = \int_p^{p_0} RT\,d(\ln p) \tag{3.15}$$

where Φ_0 is the geopotential at pressure level p_0. Thus the separation in height between two pressure surfaces will be greater where the column is warm. This occurs because warm air is less dense, so a greater column height is needed to create a given pressure difference. This can be used to explain thermal circulations.

3.2.4 Application: thermal circulations

The concept of a "*thermal circulation*" (not the same as "thermals," the small-scale rising motions driven by surface heating) is usually applied to small-scale circulations in the atmospheric boundary layer, such as the sea breeze. In a modified form, it is actually a powerful explanation of the major circulations in the tropics of Earth's climate. Heating at the Earth's surface is transferred upward through about 1 km by thermals, or, if it is strong enough, through about 10 km by deep convective motions. In either case, the effect is to make the atmosphere warm through a relatively deep layer. Consider the hydrostatic equation (with ideal gas law) in either form Eq. (3.13) or Eq. (3.15). Two pressure surfaces will have a greater separation in height if the layer of air between them is warm, as illustrated in Figure 3.5. This is simply because warm air is less dense and so a given mass (per area) occupies a greater vertical distance, relative to a neighboring region of cold air. Thus in a convective region over warm sea surface temperatures, there will tend to be a surface low, and an upper-level high (in terms of pressure at a given height), relative to a neighboring region where the air column is cooler, since it is being warmed less strongly from below. Two factors affect the strength of the circulation: the temperature gradient between the regions of course, but also the depth (in pressure or in height) of the layer in which temperature differences are occurring. Thus deep convection tends to favor stronger circulations.

In a sea breeze circulation, the air over the land warms during the day, resulting in relatively lower pressure over land near the surface (and higher pressure at about 1 km height) than over the neighboring ocean. The surface wind blows toward the low pressure, rises and returns at the top of the boundary layer. The entire Hadley cell is basically similar. The warmer part of the column in Figure 3.5 is the part heated by deep convection over warm SSTs in the tropics. The tropics have lower surface pressure than the cooler subtropics,

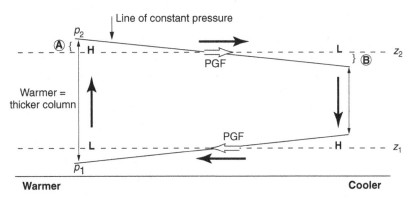

Fig. 3.5 Schematic of a thermal circulation. Solid curves give surfaces of constant pressure at pressures p_1 and p_2 with $p_1 > p_2$. Dashed curves give surfaces of constant height at lower level z_1 and upper level z_2. Temperature affects the density and thus the thickness of the column between two pressure surfaces. At A, the pressure p_2 occurs at a greater height than z_2. Since pressure increases downward, the pressure at z_2 is relatively high. Likewise at B, the increment by which p_2 lies below z_2 corresponds to a lower pressure at z_2. Relative high and low pressure at a given height are indicated by **H** and **L**.

but at upper levels, the reverse is true, resulting in the poleward flow at upper levels and equatorward flow at the surface. The main difference between the Hadley circulation and a sea breeze is that for the Hadley circulation the Coriolis force is important, so the winds turn and have a large component that runs perpendicular to the temperature gradient. This is particularly important at upper levels where there is no surface friction to slow this geostrophic component down. The subtropical jet stream is the result of this. The trade winds at the surface blow from the subtropical high pressure regions to the warmer, and hence lower pressure, tropical convergence zones, turned westward by the Coriolis force, but not running perfectly along the pressure surfaces because of the surface friction.

The monsoons and the Walker circulation have similar thermal circulation effects. As the solar heating of the land surface moves with season, the atmospheric column is warmed, and the pressure distribution follows. Thus wind patterns change with the convection zones. Similarly, during El Niño, the pattern of warm surface temperature is quickly transferred up through the tropospheric column, leading to the changes in surface pressure, and hence in the surface winds.

To summarize: a surface low (pressure) and an upper-level high will tend to occur over a region where the atmospheric column is being warmed, for instance, by warm SST. The wind pattern will be complicated by the relative importance of surface friction and the Coriolis force, but will be generally consistent with the pressure pattern.

It should be noted, however, that the pressure distribution is not completely determined by the pattern of heat sources, so the lowest pressure does not always conform perfectly to the region with warmest surface temperatures. The winds transport mass and heat, and thus affect pressure. Thus while the thermal circulation gives an excellent qualitative picture of the motions, it is also necessary to employ a thermodynamic equation and mass conservation for a complete picture.

3.2.5 Application: sea level rise due to oceanic thermal expansion

We can obtain a rough feeling for how much sea level rise to expect under global warming simply due to the thermal expansion of sea water. For an accurate calculation, one should use the full equation of state of sea water, and consider horizontal variations of temperature and other factors, but for a first estimate we just use the thermal expansion coefficient ϵ_T and approximate the warming as constant in the horizontal.

Consider a column of water of depth h. This could be the full depth of the ocean, or just a part of the depth, e.g. the deep ocean or upper ocean. The mass of water in this column is approximately $\rho h \times$ area, where ρ is the density. If the mass and area of the ocean column remain constant, then $\rho h = $ constant. Density decreases due to warming must be balanced by increases in h. For small changes $\delta \rho$ and δh, this implies

$$h\delta\rho + \rho\delta h = 0 \qquad (3.16)$$

Since ϵ_T gives the fractional decrease in density, we have $\delta\rho = -\rho\epsilon_T\delta T$, where δT is the change in temperature, yielding

$$(\delta h)/h = \epsilon_T\delta T \qquad (3.17)$$

If a 300 m deep layer of the upper ocean were to warm by 3 °C, this would imply only about a quarter of a meter rise in sea level. A 3 °C warming of a 4 km deep ocean layer would imply an almost 2 m rise in sea level, for the values of ϵ_T given in section 3.2.2.

The deep ocean warms more slowly than the upper ocean under global warming. A consequence of this is that a substantial part of sea level rise is delayed relative to surface warming. More detailed projections of sea level rise by the end of the century are discussed in Chapter 7.

3.3 Temperature equation

The *thermodynamic energy equation*, commonly referred to as the temperature equation since it gives the time rate of change of temperature, is a form of the first law of thermodynamics. It expresses the conservation of thermodynamic energy as it is converted between internal energy of the gas, work of expansion and input of heat into an air or water parcel.

3.3.1 Ocean temperature equation

In the ocean, the temperature equation simply balances the heat capacity times the rate of change of temperature of a water parcel against diabatic heating of the parcel,

$$c_w \frac{dT}{dt} = Q \qquad (3.18)$$

where c_w is the heat capacity of water, approximately $c_w = 4200 \, \mathrm{J \, kg^{-1} \, K^{-1}}$, and Q is heating in $\mathrm{J \, kg^{-1} \, s^{-1}}$.

The heating is due to heat fluxes at the surface and mixing of the heat down into the ocean. If we integrate through a surface layer of depth H, in the vertical, the relation to the fluxes is[6]

$$\rho c_w H \frac{dT}{dt} = F_{sfc}^{net} - F_{bottom}^{net} \qquad (3.19)$$

where F_{net} is net heat flux in $\mathrm{W \, m^{-2}}$. The subscript *sfc* denotes the surface of the ocean, while *bottom* here denotes the bottom of the surface layer. This form of the equation will prove useful when considering how the thermal inertia of the upper ocean can affect response to warming.

3.3.2 Temperature equation for air

The temperature equation for air is similar to the ocean temperature equation (3.18), but with one added term. This is because of the compressibility of air (i.e. pressure can change the density). The added term, $\frac{1}{\rho} \frac{dp}{dt}$, depends on rate of change of pressure with time, since it is due to the work done in expanding or compressing the air parcel when changing the pressure. The equation is

$$c_p \frac{dT}{dt} - \frac{1}{\rho} \frac{dp}{dt} = Q \tag{3.20}$$

where c_p is the heat capacity of air at constant pressure and Q is heating, discussed below. To visualize the effects of pressure, consider an imaginary column that is roughly the same temperature all the way up. Pressure is decreasing with height, according to hydrostatic balance, Eq. (3.8). When an air parcel is moved upward (to a lower pressure), the volume of the parcel must be larger according to the ideal gas law, Eq. (3.10). This expansion is done against the pressure force from surrounding air. The work of expansion must come from the internal energy of the gas, which is measured by the temperature, so the temperature must drop with each change in pressure. The heating term is made up of the various terms that we saw in the global energy budget

$$Q = Q_{solar} + Q_{IR} + Q_{convection} + Q_{mixing} \tag{3.21}$$

The heating is related to the fluxes by vertical integrals in pressure, since $dp/g =$ (increment of mass per area) in the vertical. This is the same as multiplying by density in the ocean case Eq. (3.19) and integrating in z. In the atmosphere, for instance,

$$\int_{sfc}^{top} Q_{IR} \frac{dp}{g} = F_{IR}^{top} - F_{IR}^{sfc} \tag{3.22}$$

This simply states that the heating integrated over the column is given by the difference in a given flux into the atmosphere at the top minus the flux lost out of the bottom of the atmosphere, at the surface into the ocean.

3.3.3 Application: the dry adiabatic lapse rate near the surface

As seen in Figure 2.11, temperature decreases as a function of height in the troposphere. The rate of temperature decrease with height is termed the *lapse rate*. In the atmospheric boundary layer, roughly the lowest 1 km of the atmosphere, the lapse rate is about $10\,°C\,km^{-1}$. This occurs because of the compressibility of air and the effects in dry convection (i.e. without condensation of water into clouds). The compressibility effect (the $\frac{dp}{dt}$ term in Eq. (3.20)) causes temperature to change when a parcel moves vertically, even if is not being heated or cooled. Motions without any heat exchange are termed *adiabatic*; motions with heat exchange are termed *diabatic*. Because convective motions tend to occur quickly (hours), heating rates other than condensational heating can often be neglected in determining the temperature of a small air parcel as it rises.

How does this help with lapse rate? We can define a lapse rate for an air parcel moving adiabatically, the *adiabatic lapse rate*. If no condensation occurs (i.e. no clouds form; the parcel can contain water vapor) this is called the *dry adiabatic lapse* rate,[7] which turns out to be $10\,°C\,km^{-1}$. This will not be the same as the environmental lapse rate of the large-scale atmosphere (which is referred to as the *environment* since it gives the conditions surrounding a rising convective parcel). However, the adiabatic lapse rate can affect the

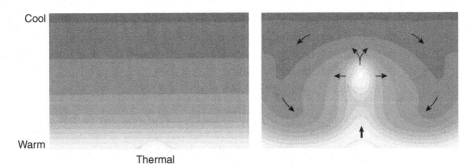

Cool

Warm

Thermal

Fig. 3.6 Schematic of dry convection (thermals). The air near the surface is warmed from below and becomes buoyant. It rises, with its temperature decreasing at the dry adiabatic lapse rate. This process tends to mix the layer toward the dry adiabatic lapse rate.

environmental lapse rate in regions where many convective parcels are moving up and down adiabatically.

One more step is needed to explain why the observed environmental lapse rate in the atmospheric boundary layer is often close to the adiabatic lapse rate. Since the atmosphere is being heated from below, air parcels warmed near the surface tend to be warmer than surrounding air. Since they are less dense they rise buoyantly as thermals, as indicated in Figure 3.6. An air parcel that is rising fairly quickly will have relatively little heat input or loss as it ascends, and so it will move approximately adiabatically. For an air parcel rising from the surface, temperature will drop at the dry adiabatic lapse rate of $10\,^{\circ}\mathrm{C}\,\mathrm{km}^{-1}$. As the parcel ascends, surrounding air parcels must descend to compensate, also at the dry adiabatic lapse rate. The parcels will eventually tend to mix with the environment, so the result of these motions is to adjust the environment toward the dry adiabatic lapse rate.

3.3.4 Application: decay of a sea surface temperature anomaly

Suppose we have an SST anomaly T_s'. A useful case to consider is one where there are negligible anomalous fluxes through the bottom of the surface layer. Subtracting the climatology of SST and fluxes from Eq. (3.18), the equation for the SST anomaly is

$$\rho c_w H \frac{dT_s'}{dt} = F_{sfc}^{net'} \tag{3.23}$$

When SST is warmer, evaporation, infrared and sensible heat fluxes out of the surface all increase so there is a negative heat flux anomaly $F_{sfc}^{net'}$ into the ocean. This negative heat flux anomaly is approximately proportional to the temperature anomaly since the warmer the SST, the more cooling occurs. This can be represented as

$$F_{sfc}^{net'} \approx -\gamma T_s' \tag{3.24}$$

Because the magnitude of the cooling by the heat fluxes depends on the warmth of the SST in a way that reduces SST anomalies, this is an example of a *negative feedback*.

The coefficient γ gives the strength of the negative feedback. In practice it depends on a number of factors including the spatial extent of the SST anomalies. For anomalies within an ocean basin a typical value would be $\gamma \approx 20\,\mathrm{W\,m^{-2}\,K^{-1}}$. For global temperature changes, the value would be much smaller.

When Eq. (3.24) is applied, the solution for SST is

$$T'_s = T'_{s0} e^{-t/\tau} \tag{3.25}$$

where T'_{s0} is the initial anomaly at $t = 0$, and the time scale $\tau = \rho c_w H/\gamma$ gives the exponential decay time. For surface layer depths on the order of 50–100 m the time scale is on the order of months. For a deeper layer, or weaker negative feedback, the time scale could be larger.

Many of the SST anomalies that one sees in the world oceans in satellite observations are produced in a very simple manner. Short-lived weather events provide an extra heat flux effect in addition to the SST-dependent part in Eq. (3.24). After passage of a weather event that has anomalously cooled or warmed the surface layer, the SST returns gradually toward normal over a period of months as described by Eq. (3.25).

3.3.5 Time derivative following the parcel

The momentum and temperature equations look fairly simple. Why are they hard to solve? And where are the heat and momentum transports that were emphasized in Chapter 2? The transports, along with many of the challenges in these equations, are hidden in the d/dt terms. The total derivative d/dt is defined *following* the air parcel as it moves. The time rate of change in a particular location, i.e. with x, y and z fixed, is given by the partial derivative $\partial/\partial t$. This is related to the total derivative following the parcel, in the case of temperature, by

$$\frac{dT}{dt} = \frac{\partial T}{\partial t} + u\frac{\partial T}{\partial x} + v\frac{\partial T}{\partial y} + w\frac{\partial T}{\partial z} \tag{3.26}$$

The terms involving velocity components multiplied by gradients (of temperature) are known as *advection* terms (from latin *vehere*, to carry), since they carry properties (in this case temperature) from one region to another.

For instance, if cold air is flowing from due west over Los Angeles, local temperature is dropping even if the temperature as measured by a balloonist travelling with the air is constant. The temperature of the air parcels being constant means $dT/dt = 0$, but the local temperature change is then given by $\partial T/\partial t = -u\,\partial T/\partial x$. Since the gradient of temperature is positive if cold air is to the west, and u is positive if the wind is from the west, the local rate of change will be negative, i.e. it will be getting colder.

The same applies to the total derivatives in the momentum equation and in the moisture equation treated below. To see how these advective effects make motions complicated, consider what happens to an air parcel in even a relatively simple wind pattern. Figure 3.7 shows a classic example of a set of adjacent air parcels, indicated in black or white, being deformed by the wind shear. Even though each parcel perfectly conserves its property (black

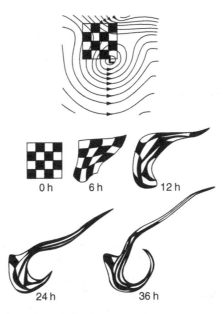

Fig. 3.7 Schematic of an initially simple set of air parcels being deformed in the flow pattern shown in the first panel. After Welander (1955).

or white), the pattern is so complex, it is difficult to distinguish boundaries. In a more complex wind pattern, the results can be even more complicated. And yet these are the air parcels that we are trying to follow with the total time derivative. If you now consider that the quantity being advected by the wind also feeds back to affect the wind (e.g. temperature affects the pressure and the pressure affects the winds), it is easy to imagine that chaotic motions will result, as indeed they do.

3.4 Continuity equation

Conservation of mass

Density is the most convenient measure of mass at a given place in the atmosphere or ocean, but if the volume changes then density changes, even though the mass of air or water is conserved. The rate of change of volume is given by the divergence, D_{3D} (measured in all three dimensions). We often talk about horizontal divergence or convergence, which considers just the part due to horizontal motions, so the subscript 3D is to recall that this includes the effects of vertical motions.

Since mass = (density × volume) is conserved, divergence reduces density according to

$$\frac{d\rho}{dt} = -\rho D_{3D} \qquad (3.27)$$

3.4.1 Oceanic continuity equation

In fact, we can use much simpler equations. In the ocean the changes of density are small enough to neglect and we can use the approximation $D_{3D} = 0$. Letting D be the *horizontal divergence*, i.e. the tendency of a parcel to expand in the two horizontal directions, gives an *oceanic continuity equation*[8]

$$D = -\frac{\partial w}{\partial z} \tag{3.28}$$

where D is defined in terms of the horizontal currents as

$$D = \frac{\partial u}{\partial x} + \frac{\partial v}{\partial y} \tag{3.29}$$

which gives the tendency of a volume to expand or contract owing to changes in u and v across it. If density remains constant then conservation of mass implies that the horizontal divergence must be balanced by suitable changes in vertical motions, as stated in Eq. (3.28). Figure 3.8b illustrates such a case: fluid is diverging in the horizontal, so additional mass is being supplied from below. In the case shown, horizontal divergence is positive and $\partial w/\partial z$ is negative to balance.

A special case of the oceanic continuity equation is the *surface height equation* which is obtained by integrating Eq. (3.28) through the full depth of the ocean.[9] Vertically integrated divergence through the depth of the ocean at a given location thus gives the rate of change of the surface height.

3.4.2 Atmospheric continuity equation

An equation of equal simplicity applies in the atmosphere if we use pressure coordinates. Let ω be the vertical velocity in pressure coordinates (i.e. the rate of rising across pressure surfaces), and let D be horizontal divergence defined along pressure surfaces. Then the *atmospheric continuity equation* is

$$D = -\frac{\partial \omega}{\partial p} \tag{3.30}$$

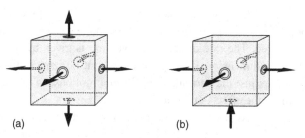

(a)   (b)

Fig. 3.8   Schematic of (a) a parcel with diverging motions (in 3D); (b) a parcel that has horizontal divergence, but for which the three-dimensional divergence is zero. This would apply in the oceans, for instance in the case of upwelling.

which looks the same as the oceanic case. Since pressure surfaces are mass surfaces according to Eq. (3.8), this should not be surprising. Any mass brought in by horizontal convergence (along pressure surfaces) must be balanced by suitable changes in the upward motion (across pressure surfaces). For instance, low-level convergence in tropical convection zones is balanced by rising motion. Note that because pressure increases downward, ω is negative for rising motions.

A special case of the atmospheric continuity equation is the *surface pressure equation* which is obtained by integrating Eq. (3.30) in pressure through the full depth of the atmosphere.[10] The rate of change of surface pressure is then given by vertically integrated divergence.

3.4.3 Application: coastal upwelling

The trade winds have a component that blows parallel to the coast of Peru. Let us idealize this to a north–south coast with a wind blowing northward along it as in Figure 3.9. We can make use of the velocity equations and the oceanic continuity equation to give some insight into what must happen.

Neglecting the d/dt term in the v equation, and assuming nothing changes very quickly in the north–south direction, Eq. (3.5) becomes

$$fu \approx F^y_{drag} \tag{3.31}$$

where F^y_{drag} contains the northward drag of the wind stress which tends to accelerate the ocean currents. It also contains a drag at the bottom of the surface layer that tends to slow the currents. Near the surface and away from the coast, driving by wind stress dominates. The wind is northward in this case, so F^y_{drag} is positive. Since the region is south of the equator, f is negative. Thus u will be negative (away from the coast). This yields a horizontal divergence out of a region near the coast (D positive). To balance this divergence there must

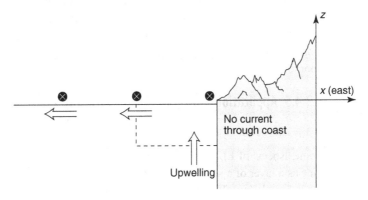

Fig. 3.9 Schematic of a northward wind component along a north–south coast (an idealized coast of Peru). We consider the divergence from the region indicated by the dashed box near the coast to get the overall view (considering the details of what happens very near the coast is more complicated). Crosses indicate flow into the plane of the diagram.

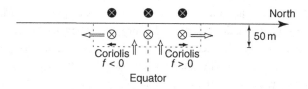

Fig. 3.10 The processes leading to equatorial upwelling. The dashed box indicates the region near the equator from which divergence occurs. Westward winds and currents are shown by crosses (into the diagram). Solid arrows indicate the Coriolis force acting on the westward current north and south of the equator.

be positive upward motion at the bottom of the surface layer. Thus there is coastal upwelling along Peru, bringing up cold water and nutrients from below the surface layer.[11]

3.4.4 Application: equatorial upwelling

In Chapter 2, the equatorial cold tongue was noted, extending across the eastern Pacific, with substantially colder SST than occurs in the western Pacific. This important climate feature sets the average conditions about which ENSO evolves. It can be understood using the momentum equations and the continuity equation, similar to the case of coastal upwelling. Figure 3.10 shows easterly winds moving over a north–south cross-section in the eastern or central Pacific. The westward wind stress exerted on the near-surface layer of the ocean results in westward acceleration of the currents. Exactly on the equator, the currents will accelerate westward, with frictional forces bringing them to a steady westward motion. Just north of the equator, the currents will also move westward, but the Coriolis force, although small, will tend to turn the current slightly to the right. There will thus be a small northward component of the flow. Just south of the equator a similar effect will occur, except that the Coriolis force turns the current to the left of its motion, and a small southward component of the current results. This will create a divergence in the surface layer (approximately the upper 50 m), which must be balanced by upward motion from below, close to the equator. The depth of the diverging currents is set by the depth over which vertical mixing is strong, since this carries the influence of the wind downward. Maximum upwelling occurs at the bottom of this diverging layer. The depth of this upwelling does not change much with the thermocline to a first approximation. Thus when the thermocline deepens, creating a deeper warm layer in the east, the upwelling brings up water that is less cold than normal.

3.4.5 Application: conservation of warm water mass in an idealized layer above the thermocline

For some aspects of El Niño theory, it is very useful to idealize the layer above the thermocline as a layer of a single temperature that is relatively warm and light compared with the cold abyss below. The thermocline is idealized as a sudden jump in temperature (and density), at a depth $-h$, as shown in Figure 3.11. The movement of the thermocline will be governed by divergence or convergence averaged over the entire layer above the thermocline. Note that this need not be the same as the divergence in the surface layer, since now

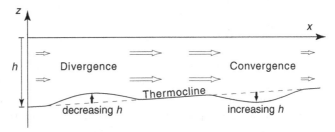

Fig. 3.11 Schematic of an idealized layer above the thermocline, showing divergence leading to a rise in the thermocline and convergence leading to a deepening thermocline. Only a longitudinal slice is shown, but convergence or divergence could occur also in north–south currents.

we are discussing a thicker layer. If we integrate the oceanic continuity equation, Eq. (3.28), through the whole layer and let $\hat{D}$ be the vertical average divergence, we have[12]

$$\frac{dh}{dt} + h\hat{D} = 0 \qquad (3.32)$$

A useful approximation to this that holds quite well for most climate purposes is to approximate h by the average depth H except where derivatives are taken and to approximate the total time derivative by the local partial derivative in time

$$\frac{\partial h}{\partial t} + H\hat{D} = 0 \qquad (3.33)$$

Note that this equation is simpler than is used in GCMs, but is useful for understanding and for intermediate complexity climate models. It was used in the Cane–Zebiak model that produced the ENSO predictions discussed in Chapter 1.

In Figure 3.11, a simple but typical configuration of currents leading to raising or lowering of the thermocline along the equator is shown. Divergence occurs where the zonal current increases to the east, convergence (i.e. negative $\hat{D}$) where it decreases eastward. Downward motion at the depth of the thermocline occurs where the vertical average current is converging. Since h measures thermocline depth (i.e. increases when the motion is downward), $\partial h/\partial t$ is positive where the current converges. Note also that divergence or convergence can occur even though the current does not change direction. For instance, divergence occurs where eastward flow increases eastward ($\partial u/\partial x$ is positive) because more flow leaves the region than enters.

3.5 Conservation of mass applied to moisture

3.5.1 Moisture equation for the atmosphere and surface

Moist processes are extremely important in the atmosphere, so weather and climate models need to keep track of water substance. In the atmosphere, the budget for water vapor is usually referred to as the moisture equation. The quantity of water vapor in the air is conveniently measured by the specific humidity

$$q = \text{(mass of water vapor)}/\text{(total mass of air parcel)}$$

If no condensation or evaporation were occurring within the parcel and no mixing were occurring, then water vapor and air would both be conserved (i.e. $dq/dt = 0$). The moisture equation in principle is thus just a budget that keeps track of sources and sinks of moisture such as evaporation, mixing and condensation. In practice, the processes that affect condensation (cloud formation, etc.) depend on small-scale motions and so they must be parameterized. Most of the complexity lies in these parameterized terms.

Using P to refer to the source/sink of moisture due to these parameterized processes, the moisture equation becomes

$$\frac{dq}{dt} = P_{convection} + P_{mixing} \tag{3.34}$$

The term $P_{convection}$ includes the loss of moisture by condensation and precipitation, as well as the part of vertical mixing that is directly associated with moist convection.[13] The term P_{mixing} includes mixing not associated with moist convection, including boundary layer turbulence. At the lowest level of the atmosphere, P_{mixing} includes surface evaporation that provides the source of moisture to the atmosphere. High-resolution climate models carry similar equations to keep track of cloud liquid water and cloud ice, in which case the term $P_{convection}$ represents conversion from water vapor to cloud condensate, and precipitation occurs as a sink term in these equations.

At the land surface, conservation of mass is again applied. Water lost from the atmosphere by precipitation serves as an input to the land surface model (discussed in section 5.3.4) or snow/ice model (discussed in section 5.3.5). Similarly, evaporation, which is a source term for the atmosphere, is a loss term for the surface budget. In land surface or snow/ice models, the equation for water substance is usually written for the mass of water per unit area within a given layer of soil, snow or ice. This has the same basic form; time rate of change of water substance balances a set of sources and sinks, some of which depend on the amount of water in the layer.

3.5.2 Sources and sinks of moisture, and latent heat

Phase changes of moisture, such as condensation or freezing, are associated with latent heat release, while evaporation or melting require a corresponding latent heat input. This implies that conservation of mass for water substance must keep track of liquid solid and vapor phases separately. It also connects the source and sink terms for each phase to the energy budget. For instance, because condensation releases latent heat, the $P_{convection}$ term in Eq. (3.34) is closely connected to the convective heating $Q_{convection}$ in Eq. (3.21). As water vapor condenses into water droplets in clouds, the latent heat release per unit mass of water condensed is given by $L = 2.5 \times 10^6 \, \text{J kg}^{-1}$. Since the specific humidity is mass of water vapor per mass of air, when humidity is reduced by an amount δq the latent heat per mass of air is given by $L\delta q$. In addition to latent heating, convective heating includes redistribution by overturning parcels which cools lower levels and warms upper levels but has zero vertical average. When vertically integrated over the column $Q_{convection}$ is therefore equal to $-L P_{convection}$. Furthermore, the vertical integral of $P_{convection}$ gives the precipitation, since that is the loss term for the condensed water. The vertical integral of P_{mixing} is just the evaporation, since that is the source of water at the bottom of the column, and there is no loss from the top of the atmosphere. During evaporation, the same latent heat per unit mass of water evaporated cools the surface, and must be included in the heat budget of the surface. Likewise, the latent heat of fusion per unit mass, $L_f = 3.34 \times 10^5 \, \text{J kg}^{-1}$, is required to melt snow or ice, or is released when liquid water freezes.

3.5.3 Application: surface melting on an ice sheet

The coupling between conservation of mass for water/ice and energy balance that includes the latent heat of melting is important in snow and ice surface models, discussed in more detail in Chapter 5. A simple calculation provides a sense for the rate at which an ice sheet can be melted by a small perturbation in the surface energy budget. Suppose that during the melting season the surface temperature for a region on an ice sheet remains at freezing so that an additional increment of downward heat flux from the atmosphere associated with global warming, say 5 W m^{-2}, is entirely used for melting ice. The rate of mass loss per unit area is thus 5 W m$^{-2}/L_f$. We can divide by the density of the ice, $\rho_i = 0.9 \times 10^3 \, \text{kg m}^{-2}$, to convert this to the rate at which the thickness of the ice decreases owing to local surface melting, and use $365 \times 86\,400$ seconds per year to convert this to meters per year. The result is 5 W m$^{-2}(L_f \rho_i)^{-1} \, (365 \times 86\,400) \approx 0.5 \, \text{m yr}^{-1}$. Considering that the Greenland ice sheet is roughly 1.5 km thick, this provides an indication that melting occurs on millennial timescales. Of course, this calculation applies only during the melting season, the radiative forcing changes as a function of time, and ice flow is important to the mass balance of an ice sheet, so this should be taken only as a rough indication of the balances involved.

3.5.4 Salinity equation for the ocean

Salt affects ocean density, so ocean models must keep track of salinity. The salinity equation is analogous to the atmospheric moisture equation, and is simply an expression of the conservation of mass. In fact, because the salt remains in the ocean, it is basically another expression of conservation of mass of water substance as it moves among components of the climate system. The quantity of salt in the water is measured by the salinity

$$s = (\text{mass of salt})/(\text{total mass of water parcel})$$

Source/sink terms for salinity occur in the interior of the ocean because of small-scale mixing between a water parcel and neighboring parcels. At the surface of the ocean, evaporation removes water and thus increases salinity, while precipitation adds fresh water to the ocean surface thus reducing salinity. River inflow and sea ice freezing or melting create additional source or sink terms at the surface in some locations.

The mixing depends on small-scale motions that must be parameterized. Using P^s_{mixing} to refer to the change of salinity due to these parameterized processes, the salinity equation is thus

$$\frac{ds}{dt} = P^s_{mixing} \tag{3.35}$$

If no mixing is occurring, then water and salt will both be conserved, so $ds/dt = 0$. At the surface of the ocean, P^s_{mixing} depends on evaporation and precipitation so that it matches the net flux of water lost or gained from the atmosphere at every point. These surface effects are then mixed down into the water below. If sea ice is present, melting or freezing can likewise make the surface water fresher or more saline.

3.6 Moist processes

While the moisture equation itself is very simple, a number of physical processes must be taken into account in order to calculate the sources and sinks of moisture. When considered in full detail, these processes can include such complex effects as the mixing of surrounding air with rising plumes of cloudy air, the growth of ice crystals in clouds, the fall rate of raindrops, the tendency of raindrops to aggregate when falling, and their re-evaporation as they reach warmer dryer levels below. Many of the above processes are represented in recent climate models but not all are of equal importance. The leading effects of moisture can be understood from just a few basic moist processes.

3.6.1 Saturation

If water vapor is continually added to air, for instance by evaporation, at a certain point the air becomes saturated. Evaporation tends to stop, and any additional water vapor simply condenses back out. This equilibrium occurs for a certain value of moisture content. This *saturation* value is most easily given in terms of the saturation vapor pressure (which can be converted into a saturation value of the specific humidity, q, if given the total pressure of dry air plus water vapor). Consider the equilibrium of saturated water vapor over liquid water for various temperatures. At warmer temperatures, the molecules in the liquid water have more kinetic energy and are more likely to escape into the air and less likely to condense out. Warm air can thus hold more water vapor. The saturation vapor pressure is a precise function of temperature, shown in Figure 3.12. This saturation value separates values of temperature and vapor pressure where condensation tends to occur from values that are subsaturated. *Relative humidity* is given by the actual vapor pressure of an air parcel divided by the saturation value at that temperature. This provides a measure of the degree of subsaturation. Depending on its history of moisture sources and sinks and heating and cooling, an air parcel can have values of water vapor content and temperature anywhere below the saturation line in Figure 3.12. If any process occurs that would yield values

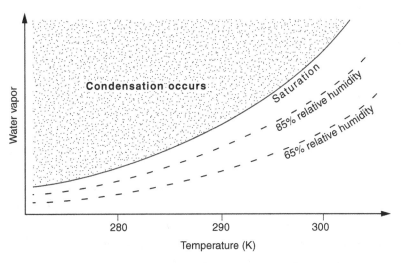

Fig. 3.12 Values of water vapor content (measured by vapor pressure) versus temperature. The saturation vapor pressure as a function of temperature is given by the solid curve. Dashed curves show selected values of the relative humidity as a function of temperature.

above the saturation line, condensation will occur. This will continue until the parcel has temperature and moisture values that lie on the saturation curve in Figure 3.12 (how quickly this happens depends on other factors, such as the presence of particles on which droplets can initially form).

An unsaturated air parcel tends to conserve its water vapor concentration (if not being strongly mixed), so if temperature decreases it will eventually saturate and condensation will begin to occur. Fog, for instance, can form by air cooling at constant pressure and water vapor concentration, which in Figure 3.12 would be a simple movement from right to left at constant water vapor pressure. If similar condensation occurs for a large air mass, it is simple to represent in a climate model since the saturation vapor pressure is easy to calculate. However, much condensation occurs in small-scale motions associated with convection which requires additional considerations.

3.6.2 Saturation in convection; lifting condensation level

In section 3.3, we saw how temperature drops with increasing height (decreasing pressure) as a small air parcel rises adiabatically in a thermal. A parcel that initially contains moisture will conserve it (keeping constant specific humidity q) during this process. Because the temperature is dropping, the saturation value of the specific humidity is decreasing until eventually the air becomes cold enough that excess water vapor must begin to condense. The pressure level at which this occurs is called the *lifting condensation level*, and it gives cloud base for convective clouds. This can be calculated using a combination of Eq. (3.20) and the saturation curve.

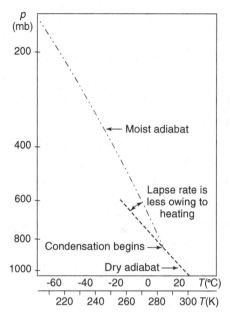

Fig. 3.13 The moist adiabat as applied to a rising air parcel containing moisture. It initially follows the dry adiabat (dashed curve) until condensation begins, then follows the moist adiabat (dot-dashed curve).

3.6.3 The moist adiabat and lapse rate in convective regions

Now consider what happens when a saturated parcel continues to rise beyond the point where condensation begins, as seen in Figure 3.13. As the parcel moves to a slightly lower pressure, the temperature will tend to drop. This will cause additional water vapor to condense, which will release latent heat into the parcel. The effect of this is simply that the temperature drops less than for an unsaturated parcel (for a given reduction in pressure). Thus the process is still adiabatic in the sense that the parcel is not exchanging heat or mass with its environment. But the condensation of moisture within the parcel is important, so the term *moist adiabatic process* is used. The rate of reduction of temperature with height is the *moist adiabatic lapse rate* which is less than the dry adiabatic lapse rate because of the release of heat. A typical value would be roughly $6 \, \text{K km}^{-1}$ (as opposed to $10 \, \text{K km}^{-1}$ for the dry adiabat). The moist adiabatic lapse rate changes with height simply because at upper levels most of the moisture has been condensed out and it reverts to almost the dry adiabat. The curve followed by this moist parcel is called the *moist adiabat*. It is a function of the temperature and moisture of the parcel when it starts out in the boundary layer. The equation is ugly, but it is easily programmed.[14] The shape of the curve in Figure 3.13 is sufficient to understand the effects.

3.6.4 Moist convection

A cloud parcel rising according to the moist adiabat has a temperature given by the temperature and moisture of the boundary layer. If this parcel is warmer and therefore less dense than the surrounding air, it will continue to rise until it reaches a level where it is no longer buoyant. The overall effect of such small-scale convective motions with rising warm air parcels will thus be to warm the troposphere through a deep layer. This is why we were able to treat the Hadley and Walker circulations as approximately large, deep thermal circulations. While the first approximation to effects of convection is simple, climate models require a considerably more detailed representation. Parameterizations of moist convection begin with the basics presented here and add representations of such effects as how mixing affect rising cloud plumes, how ensembles of different clouds interact with the large scale, and so on. Examples are discussed in section 5.3.

To summarize moist processes, we have a good understanding of the equations that govern temperature and moisture at large scales, since these are essentially budgets of conservation of energy and mass. We have a set of equations governing very small-scale parcels making a fast ascent that can be approximated as adiabatic. There is a gray area in between that involves the net effect of many small-scale motions on the large scale. We have some insight into how these should work from our knowledge of the individual small-scale parcels, but it is approximate.

3.7 Wave processes in the atmosphere and ocean

Several types of wave motions play crucial roles in the climate system through their effects on ocean and atmosphere dynamics. The main wave types are gravity waves, Rossby waves and Kelvin waves.

3.7.1 Gravity waves

Gravity waves are due to the effects of gravity acting on perturbations to a density gradient in the vertical. A familiar example is the surface gravity waves on the surface of the ocean. In this case, the density difference is the sudden jump between the density of the ocean below and that of the much less dense atmosphere above. If a region of the dense fluid (water) is raised, it has negative buoyancy relative to the surrounding fluid (air). Gravity tends to pull it back down, accelerating currents that raise neighboring regions of the dense fluid, causing the wave to propagate. Exactly the same principle holds for the smaller vertical density changes in the atmosphere and ocean. Because the density changes continuously in the vertical, the gravity waves can have various vertical structures and their speed increases for waves that have larger depth scale. In the atmosphere, gravity waves that extend through the depth of the troposphere propagate horizontally at speeds on the order of $60\,\mathrm{m\,s^{-1}}$.

Gravity waves have a behind-the-scenes role in the climate system. If a process such as heating on a small scale by a convective cloud tends to produce horizontal density and pressure gradients, gravity waves will act to reduce these gradients, rapidly spreading the increase of temperature over a region of hundreds of kilometers around the heat source. This is analogous to what happens if you add a bucket of water to the middle of a pond. The depth of the pond does not increase just at the point where water was added. Rather, gravity waves propagate out until, after a period of adjustment, the depth of the pond is increased evenly, with no remaining pressure gradients associated with differences in surface height. The time for this adjustment to be completed depends on how long it takes the waves to cross the pond.

In the atmosphere and ocean, gravity waves tend to reduce pressure gradients for scales smaller than a scale where the Coriolis force becomes important. For midlatitudes, the distance that a gravity wave can travel in a time $1/f$, where f is the Coriolis parameter, is known as the Rossby radius of deformation. The value of f at $45°$ N is about 1.03×10^{-4} s^{-1}, so for an atmospheric gravity wave speed of 60 m s^{-1}, the Rossby radius is about 600 km. In our discussion of atmospheric climate processes, we often take the winds and pressure gradients to be approximately in geostrophic balance. This holds because any portions of the winds and pressure gradients that are not in balance are quickly adjusted by gravity waves.

3.7.2 Kelvin waves

In the tropics, where f varies from zero at the equator to substantial values in the subtropics, an interesting wave type occurs that acts like a gravity wave in the longitudinal direction, along the equator. In the north–south direction, however, it obeys geostrophic balance. This is known as the *Kelvin wave*. It is derived in section 4.6.2 for the oceanic case, which is relevant to El Niño. It has the property of traveling at a gravity wave speed, eastward along the equator. When it encounters a coast, it can travel up the coast with the along-coast balances acting like a gravity wave while being in geostrophic balance in the direction perpendicular to the coast. The width of the Kelvin wave is set by a Rossby radius. For the equatorial case, where f varies strongly, an equatorial Rossby radius must be used (derived in section 4.6.3) that turns out to be about 20 degrees of latitude in that atmosphere and 3 degrees of latitude in the ocean.

Like gravity waves, Kelvin waves tend to do their job quickly relative to most climate processes. However, their spatial form is so characteristic that often the steady state result that is left by a Kelvin wave passage is still referred to as a Kelvin wave. Thus while gravity waves are seldom explicitly discussed in climate processes, Kelvin waves come up more often.

3.7.3 Rossby waves

Rossby waves are the wave type most directly important to climate processes such as teleconnections, communicating influences from one part of the climate system to regions much further away. A brief summary of Rossby wave properties is given here to provide background for discussion of El Niño teleconnections in Chapter 4.

Rossby waves depend fundamentally upon the variation of the Coriolis parameter with latitude, known as the beta effect, as discussed following Eq. (3.3). Figure 3.14 shows a simplified derivation of how this occurs for a northern hemisphere case. In this case, a sinusoidal pattern is assumed in pressure, although the diagram focuses on one low pressure region. The pressure gradient is in balance with the Coriolis force at both the northern and southern sides of the low. Because f increases with latitude, it is larger on the northern side

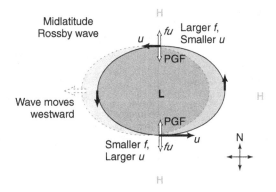

Fig. 3.14 Schematic of Rossby wave westward propagation for the case where the mean wind is zero, shown for the northern hemisphere midlatitudes.

of the low than on the southern side. In order to have the same Coriolis force balancing the pressure gradient, this implies that the zonal wind u must be smaller on the north side than the south side. This in turn means that more mass is transported eastward than westward. Because the pressure is related to the amount of mass in the column, the pressure will tend to increase on the eastern side of the low and decrease on the western side. The low will thus move west and be replaced by a high on the eastern side. Since a similar process holds for the high, the whole pattern will tend to move westward. The speed of propagation of the highs and lows for a sinusoidal wave is known as the phase speed.

The case shown in Figure 3.14 is in the absence of a large-scale mean flow. This is quite relevant for the oceans where mean currents are smaller than the typical oceanic Rossby wave phase speed. In the atmosphere, Rossby waves at midlatitudes are strongly affected by westerly climatological winds. The processes in Figure 3.14 are crucial to the existence of a Rossby wave but the propagation is modified. While the inherent wave processes tend to cause westward phase speed, the wind is carrying the entire pattern eastward so the net phase speed can be either eastward or westward. For climate applications, the most common Rossby waves to encounter are those that have zero phase speed, since these can remain locked to a stationary source that maintains the wave, like flow of the mean wind over a mountain or convective heating. An example of these stationary Rossby waves is shown in Figure 3.15. They have a particular wavelength where the westward propagation due to the beta effect balances the wind, namely $2\pi (U/\beta)^{1/2}$ where U is the mean wind. This stationary Rossby wavelength is several thousand kilometers for typical values of the mean wind. The waves appear on the downstream side of the source, i.e. to the east. For a source that is compact in latitude, typically there will be two *wavetrains*, one propagating northeastward and one southeastward, as seen in Figure 3.15. This is because both have the same properties initially so both are excited by the mountain (the direction is determined by the north–south extent of the source compared with the east–west extent).

Far from the source, the waves no longer depend on the properties of the source, but rather on the regions through which they have traveled. Because the wavelength depends on $(U/\beta^{1/2})$ and β goes to zero at the pole, the waves can never reach the pole, but rather reach a maximum latitude, then turn and propagate back southeastward. As a by-product

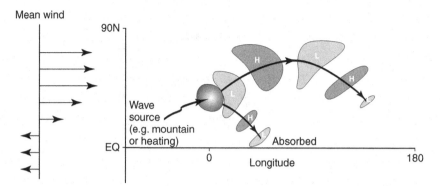

Fig. 3.15 Typical Rossby wave pattern excited by a stationary source, such as a region of sustained convective heating or a mountain perturbing the climatological mean wind. A simplified latitudinal pattern of zonal mean wind is shown at left for reference. The oscillatory Rossby wavetrain of highs and lows can only exist where there are mean westerly winds. Thus the southeastward propagating wavetrain is absorbed in the tropics. Curves with arrows show the path of the wavetrains as they are set up when the source is turned on. The northeastward propagating wavetrain reaches a maximum latitude and then returns southeastward.

of this process, the amplitude increases toward the north in the wavetrain. The result is the typical low–high–low pattern seen in the figure, with a spatial extent somewhat larger than North America.

If U is negative, as occurs in the tropics, then the wavelength, which depends on the square root of U, becomes imaginary. This corresponds to waves that decay in space instead of oscillating. The southeastern propagating wavetrain tends to be strongly absorbed as it enters the easterly wind region. Furthermore, because f is becoming small, the size of the pressure perturbation becomes small (the winds would tend to remain of similar size if it were not for tropospheric damping, and the pressure is approximately proportional to f times the wind). Thus the southward propagating wavetrain is typically scarcely seen in observations.

If the source of the wave is within the tropics, typically because of a convective heating anomaly, then a pattern essentially the same as the northward propagating wavetrain emerges. The vertical structure of the wave near the source depends on whether the source is a mountain or convective heating, but the wavetrain far from the source does not. It has a deep structure, with pressure perturbations of the same sign through the depth of the troposphere.

3.8 Overview

A summary of some of the most important concepts from this chapter:

Section 3.1

- An approximate balance between the Coriolis force and the pressure gradient force holds for winds and currents in many applications (geostrophic balance). See Figure 3.4.

- The Coriolis force tends to turn a flow to the right of its motion in the northern hemisphere (left in the southern hemisphere); the pressure gradient force acts from high toward low pressure.
- The Coriolis parameter f varies with latitude (zero at the equator, increasing to the north, negative to the south); this is called the beta effect (β = rate of change of f with latitude).
- In the vertical direction, the pressure gradient force balances gravity (hydrostatic balance). This allows us to use pressure as a vertical coordinate. Pressure is proportional to the mass above in the atmospheric or oceanic column.

Section 3.2

- The ideal gas law gives the relationship of density to pressure and temperature in the atmosphere.
- The density in the ocean depends on temperature (warmer = less dense) and salinity (saltier = more dense).
- A warm atmospheric column, since it is less dense, has a greater thickness between two pressure surfaces than a cold column.
- This explains thermal circulations (see Figure 3.5).

Section 3.3

- In the ocean, the time rate of change of temperature of a water parcel is given by the heating (for a surface layer this is the net surface heat flux from the atmosphere minus the flux out the bottom by mixing).
- The temperature equation is similar for the atmosphere but when an air parcel rises, the temperature decreases because the parcel expands as it goes to lower pressure.
- For an air parcel rising quickly (e.g. in thermals) so little heat is exchanged, temperature decreases at $10\,°C\,km^{-1}$ (the dry adiabatic lapse rate).
- The time derivatives following the parcel hide some of the complexity of the system because the parcels themselves tend to deform in complex ways if followed for a long time. This is particularly important for weather motions and results in the loss of predictability for weather. The time derivative for temperature at a fixed point is obtained by expanding the time derivative for the parcel in terms of velocity times the gradients of temperature (advection). A similar procedure applies in other equations.

Section 3.4

- Conservation of mass can be expressed as horizontal divergence or convergence (of currents or winds) being balanced by changes in the vertical motion. In the atmosphere this holds in pressure coordinates.

- Equatorial upwelling results from divergence of ocean surface currents away from the equator. Water must rise from below to compensate. The divergence at the equator is due to effects of easterly winds and the Coriolis force (see Figure 3.10).
- For the layer of warm water above the thermocline, convergence of upper ocean currents implies a deepening of the thermocline (see Figure 3.11).

Section 3.5

- Conservation of mass gives equations for water vapor (atmosphere) and salinity (ocean).
- The main sinks of water vapor are due to moist convection (resulting in precipitation) which at the same time produces convective heating in the temperature equation (from condensation in clouds). The source of water vapor is evaporation at the surface. These processes involve small-scale motions and must be parameterized.
- Salinity at the ocean surface is increased by evaporation and decreased by precipitation.

Section 3.6

- Saturation of moist air depends on temperature according to Figure 3.12. Relative humidity gives the water vapor relative to the saturation value.
- A rising parcel in moist convection decreases in temperate according to the dry adiabatic lapse rate until it saturates, then has a smaller moist adiabatic lapse rate. The temperature curve in Figure 3.13 (the moist adiabat) depends only on the surface temperature and humidity where the parcel started.
- If this curve is warmer than the temperature at upper levels, convection typically occurs.

Section 3.7

- Waves play an important role in communicating effects from one part of the atmosphere to another.
- Rossby waves depend on the beta effect. Their inherent phase speed is westward. In a westerly mean flow, stationary Rossby waves can occur in which the eastward motion of the flow balances the westward propagation. Stationary perturbations, such as convective heating anomalies during El Niño, tend to excite wavetrains of stationary Rossby waves.

In this chapter, we have assembled essentially all the equations and most of the processes that govern the atmosphere and ocean. Furthermore, we have several examples of applications and a taste of some of the ways these can be used for quantitative solution. Almost all phenomena that you see in meteorology, oceanography, and climate dynamics are contained in these equations. Perhaps not surprisingly, that is why the full set is rather challenging to solve. How these equations are implemented on the computer is the subject of Chapter 5. An equally important task in unraveling climate interactions is to focus in on particular aspects of these equations, using approximations that make the effects more understandable. Some of these approximate versions will be used to assist in understanding El Niño dynamics in Chapter 4.

Notes

1 Another apparent force, the centrifugal force, applies to bodies in the rotating frame, but on Earth it appears merely as a tiny modification to the effective local value of the gravitational force, with the shape of the Earth compensating so there is no horizontal component by having a slight bulge at the equator relative to the poles. The Coriolis force due to eastward/westward motions is essentially an increase/decrease in the centrifugal force relative to what is balanced by the shape of the Earth.

2 The mixing term F^x_{drag} in the momentum equation is related to stress in the x direction τ_x by $F^x_{drag} = \frac{1}{\rho}\frac{\partial \tau^x}{\partial z} + F^x_{HM}$, where F^x_{HM} is a contribution due to horizontal mixing which is often smaller than the vertical mixing term associated with τ_x. A similar equation holds for F^y_{drag}, relating it to stress in the y direction, τ^y. Averaged through an ocean surface layer of depth H and constant density, one obtains $\frac{1}{H}\int_{-H}^{0} F^x_{drag}dz = (\rho H)^{-1}(\tau^x_s - \tau^x_{-H}) + F^x_{HM}$, where τ^x_s is the surface stress input by the wind that drives the ocean currents. The stress at the bottom of the layer τ^x_{-H} acts to slow surface currents, as does F^x_{HM} (here approximated as vertically constant).

3 Integrating Eq. (3.8) in the vertical from the top of the atmosphere to a level z, using $p = 0$ at the top of the atmosphere, gives $p(z) = \int_{\infty}^{z} \rho(z)g dz$. Since g does not vary significantly with z within the depth of the atmosphere where ρ is significant, $p(z)$ is equal to g times the mass of the atmosphere above level z.

4 The derivation of the geopotential gradient term that replaces the PGF in pressure coordinates involves the identity for partial derivatives (for any three variables, in this case x, z and p) $\frac{\partial p}{\partial x}\frac{\partial x}{\partial z}\frac{\partial z}{\partial p} = -1$, so $\frac{1}{\rho}\frac{\partial p}{\partial x} = -\frac{\partial z}{\partial x}(\frac{1}{\rho}\frac{\partial p}{\partial z}) = \frac{\partial gz}{\partial x}$.

5 The ideal gas constant R for air (expressed per kg) depends on the composition of air and the molecular weights of each constituent. The value given is for dry air; R can change slightly when humidity is high, an effect that is included in climate models.

6 In Eq. (3.19), T has been assumed constant through the surface layer, since it is well mixed. The more general relation, which holds for any layer, is $\int_{bottom}^{top} \rho c_w \frac{dT}{dt} dz = F^{net}_{top} - F^{net}_{bottom}$. There is also a smaller contribution due to horizontal mixing that is omitted for clarity.

7 Derivation of the dry adiabatic lapse rate. For parcels moving up or down adiabatically, the solution to Eq. (3.20) (with $Q = 0$) is $T = T_0(p/p_0)^{R/c_p}$ for a parcel starting at p_0 with temperature T_0. This *dry adiabat* is slightly curved for T as a function of p but has a constant lapse rate as a function of height z. If we take the derivative with respect to z to get the lapse rate, and use the hydrostatic equation Eq. (3.13), we obtain $-\frac{\partial T}{\partial z} = \frac{g}{c_p} = 10\frac{C}{km}$.

8 $D_{3D} = \frac{\partial u}{\partial x} + \frac{\partial v}{\partial y} + \frac{\partial w}{\partial z}$. Another commonly used approximation (slightly more accurate than Eq. (3.28)) is to neglect $\frac{\partial \rho}{\partial t}$ in Eq. (3.27) but retain the advection terms acting on density. This yields $\frac{\partial(\rho u)}{\partial x} + \frac{\partial(\rho v)}{\partial y} + \frac{\partial(\rho w)}{\partial z} = 0$.

9 The vertical velocity at the ocean surface is just the time rate of change of the surface height η, so vertically integrating Eq. (3.28) gives the surface height equation (here using the approximation that neglects density variations) $\frac{d\eta}{dt} = -\int_{column}(\nabla \cdot v)dz$.

10 The vertical pressure velocity at the surface is just dp_s/dt, where p_s is the surface pressure. Thus vertically integrating Eq. (3.30) gives the surface pressure equation $\frac{dp_s}{dt} = -\int_{column}(\nabla \cdot v)dp$.

11 The example of coastal upwelling is simple to treat in qualitative terms as done in Figure 3.8 and related text. However, if you try to evaluate the exact value of the upwelling, you end up trying to take the x derivative of u. You also find that Eq. (3.31) can only hold if F^y_{drag} goes to zero at the coast (since u must be zero there; no current flows through the coast). Since the wind is not zero at the coast, we cannot neglect effects of friction on v, hidden in F^y_{drag}. These effects enter in the region close to the coast to give the details of the solution there. It is typical of interesting

climate problems that messy effects like this tend to come up, requiring some treatment of the parameterized terms.

12 Equation (3.32) is derived by integrating Eq. (3.28) to give $\int_{-h}^{0} D dz = -(w(0) - w(-h))$. The vertical velocity at the thermocline is $w(-h) = -dh/dt$. At the surface, $w(0) = 0$.

13 Most climate models further subdivide the term denoted $P_{convection}$ in Eq. (3.34) into several parts: part associated with deep convection, part associated with shallow convection and part associated with condensation during large-scale uplift of an air mass, such as can occur in frontal systems. Our discussion applies to the sum of these. The subdivision becomes important when parameterizing the detailed dependence on other large-scale variables.

The value of L varies slightly with temperature. The value given in the text applies at $0\,°C$. For deposition of water vapor directly onto ice crystals (as occurs at upper levels) or sublimation of ice directly to water vapor, a latent heat of sublimation is used (about 13% larger than L).

14 For further treatment of moist adiabatic processes see Bohren and Albrecht (1998).

El Niño and year-to-year climate prediction

4.1 Recap of El Niño basics

Chapter 1 provided a view of the essential features of the El Niño/Southern Oscillation phenomenon, or ENSO, as observed in the tropical Pacific Ocean. ENSO involves variations about the average climate (termed anomalies) in the sea surface temperature (SST), convection zones, atmospheric pressure gradients and winds, thermocline depth and upper ocean currents. The relation between these variables indicate that El Niño and the Southern Oscillation are connected phenomena. There are indications of oscillatory behavior with a 3- to 5-year preferred time scale, although also considerable irregularity in the behavior. In this chapter, the physical mechanisms relating these variables are discussed in more detail using some of the tools developed in Chapter 3. First the mechanisms for maintaining a fully developed El Niño (warm) or La Niña (cold) phase are discussed, followed by the mechanism for alternation between the extreme phases.

While the heart of ENSO lies in the tropical Pacific, impacts in other parts of the world are also important. Such remote effects of El Niño or La Niña are termed *teleconnections*. These are essentially by-products of the changes in the tropical Pacific. This would include changes in temperature and rainfall over the USA, South America, Australia, Southeast Asia and India, and Africa. Although these are effects we are interested in, prediction of these effects is often less reliable than that of the El Niño itself. After treating the dynamics and basis for predictability of the main part of ENSO, we discuss mechanisms for teleconnections in section 4.8. Selected further interannual and decadal time scale phenomena in the climate system are then presented.

4.1.1 The Bjerknes hypothesis

As noted in the overview of ENSO studies in Chapter 1, Bjerknes first hypothesized in 1969 that interaction between the ocean and atmosphere was essential to ENSO. During the 1980s and 1990s many investigators contributed to making this conjecture a full fledged theory. Modestly on the part of these scientists, "the Bjerknes hypothesis" is used as the name for this theory. In essence, ENSO arises as a coupled cycle in which anomalies of SST in the Pacific cause the trade winds to strengthen or slacken, and this in turn drives the ocean circulation changes that produce anomalous SST.

In Bjerknes' own words:

> It is the gradient of SST along the equator which is the cause of [...] the Walker circulation.
> An increase in equatorial easterly winds [is associated with] an increase in upwelling and
> an increase in the east–west temperature contrast that is the cause of the Walker circulation
> in the first place. [...] On the other hand, a case can also be made for a trend of decreasing
> speed [...] There is thus ample reason for a never-ending succession of alternating trends
> by air–sea interaction in the equatorial belt, but just how the turnabout between trends
> takes place is not yet clear.

Although Bjerknes did not guess the essential role now thought to be played by thermocline depth anomalies, this certainly was a remarkable conjecture, since the essence of it still
stands as the basis of modern work. As we shall see, the dynamics of "how the turnabout"
takes place involves subsurface ocean dynamics. At the time of Bjerknes' work, neither the
theory nor the observational system for this aspect had been developed.

4.2 Tropical Pacific climatology

Before examining the feedbacks involved in El Niño in more detail, an overview of the
tropical Pacific climatology of both atmosphere and ocean is in order. Recall that climatology refers to long-term average conditions, which in this case will typically be over the past
several decades, for which good data are available. A span of a few decades is significantly
longer than the typical time scale of the ENSO cycle, so the climatology may be taken
as fixed during the evolution of an El Niño event. Qualitatively, the tropical climatology
has been the same during the entire century (and probably longer). There can, however,
be variations in details from one decade to another and the climatology may change in the
future because of global warming.

Recall from Chapter 2 that there are two aspects of the atmospheric circulation that
contribute to the trade winds blowing from east to west across the Pacific: the Hadley
circulation and the Walker circulation. Both are basically large thermal circulations associated with warming of the atmospheric column over warm SSTs (Figure 2.16), and the rising
branches are associated with strong precipitation over these warm regions (Figure 2.13). The
Hadley circulation involves the longitudinal-average circulation in the north–south direction, whereas the Walker circulation is associated with east–west contrasts. The Hadley
circulation (Figure 2.12) contributes easterly winds across the entire tropics as air flows
back toward the equator at low levels and is turned by the Coriolis force. Over the Pacific, the
winds associated with the Walker circulation (Figure 2.14) are actually somewhat stronger,
but they are in the same easterly direction. They blow from the relatively cold eastern Pacific
to the western Pacific warm pool.

Figure 4.1 shows a schematic of the Pacific climatology which includes both atmospheric
and oceanic circulation. On the atmospheric side, the Walker circulation is emphasized. The
trade winds blow across the Pacific and the air rises in the convergence zone over the warm
SSTs in the west. It diverges at upper levels (not shown). Note that scales are different for
various features in the schematic: the depth of the deep convection in the western Pacific is

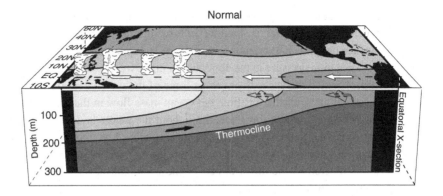

Fig. 4.1 Schematic of the Pacific in three dimensions. Sea surface temperature and subsurface temperature (in longitude–depth cross-section at the equator) are shown by shading (warmest at the surface in the western Pacific). The regions of strong convection are indicated with deep convective clouds (not to scale). Surface winds along the equator are indicated by white surface arrows. Near-equatorial currents are shown in the ocean surface layer, including the diverging surface currents moving poleward from the equator at a few degrees of latitude. The equatorial undercurrent is indicated by bold arrow just above the thermocline.

about 10 km, while the ocean vertical scale is only 300 m. The horizontal scale across the basin is about 15 000 km, while individual convective clouds are only on the order of 1 km.

On the oceanic side, the temperature is shown both at the surface and in a slice along the equator, indicating the connection of cold surface temperatures in the east to the subsurface temperature. The thermocline corresponds to the temperature difference between the deepest layer and the less cold water above. This temperature difference is larger than the temperature differences between the other shaded regions in Figure 4.1 which indicate subtler gradations of temperature in the upper ocean above the thermocline. The temperature boundaries correspond roughly to 28, 24, and 20 °C, but the water below the thermocline has typical temperatures of 15 °C or colder.

The dynamics of equatorial and coastal upwelling has been discussed in Chapter 3. Surface currents exactly on the equator tend to flow westward, accelerated by the wind stress on the ocean surface due to frictional effects. On either side of the equator they tend to diverge slightly under the effects of the Coriolis force. Although the divergent component is not large, it is very important in bringing up cold water from below (vertical arrow, indicating upwelling). The equatorial undercurrent is a swift, narrow current along the equator slightly above the thermocline that returns water eastward. The slope of the thermocline is important both in the climatology and in ENSO dynamics. In the climatology, the shallow thermocline in the east allows the upwelling near the surface to bring up colder water. If the eastern Pacific thermocline were as deep as in the west, surface waters would not be nearly so cold despite the presence of upwelling.

The thermocline slope is associated with a slope in the surface height of the ocean (too small to be seen in the diagram). The western Pacific is about 40 cm higher than the eastern Pacific, while the thermocline is almost 100 m deeper. Surface height and thermocline depth are related to each other because where the thermocline is deep, the column of water is on average warmer and thus slightly less dense. It thus requires a slightly taller column to have the same mass per unit area, leading to a higher sea surface height than for a column where the thermocline is shallower. This is discussed in more detail in section 4.4. Where surface height is higher, pressure in the ocean upper layer is higher. The pressure gradient between high pressure in the upper ocean in the west Pacific and lower pressures in the east approximately balances the average force of the wind stress on the ocean layer above the thermocline. Thus there is little net ocean mass flow in the east–west direction. However, the pressure gradient force acts throughout the whole layer above the thermocline, while the force due to the wind stress is concentrated in the upper layers. This is why there are westward currents at the surface and a returning undercurrent below. Surface currents are more affected by the direct effects of the wind and the subsurface currents more affected by the eastward pressure gradient force.

In the climatology, the pressure gradient caused by the slope of the thermocline acts to balance the wind stress. If the winds change, however, the pressure gradient force in the ocean will no longer balance the force of the wind stress and adjustment will have to occur, as happens during ENSO.

4.3 ENSO mechanisms I: extreme phases

The extreme phases of the ENSO cycle are El Niño (warm phase) and La Niña (cold phase). We deal first with these phases because they have the most spectacular effects and contain the ocean–atmosphere feedbacks associated with the Bjerknes hypothesis (referred to as the Bjerknes feedbacks for brevity). In section 4.5 we examine the more complicated part: how "the turnabout" between warm and cold phases takes place, and what occurs in the transition phases of the cycle.

Figure 4.2 shows the state of the atmosphere and ocean in the Pacific during El Niño and La Niña. While we often show anomalies for these conditions, it is useful to remember what the total fields look like. For instance, during El Niño, there are large positive anomalies of precipitation in the central and eastern Pacific (Figure 1.8) and negative anomalies in the western Pacific. In terms of the total precipitation (climatology plus anomaly, the actual precipitation), the rainfall has simply spread out across the Pacific. This results in more rainfall than normal in the east and less than normal in the west. The convection is more widespread because the SST is warm over a greater part of the basin. Since the Walker circulation is basically a thermal circulation, pressure gradients across the basin are smaller, and the trade winds weaken. In terms of anomalies, this corresponds to a westerly anomaly, even though the total winds may still be easterly.

In the ocean, the warm water above the thermocline tends to slosh back eastward, since the wind no longer balances the pressure gradient force. The thermocline thus deepens in the east. Equatorial upwelling still brings up water from approximately the same depth as in normal conditions, but because the thermocline is deeper during El Niño, this water comes from well above the thermocline, so is not as cold as normal. This is the main effect on SST.[1] These features of the modern version of the Bjerknes hypothesis are summarized in Figure 4.3. This is a positive feedback loop; that is, if one begins with a warm

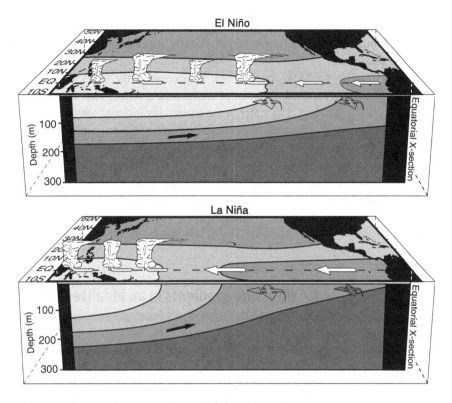

Fig. 4.2   Schematic of the Pacific for El Niño and La Niña conditions. Conventions are the same as for Figure 4.1.

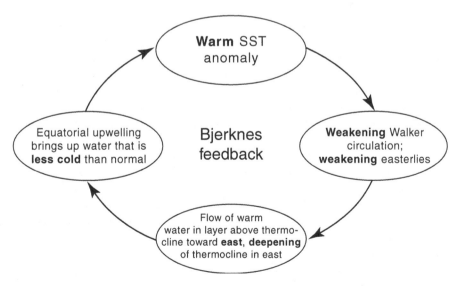

Fig. 4.3   The feedbacks operating in the Bjerknes hypothesis, shown for the case of strengthening El Niño warm conditions.

SST anomaly, the changes in atmospheric and oceanic circulation tend to reinforce the initial SST anomaly. An initial westerly wind anomaly will lead to the same chain of processes.

During La Niña, the feedback loop operates the same way but with the signs of the anomalies reversed. In terms of total fields, essentially all the features in the climatology are strengthened. The water in the eastern Pacific cold tongue along the equator is colder than normal. The surface pressure gradient in the atmosphere and trade winds increase. The convection is more concentrated in the western Pacific than normal. This implies that negative rainfall anomalies occur in the central and eastern Pacific where rainfall is disfavored over cold SST. In response to the increased easterly winds, the thermocline slope increases so there is a greater pressure gradient across the basin in the ocean as well. The thermocline in the east thus *shoals* (becomes shallower), so upwelling brings up water that is even colder than normal. This reinforces the anomalously cold SST.

4.4 Pressure gradients in an idealized upper layer

Since the interplay between wind stress force, pressure gradient force and Coriolis force is so important to ENSO dynamics, it is useful to have a relatively simple model in which to discuss them. Figure 4.4 shows a model of the upper layer of the ocean that is used for understanding the movement of warm water in the layer of the ocean above the thermocline,

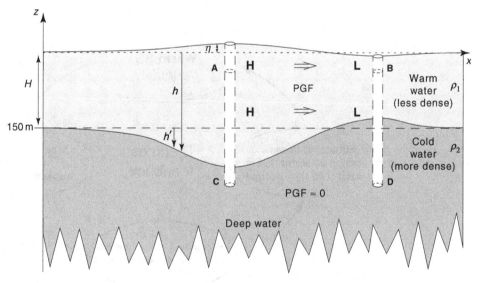

Fig. 4.4 A model of the upper layer of the ocean above the thermocline. The thermocline is idealized as a sharp transition between warm, light water and cold, dense water at a depth h. The mean value of the thermocline depth is denoted H, while H and L denote relatively high or low pressure at a given depth, and PGF denotes the pressure gradient force.

as seen in Chapter 3, and the distribution of pressure gradients that contribute to the balance of forces on these currents.

The upper layer is idealized as having a constant density, which is lighter than the layer below in the deep ocean since the water above the thermocline is considerably warmer than the layer below. When discussing effects of subsurface temperature on SST, we need to keep track of more subtle variations of temperature, but for the purpose of calculating pressure gradients, this approximation (warm above, cold below) is sufficient. To find the distribution of high and low pressure regions in the upper layer, we note the following:

- Pressure is proportional to the mass above a unit area; considering columns of water as shown, the relative pressure can thus be evaluated.
- The surface is higher at location A than at B, and the density is the same, so there is more mass above A. Hence there is a pressure gradient force from A toward B.
- Points C and D are in the deep ocean, well below the thermocline. Above C there is additional mass where the surface height is raised, but the column contains more of the less dense water. Above D, the column is shorter, but there is more of the dense water below the thermocline because the thermocline is raised. They add up to very close to the same mass so the pressure gradient force is small in the deep ocean.
- The pressure gradient force (PGF) is small in the deep ocean because small currents through the very deep layer can quickly adjust the surface height to balance whatever the thermocline is doing.[2]
- Because of this adjustment to keep the PGF small in the deep ocean, the surface height η is proportional to the thermocline depth (on time scales of weeks or longer). The surface elevation is high where the thermocline is deep, and thus pressure in the upper layer is high.
- The surface height is given by $\eta = \rho_1^{-1} \Delta \rho h'$ where $\Delta \rho = \rho_2 - \rho_1$ is the difference in density between upper ocean and deep ocean, which is much smaller than the total density. Thus the sea surface height varies by only tens of centimeters when the thermocline varies by tens of meters.

In summary: the sea surface height and pressure in the upper ocean layer are high where the thermocline is deep and are low where the thermocline is shallow.

4.4.1 Subsurface temperature anomalies in an idealized upper layer

This simplified upper layer can also be used to understand what subsurface temperature anomalies look like before examining them in observations. Figure 4.5 shows an example where the thermocline is deeper than normal in the east and shallower than normal in the west (as occurs during El Niño). Where the thermocline is deeper than normal, warm upper layer waters occur where normally there is cold water. Thus there are warm anomalies between the thermocline and the depth of the normal thermocline. Where the thermocline is shallower than the normal thermocline, there is cold water in regions that are normally warm, so there is a cold anomaly. Above or below these regions, the water temperature is the same as it is under normal conditions, so there is no anomaly.

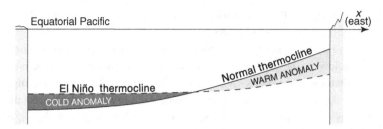

Fig. 4.5 Two positions of the thermocline, indicating regions of warm and cold thermocline anomalies. The climatological thermocline (solid line) and an example of the thermocline in a particular month during an El Niño (dashed line) are shown.

4.5 Transition into the 1997–98 El Niño

4.5.1 Subsurface temperature measurements

The movements of the thermocline are very important in El Niño, and yet measurements of subsurface ocean temperature have been very sparse in the past. One source of subsurface measurements is to have a research vessel cruise across the area, lowering instruments on a line, or dropping instrument packages known as expendable bathythermographs (XBTs) into the ocean as they go. In earlier decades, this was the primary source of subsurface data, which were quite sparse and did not provide a continuous picture in space or time.

With the 1997–98 El Niño, there was an unprecedented chance to get accurate views of the subsurface temperature structure because of a measurement system called the TAO array (Tropical Ocean–Atmosphere array). This is a system of moored buoys that had been put in place over the previous decade as part of a concentrated research program called the Tropical Ocean–Global Atmosphere (TOGA) program.[3] These buoys (like the one in Figure 4.6) are anchored to the bottom of the ocean and are instrumented to measure oceanic temperature over the upper few hundred meters, as well as surface wind, humidity and temperature in the atmosphere. Some of the buoys on the equator also have current meters. This required solving a considerable engineering challenge because the surface buoy must be tautly moored to the ocean floor 4 km below. Any slack in the cable would degrade the accuracy of the current measurement.[4]

The TAO array moorings are distributed across the Pacific as shown in Figure 4.7, in lines straddling the equator so as to optimally capture surface and upper ocean measurements in the main ENSO region. Initial success in the Pacific led to subsequent deployment of similar buoys in the Atlantic, and ongoing deployment in the Indian Ocean. The observations are transmitted via satellite to data centers, where they are compiled and made available to researchers and forecast centers almost immediately. The buoys require periodic servicing by research vessels, and so this observational system has a maintenance cost that is substantial by the standards of climate research programs, although less than the cost of a typical satellite observation program. This array is just one component of the global ocean

Fig. 4.6 A buoy from the TAO array. Courtesy of the Pacific Marine Environmental Laboratory.

observing system, which also includes drifting buoys, automated floats that move up and down in the water column, and instruments dropped from ships.[5]

4.5.2 Subsurface temperature anomalies during the onset of El Niño

The observations of subsurface temperature during the transition from slightly cold conditions in 1996 into the 1997–98 El Niño warm phase provide a good example of what happens during the "turnabout" between warm and cold phases that Bjerknes could not explain. Even when there are no anomalies at the surface and one might think that the

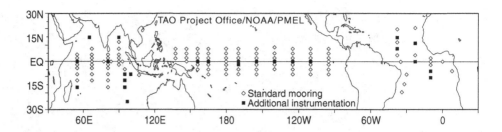

Fig. 4.7 The global tropical moored buoy array, with the original TAO array in the Pacific augmented by subsequent programs (Indian Ocean locations include planned deployments). Diamonds mark locations of buoys that measure subsurface temperature and surface atmospheric variables. Squares mark locations of buoys that have additional instrumentation including current meters or measurements of air-sea fluxes. Adapted from a figure courtesy of the Pacific Marine Environmental Laboratory.

Pacific climate system is in its normal state, anomalies beneath the surface can be slowly evolving, moving the system into the next phase of the cycle. These observations fit very nicely with what had been inferred earlier from model simulations of ENSO.

They are also consistent with Wyrtki's observation in the 1970s that sea level height in the western Pacific increases (related to the thermocline deepening) prior to El Niño. They also indicate why it is possible to predict El Niño in advance: since subsurface anomalies are present, a model that can predict slow subsurface evolution can have some skill at predicting the later impact at the surface.

Figure 4.8 gives a three-dimensional view of the temperature anomalies in the Pacific, showing simultaneously SST anomalies north of $5°$ S, and a vertical cross-section of subsurface temperature anomalies down to 300 m depth along the equator. As illustrated in Figure 4.5, temperature anomalies near the depth of the thermocline (about 150–200 m in the west, 100 m in the east) correspond to deepening (positive temperature anomalies) or shallowing (negative anomalies) of the thermocline. Even a year in advance of the maximum surface warming, the precursors of El Niño are visible subsurface in the western Pacific. From January 1997, climate models were already predicting an El Niño for the end of the year, based on these initial conditions. The depression of the thermocline extends slowly from west to east along the equator. When warm subsurface temperature anomalies caused by the thermocline deepening reach the east, they are carried by equatorial upwelling to the surface. Once SST becomes anomalously warm, as seen in Figure 4.8c, the feedback described by the Bjerknes hypothesis begins. Westerly wind anomalies in the central Pacific cause the eastern Pacific thermocline to deepen still further, causing additional warming, and so on. However, the eastward transfer of warm water above the thermocline that is causing the thermocline to deepen in the east has the opposite effect in the west, causing the thermocline to become shallower. The shallower thermocline is seen as subsurface cold anomalies. Even during the onset of El Niño at the surface, the seeds of its destruction are being sown in the western Pacific. In Figure 4.8c and d, cold subsurface anomalies may already be seen in the west. The eastward extension of the cold anomalies brings the gradual erosion of the surface warm anomalies. This initiates a reversal of the chain of feedbacks in the Bjerknes hypothesis, which acts to drive the system into La Niña.

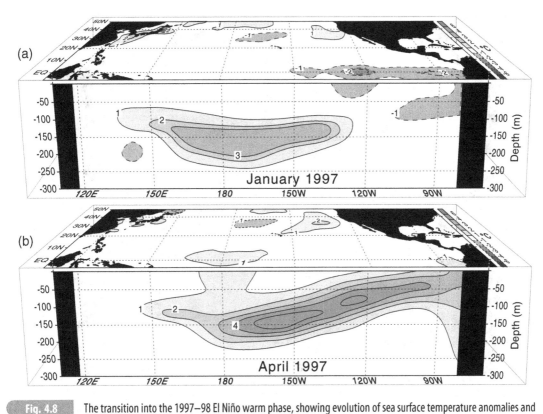

Fig. 4.8 The transition into the 1997–98 El Niño warm phase, showing evolution of sea surface temperature anomalies and subsurface temperature anomalies along the equator. Note the surface temperature anomaly plot extends slightly south of the equator. (a) In January 1997, the surface temperature in the east is still colder than normal, but in the west, warm subsurface temperature anomalies associated with a lower thermocline have already existed for some time, partly owing to movement of warm water onto the equator from off-equatorial regions to the north and south. Since pressure gradients associated with these anomalies are not in balance with the wind stress, they evolve in time, spreading eastward along the equator. (b) By April, the thermocline has deepened in the east, but effects of subsurface warm anomalies are not yet seen in surface temperatures. The transition into the 1997–98 El Niño warm phase, showing evolution of sea surface temperature anomalies and subsurface temperature anomalies along the equator.

Although the subsurface temperature anomalies migrate along the equator from west to east, they are not simply being carried along passively by a fixed current, nor are they propagating freely like a wave. For instance, when the cold anomalies are extending eastward during the onset of La Niña, anomalous ocean currents are toward the west, moving warm water in the layer above the thermocline toward the west, and thus causing it to shoal in the east. Furthermore, the equatorial subsurface temperature anomalies in the west are associated with off-equatorial anomalies of the same sign that tend to provide a reservoir of warm or cold water in the transition phase. These motions involve an interplay between wind stress anomalies, the pressure gradients associated with the thermocline anomalies, the anomalous currents, and the Coriolis force. These can be understood using the tools developed in Chapter 3, as discussed in section 4.6.

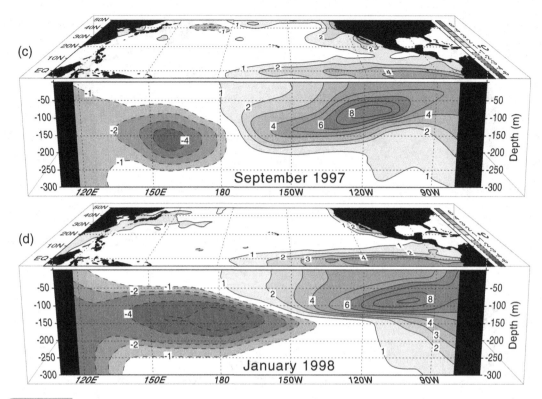

Fig. 4.8 (c) By September, the warm anomalies have been communicated to the surface, setting in motion the Bjerknes feedbacks. In response to reduced easterly winds, the thermocline slope has further reduced, intensifying warm anomalies in the east. In the west, the shoaling of the thermocline has produced cold anomalies, and even more so off the equator. (d) By January 1998, this process has produced large anomalies both at the surface and subsurface in the east. The slower adjustment process involving off-equatorial anomalies in the west has produced growing subsurface cold anomalies. The subsequent eastward extension of these continued the cycle, swinging the system into the La Niña cold phase in 1999. The La Niña phase develops much like the anomalies shown for El Niño, but with opposite sign. After figures courtesy of David Pierce, Scripps Institution of Oceanography. [6]

This process involves not only the anomalies on the equator shown in Figure 4.8, but also subsurface temperature anomalies off the equator. The anomalies of sea surface height in Figure 1.10 show these off-equatorial thermocline anomalies at the peak of the 1997–98 El Niño. The high sea surface height regions along the equator in the east are associated with the deepened thermocline and warm subsurface temperature anomalies. The low sea surface height regions in the west are regions where the thermocline is shallower than normal. The cold subsurface anomalies in the west on the equator in Figure 4.8d correspond to this. In these regions, the mass of warm water above the thermocline has been depleted by sloshing over to the eastern part of the basin. Remember that the warm and cold anomalies are mainly associated with rearranging the amount of warm water above the thermocline, and not with actual heating or cooling. The depletion of warm water extends off the equator in the western Pacific, forming a sort of horse-shoe shape. The sea surface height is lower (thermocline shallower) slightly off the equator than right on the equator. In the next section we will examine the dynamics that produce this, and see how it affects the transition into the next phase of the cycle. Basically the cold anomalies along and off the equator in the west are a La Niña waiting to happen. This allowed model predictions of the 1998–99 La Niña, just as the subsurface warm anomalies allowed predictions of the preceding El Niño.

4.5.3 Subsurface temperature anomalies during the transition to La Niña

Figure 4.9 provides a view in the same format as Figure 4.8 of the transition from the 1997–98 El Niño into the La Niña of 1998–99. The cold subsurface temperature anomalies in the

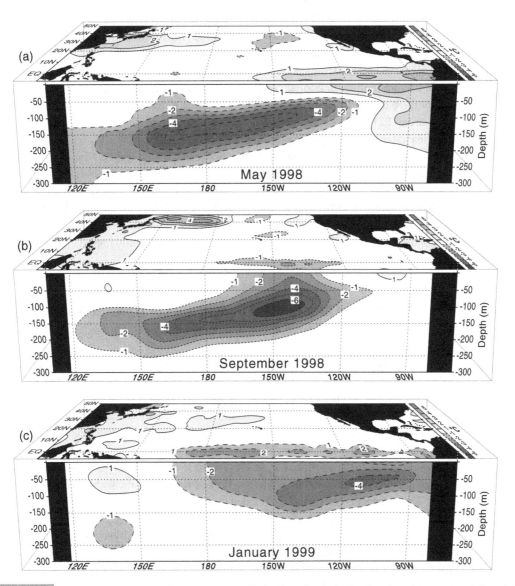

Fig. 4.9 Three-dimensional view of temperature anomalies at the surface and subsurface along the equator, as in Fig. 4.8, but during the transition into La Niña in 1998–99. After figures courtesy of David Pierce, Scripps Institution of Oceanography.

west extend slowly across the equator, eroding the warm anomalies in the east (Figure 4.9a). By September (Figure 4.9b), upwelling in the east-central Pacific has carried the effects of the subsurface cold anomalies to the surface. The cold SST anomalies are affecting the atmosphere, strengthening the trade winds in the La Niña version of the Bjerknes feedbacks. Through the fall and winter the cold SST anomalies strengthen and extend across the east Pacific, with further shallowing of the thermocline in the east, as seen in Figure 4.9c for January 1999.

While the La Niña anomalies are, to a first approximation, just the negative of the El Niño anomalies, some differences may be noted. The La Niña SST anomalies are slightly stronger toward the central Pacific, as is common for La Niña events. This is partly because of atmospheric effects: since the eastern Pacific has little rainfall during normal conditions the La Niña reduction in convection is concentrated further westward than the El Niño enhancement. On the oceanic side: because the thermocline on the easternmost part of the basin is quite shallow in normal conditions, it is hard to make the cold tongue very much colder there during La Niña. Thus it is less common to have a very strong La Niña than a very strong El Niño. Overall, however, the similarities between the pattern of El Niño and La Niña anomalies are more important than the differences.

It is also worth noting that the succession between El Niño and La Niña is not always so rapid and can be more complex than the example provided by the 1998 case. The 1998–99 La Niña continued through the following year, albeit with some variations in intensity. The irregularity of the ENSO cycle is an important feature that is strongly affected by weather events, as described in section 4.7. But the cyclic aspect and the ability to anticipate future SST anomalies from existing subsurface anomalies are sufficiently important that we should further examine the dynamics behind the transition phase.

4.6 El Niño mechanisms II: dynamics of transition phases

Consider the idealized upper ocean layer in Figure 4.4. Pressure gradients are proportional to the sea surface height (or thermocline depth). These in turn can be changed by currents moving mass around in the layer, for instance from east to west. If the Earth were not rotating, these motions would be rather similar to the sloshing of shallow water in a bathtub (in fact, they would be mathematically equivalent). Such models are thus called "shallow water models." The discussion below will be phrased in terms of the thermocline depth, but of course we could equally use sea surface height. It is easiest to use the concepts from a shallow water model, and discuss the dynamics in terms of mass transfer of the warm water in the upper layer. These concepts can easily be extended to a more general model, with deep thermocline corresponding to warm subsurface temperature anomalies, as discussed in section 4.4.

Water sloshing in a non-rotating bathtub behaves symmetrically to the "east" and "west" of a forcing. An important feature of equatorial ocean dynamics is that this is *not* true. Behavior to the east of a wind stress anomaly is very different than to the west. This was seen in observations of sea level height in Figure 1.10 where sea level height was anomalously high (and the thermocline was deeper than normal) in the eastern Pacific along the equator during El Niño. The deepening in the east was compensated by a shallower than normal thermocline in the western Pacific, but the maximum shallowing (most negative sea level height anomalies) actually occurred slightly off the equator to the north and south. Surprisingly, this difference in behavior between west and east sides is because of the beta effect: the change of the Coriolis force with latitude. It is not immediately obvious why a change of the Coriolis force in a northward direction should produce asymmetric behavior between east and west, but this will become clear as we examine the behavior of an equatorial jet in section 4.6.1. Section 4.6.4 then shows what this implies for the response of the thermocline and current anomalies to a wind anomaly, and how mass transfer associated with equatorial jets is key to El Niño dynamics.

4.6.1 Equatorial jets and the Kelvin wave

It makes intuitive sense that because the Coriolis force is zero at the equator, currents should be able to flow east or west along the equator. Such currents confined to a narrow band along the equator are termed equatorial jets and these have interesting properties. For instance, a steady (constant in time) wind stress anomaly in the western Pacific can produce a continuous equatorial jet that extends along the equator far to the east. Furthermore, the current anomalies in an equatorial jet must be accompanied by thermocline depth anomalies along the equator.

The relationship between currents and thermocline depth in an equatorial jet arises because even a small distance off the equator the Coriolis force affects the currents. The currents can continue to flow east or west because the jet sets up thermocline depth anomalies that produce a north–south pressure gradient force (PGF). This can give rise to a self-consistent jet, whose latitudinal dependence and balance of forces are shown in Figure 4.10. Consider north–south forces on a current moving eastward. Just north of the equator, the Coriolis force will be southward, toward the equator. South of the equator, it will be northward. This can be balanced by a pressure gradient force outward from the equator, which would occur if the thermocline is deep (sea surface height is high). The pressure in the upper ocean is high where the thermocline depth h is large. Because the pressure reaches a

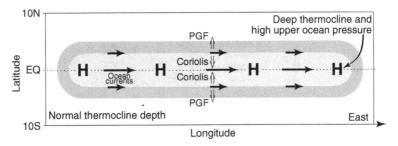

Fig. 4.10 Schematic of an equatorial jet. Eastward currents (black arrows) are maximum at the equator. Thermocline depth is shown by shaded regions, with larger values along the equator, decreasing to the normal value of thermocline depth away from the equator. Sea surface height has the same shape as thermocline depth (though the amplitude is much smaller). Open arrows show the balance of forces between the pressure gradient force (PGF) and Coriolis force at two locations. At the equator the north–south forces are zero even though the current is largest there.

maximum on the equator, there is no gradient in a small region straddling the equator. The PGF is thus zero at the equator, as is the Coriolis force. Slightly north of the equator, the change in thermocline depth with latitude yields a PGF directed northward from the high pressure region on the equator to the region with normal pressure away from the equator. This PGF balances the Coriolis force on the eastward current. South of the equator, the PGF is again away from the equator, balancing the Coriolis force toward the equator. Farther away from the equator, the PGF becomes small and the Coriolis parameter becomes large so the current needed to produce a balancing Coriolis force decreases to zero. Notice that β, the change of the Coriolis parameter with latitude, is crucial to this picture. Without the change in sign of the Coriolis parameter at the equator (and its increasing magnitude away from the equator), the equatorial jet could not exist.

If you consider westward velocity at the equator, then the Coriolis force will be outward from the equator, and a similar argument will work but with low pressure, i.e. shallow thermocline, at the equator. The jet will be the negative of what is shown in Figure 4.10, with currents reversed and a low along the equator. Thus a westward jet with shallow thermocline at the equator is also possible.

So far this seems symmetric for eastward or westward motion, but we have only considered the balance of forces. Now consider conservation of mass in the layer. An eastward current can add mass to the east, and thus extend the region of equatorial high pressure (deep thermocline) eastward, as shown in Figure 4.11a. The eastward jet can thus self-consistently extend itself to the east, adding mass as it goes. A westward jet removes mass from its eastern

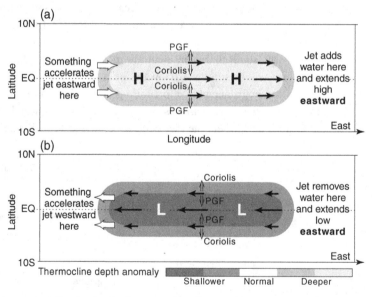

Fig. 4.11 Schematic of an equatorial jet showing that it can extend itself eastward but not westward. Shading shows thermocline depth; arrows show east–west currents south of, at, and north of the equator. The balance of forces is also shown. (a) For an eastward jet, mass carried by the currents is added to the eastern edge, and so the high on the equator is extended eastward. (b) For a westward jet, mass is removed from the east, so the low on the equator is also extended eastward. In either case, the jet extends eastward and a wind stress (open arrows) or other force is required on the western side to maintain the jet.

edge (Figure 4.11b). This is just what is needed to extend the low pressure (low sea surface height, shallow thermocline) toward the *east*. So in either case, the jet tends to expand *eastward*. On the western side of the jet, no such self-extending balance is possible: e.g. for an eastward jet, mass is being removed where the jet requires an equatorial high. Thus something else, such as a wind stress or a pre-existing pressure gradient, is required to accelerate the jet on the western side. The amplitude of this acceleration (e.g. the strength of a wind anomaly) is what sets the amplitude of the jet. In a loose analogy, the acceleration acts on the eastward jet like a leaf-blower but acts on the westward jet like a vacuum cleaner. In either case, the balance of the currents and thermocline depth within the jet permits it to travel coherently on the order of ten thousand miles eastward across the Pacific along the equator.

The eastward edge of the jet is known as a Kelvin wave. Its dynamics are very much like the waves in a sloshing bathtub, except that the demands of balancing the PGF with the Coriolis force imply that it can only propagate eastward and that it is trapped near the equator. The speed at which the Kelvin wave propagates is independent of the speed of the currents. Consider the case of a very small amplitude jet with weak currents. The thermocline depth change that accompanies these will likewise be small. At the eastern edge only a small amount of mass is needed to deepen the thermocline by the small amount needed to extend the jet. Thus weak currents can extend the jet eastward just as efficiently as strong currents. More detail on the derivation of the Kelvin wave and the width of the equatorial jet is available in sections 4.6.2 and 4.6.3.

Since water is being transfered east or west along the equator by the jet, something must also happen to the thermocline to the west of where the wind stress accelerates the jet. This is important to the change between El Niño and La Niña, as seen in section 4.6.4.

4.6.2 The Kelvin wave speed

A derivation of the Kelvin wave speed is provided by combining conservation of mass and the zonal velocity equation. Consider a jet with a small velocity anomaly u' and a thermocline anomaly h' that is much less than the mean depth of the thermocline H. At the eastern edge of the jet we want to find the increment of longitude δx by which the jet advances in an increment of time δt. Conservation of mass gives (either by directly considering Figure 4.12 or from Eq. (3.33))

$$\frac{h'}{\delta t} = H\frac{u'}{\delta x} \tag{4.1}$$

Now the eastward pressure gradient is $\frac{g\eta}{\delta x}$ where the surface elevation η is related to the thermocline depth anomaly by $\eta = \frac{\Delta\rho}{\rho}h'$. Thus the velocity equation, balancing acceleration against the PGF (by directly considering balances or from Eq. (3.4)), is

$$\frac{u'}{\delta t} = \frac{\bar{g}h'}{\delta x} \tag{4.2}$$

where we have defined a *reduced gravity* $\bar{g} = g\frac{\Delta\rho}{\rho}$ that is associated with the small difference in density between the upper layer and the deep water.

Combining Eq. (4.1) and Eq. (4.2) gives the phase speed $c = \frac{\delta x}{\delta t}$ as

$$c = (\bar{g}H)^{\frac{1}{2}} \tag{4.3}$$

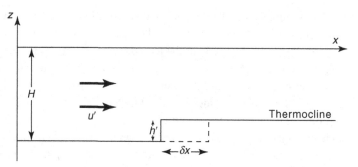

Fig. 4.12 Schematic of a Kelvin wave front at the eastern edge of an equatorial jet. In a time δt it advances a distance δx, deepening the thermocline by an amount h'. This depth change is in balance with the velocity u' according to the velocity equation and mass conservation for the layer. This yields a relation between δx and δt giving a propagation speed for the wave.

For typical density differences and thermocline depth, this is about 2.5 m s^{-1}. In other words, it crosses the Pacific basin in about 2 months. This is quite fast compared with El Niño evolution, so the Kelvin wave front itself does not play a strong role in setting El Niño time scales. Rather, the equatorial jet that it sets up plays an important role in transferring mass in the warm upper layer across the basin, since the jet continues even once the wave front has passed. Kelvin waves rocket across the Pacific frequently in response to weather fluctuations of wind. The effects of weather on El Niño can indeed be important, as is discussed in section 4.7. However, there can be confusion between the arrival of a Kelvin wave and the start of El Niño because the short-term response to a weather event is often more visible than the gradual process that occurs in the cyclic transition to El Niño.

4.6.3 What sets the width of the Kelvin wave and equatorial jet?

Here is a quick way of getting the latitudinal scale of the width, about the equator, of the Kelvin wave and the equatorial jet it leaves behind. A more complex derivation gives the same answer. Imagine that a wave motion similar to the Kelvin wave would start to move poleward from the equator at the speed c given in Eq. (4.3). After traveling a certain distance, it would be modified by the Coriolis force and would no longer be able to continue. This distance would be characterized by the length scale $L = c/f$. Note f is in s^{-1} and c in m s^{-1} giving units of length. This is the distance traveled at speed c in a time $1/f$. But what is a suitable value for f? Since f is zero at the equator and increases as $f \approx \beta$, the value of f reached is $f \approx \beta L$. Combining the two expressions gives

$$L = \sqrt{\frac{c}{\beta}} \tag{4.4}$$

This is known as the *equatorial radius of deformation*. The Kelvin wave has a Gaussian shape with this length scale (like a bell curve in latitude) and so does the equatorial jet (to a first approximation), as seen in Figure 4.10. The equatorial jet width is additionally affected by damping effects, but the scale is similar.

4.6.4 Response of the ocean to a wind anomaly

Now consider how the ocean responds to a wind anomaly, using the example of a small patch of wind in the region of the main ENSO wind anomalies: on the equator, near the International Date Line (180 degrees longitude). For simplicity, we will assume that the wind stress suddenly appears and remains constant for about half a year. This allows us to focus on the ocean response, without worrying about how the wind is changing (as it would during an actual El Niño). We will then consider what happens if the wind stress weakens. This will give us a prototype for what the ocean does at the end of an El Niño, to bring it toward La Niña. We choose a westerly wind anomaly, appropriate to an El Niño phase and illustrate the evolution of thermocline depth anomalies and ocean upper layer current anomalies in Figure 4.13. The behavior would be the same for an easterly wind anomaly, but with the signs reversed (deeper thermocline replaced by shallower thermocline, and current direction reversed). In considering the horizontal pressure gradients in Figure 4.13, recall that deeper than normal thermocline implies high pressure anomalies in the upper ocean, and shallower than normal thermocline implies low pressure.

In Figure 4.13a, the eastward acceleration by the wind has produced an eastward equatorial jet. This is moving mass to the east, deepening the thermocline as it goes (setting up a high pressure region along the equator that is in balance with the currents). The mass is being taken from the region to the west of the stress, which should cause a low pressure region. But low pressure along the equator cannot balance an eastward current. The type of self-consistent balance in the equatorial jet above cannot work in the same way, with the anomalies confined to the equator. How can one get eastward currents in a region that must have low pressure due to the depleted upper layer mass? To balance the Coriolis force on eastward currents the pressure gradient force must be away from the equator, and thus the pressure must be lower off the equator than on it. This balance is accomplished by removing more upper-layer mass a few degrees off the equator than is removed on the equator. Thus to the west of the wind, the shallowing of the thermocline is strongest off the equator. This is seen in Figure 4.13b. There is still an eastward equatorial jet, but it is in balance with a thermocline that is even shallower to the north and south than it is on the equator. Some water must recirculate around these lows, further from the equator (approximately in geostrophic balance). At the western edge, this pattern is expanding westward, taking mass from the undisturbed region to the west. This leading edge is known as a *Rossby wave* front.[7]

Some other features may be noted in Figure 4.13b: on the eastern side of the Pacific, the deepening that began on the equator is spreading up the coast, where it can eventually reach past California, and even as far as Alaska. On the equator, the deepened thermocline is in El Niño conditions. In a coupled system, the deepened thermocline along the equator to the east of the wind anomaly would produce warm SST anomalies, which would sustain the westerly wind anomaly, as in the Bjerknes hypothesis. In the example shown here, we began by imposing a fairly strong wind anomaly on the ocean, and at this stage this wind anomaly is consistent with what the warm SST would produce, so we simply hold it constant into the next panel, to see how the ocean component evolves.

One important thing to recall from the equatorial jet discussion above is that the deepened thermocline along the equator can only be sustained in balance with the eastward

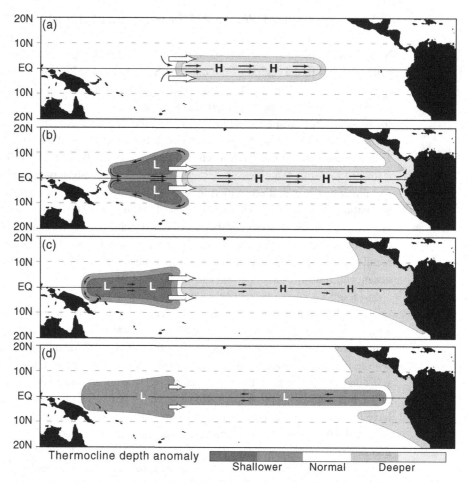

Fig. 4.13 Response of the ocean to a wind anomaly. A specified westerly wind anomaly (broad arrows) is suddenly turned on and then held constant until the last panel, when it is weakened. Shading shows the thermocline depth anomaly. Regions with deeper thermocline have higher pressure in the upper ocean, denoted H, relative to surrounding regions of normal thermocline depth. Likewise shallow thermocline regions have low pressure (L) relative to normal. Thin arrows show the ocean currents in the layer above the thermocline. (a) After a few weeks. To the east of the wind anomaly there is an equatorial jet whose eastern edge is extending rapidly as a Kelvin wave. (b) After a couple of months. On the western side a different balance is required to supply mass to the equatorial jet (see text). (c) After about half a year. The depletion of upper-layer mass on the western side has extended to the western boundary, and the balances at the leading edge have changed. Currents in the equatorial jet have thus weakened, and the deep thermocline anomaly in the east has been reduced. (d) The wind has been weakened (similar to what would happen through Bjerknes feedbacks). Imbalance of the east–west pressure gradient with the weaker wind has led to a reversed equatorial jet moving upper-layer mass to the west, and a shoaling thermocline extending eastward.

equatorial jet. However, this current requires warm water mass to be continually removed from the west. One might say that the El Niño is living on borrowed mass, and, even worse, depends on being able to continue to borrow at a given rate. This analogy suggests that the warm conditions are unlikely to continue indefinitely, and a turnabout must come.

As the jet reaches the western boundary, it can no longer "borrow" warm water mass from the west. This process may occur gradually and can be complicated by the geography consisting of many islands, but the overall effect is quite robust when considered in terms of no longer being able to supply mass to the jet. The anomalous current must therefore be reduced. Since the thermocline depth is in balance with the current, the thermocline depth anomaly in the east must also decrease. This is pictured in Figure 4.13c. If the wind stress remained constant, the ocean would come at last to equilibrium in this state.

Now we must consider the effects of the feedbacks in order to understand the next step. Since the thermocline depth in the east is less deep, the SST warm anomaly must begin to decline relative to the previous stage. Thus the winds would also weaken. In Figure 4.13d, we show a case where the winds have been weakened and consider the ocean response. The pressure gradient that had finally come into balance with the previous wind (Figure 4.13c) with high pressure (deep thermocline) in the east and low pressure in the west is now greater than the wind force. So warm water begins to flow back westward across the basin. This occurs in a westward equatorial jet, like the one from Figure 4.11b. Recall that even though the jet flows westward, it extends eastward, so the anomalies of shallow thermocline that were previously in the west extend out toward the east. These correspond to subsurface cold anomalies that we saw built up in the west in Figure 4.8c and beginning to extend eastward in Figure 4.8d and Figure 4.9 to end the 1997–98 El Niño.

At this point the thermocline is shallower in the east than normal. Upwelling carries colder water to the surface and sends the system into La Niña as was seen in Figure 4.9b and c. As the wind anomalies reverse and become easterly, the thermocline shallows still more. The anomalies then look like Figure 4.13b with the signs reversed: shallow thermocline anomalies instead of deep, and easterly wind anomalies instead of westerly. The thermocline anomaly in the east along the equator must have westward currents in balance with it. Just as in the case of El Niño, the equatorial jet proceeds to transfer mass across the basin, but in the case of La Niña, it moves warm water mass toward the west. This transfer continues to deepen the thermocline in the west, especially off the equator, until the Rossby wave front reaches the western boundary of the Pacific. Then the westward jet must weaken (as in Figure 4.13 c and d but with the signs reversed) and deep thermocline anomalies develop along the equator, setting the stage for the development of the next El Niño. And so on, for evermore.

4.6.5 The delayed oscillator model and the recharge oscillator model

In reading the literature on El Niño, you may encounter references to the *delayed oscillator model* or the *recharge oscillator model* of ENSO. These are simple models of ENSO that aim to capture its essence in one or two differential equations for a single variable. The delayed oscillator model makes approximations that emphasize reflection of a Rossby wave front from the western boundary and its delayed negative feedback on SST via a Kelvin wave. The recharge oscillator model emphasizes the measure of mass between the region off the equator in the west and the equatorial region with subsequent influence on SST in the east. The explanation given above is based on the features of both these models and more detailed evolution seen in other ENSO models. The delayed oscillator model, developed in 1988–89 by Suarez, Schopf, Battisti and Hirst, was important in first illustrating the role of thermocline evolution in setting the slow evolution of ENSO.[8]

4.6.6 ENSO transition mechanism in brief

To summarize: the Bjerknes hypothesis gives the feedback mechanism for intensifying an El Niño or La Niña phase. In an El Niño, westerly wind anomalies, deep thermocline anomalies and warm SST anomalies develop in cooperation. In a La Niña, easterly wind anomalies, shallow thermocline anomalies and cool SST anomalies develop in cooperation. The transition comes about because, considering the El Niño case:

- The ocean adjusts differently to the east of a wind stress than to the west. To the east, the change of Coriolis force with latitude gives a balance between the Coriolis force and the pressure gradient force for an equatorial jet that extends rapidly to the east along the equator. To the west, more complex motions must occur that move water from off the equator.
- The deepening of the thermocline to the *east* of the wind stress during the early stages of El Niño is thus greater than can be sustained in the long term. Subsurface conditions evolving slowly on the western side of the basin are not yet in balance.
- The deepened thermocline in the east is in balance with an equatorial jet that transfers warm water mass (in the layer above the thermocline) from west to east. The western edge of the jet (a Rossby wave front) progresses westward.
- This progression runs into the western boundary and stops. The jet must then weaken.
- When the jet weakens, the thermocline in the east shallows, so the warm SST and westerly wind anomalies are reduced.
- The pressure gradient is then greater than the wind force, so shallow thermocline anomalies progress eastward. In the east, they cool SST, beginning the La Niña half of the cycle.
- Repeat with signs switched, for La Niña.

Figure 4.14 illustrates the full cycle. Note that the positive feedback loops at the top and bottom of the figure are the same mechanisms but for cold and warm phase respectively. These feedback loops are exactly as in Figure 4.3, which showed the El Niño case. Likewise the mechanisms in the transition phases on left and right sides of the figure are the same, but with anomalies of opposite sign.

In even briefer terms: the thermocline depth anomalies (i.e. subsurface temperature anomalies) and ocean current anomalies in the western part of the basin are more slowly evolving than in the east. When they finally come toward balance with El Niño winds, they end the positive feedback of thermocline depth in the east onto SST and wind anomalies, causing a reduction in their amplitude. This in turn puts the thermocline depth and currents out of balance again and causes a shallowing of the thermocline in the east, beginning La Niña.

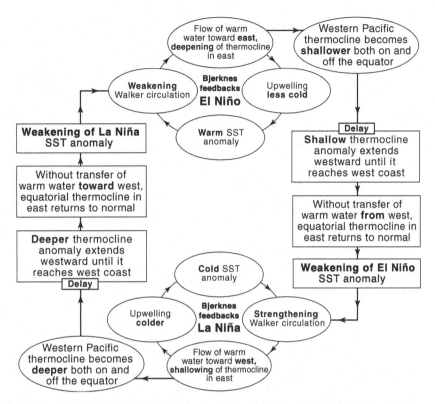

Fig. 4.14   Schematic of the mechanism for ENSO cyclic behavior. The feedback loops marked "Bjerknes feedbacks" at the top and bottom (ovals) indicate the sequence of ocean–atmosphere feedbacks that reinforce El Niño and La Niña extreme phases respectively. An additional oval in each phase gives the thermocline behavior in the western Pacific as a consequence of deepening or shallowing in the east. Initially the western Pacific thermocline has no effect on the feedbacks but it eventually has a delayed effect leading to the steps of the transition phase (rectangles). The first two steps involve subsurface processes not visible at the surface. Bold type denotes aspects that differ between phases.

4.7 El Niño prediction

Any phenomenon that evolves slowly or, even better, tends to be cyclic has the potential to have some degree of predictability. If El Niño conditions are present in September, it is reasonable to predict that they will persist through the winter. It might even be tempting to predict a La Niña for two years after the El Niño, and vice versa, since that tends to be the most common time interval. However, these predictions might not be very skillful. Initially, statistical models were developed to improve these considerations by looking for lagged relations in time. For instance, given a long time series of SST, wind and/or sea level pressure data one might seek a statistical regression coefficient of SST in a given month with variables from several months earlier. Such models were of some use, even before the dynamics of El Niño was understood, and some statistical prediction schemes are still in use and show considerable skill. The current phase of El Niño forecasting came about with the development of coupled ocean–atmosphere models with predictive capability.

Because of the slow time scales of evolution of subsurface temperature anomalies, when initial information about these anomalies is inserted into the ocean component of the model, and the atmospheric component captures the feedbacks essential to the development, one can hope to make skillful predictions of the next El Niño or La Niña some time in advance. This insertion of information must be done in a balanced way or huge errors will result. For example, imagine putting in the subsurface temperature anomalies during the onset of La Niña but erroneously omitting the wind anomalies. Without the right balance the model would slosh into an erroneously early El Niño. Other factors that must be taken into account include the following data are available only at certain locations, whereas the model must have initial conditions at all grid points, and data have errors, as do the models. The overall process of data interpolation and balancing with the model dynamics, while accounting for the error bars on both, is called *data assimilation*. This procedure can be more costly in computer time than the model simulation itself. Although forecast capability was initially developed in the universities and other research institutions on an experimental basis, the effort of sustaining operational predictions with detailed data assimilation and large models is being taken over by specialized forecast centers. These include the US National Center for Environmental Prediction (NCEP), the European Centre for Medium-range Weather Forecasting (ECMWF), and the International Research Institute for Seasonal-to-Interannual Climate Prediction (IRI).

Figure 4.15 shows a forecast of the development of the 1997–98 El Niño made from March of 1997. The forecast is one of several successful forecasts of this event, in this case made by NCEP. The NCEP model was predicting an El Niño for the following winter from as early as January 1997. This forecast, initialized with data from March, predicted a warm event growing through the summer and into the winter, and this forecast was made prior to any sign of warm SST anomalies in the observations. However, the March forecast predicted an El Niño of normal magnitude. By June, a number of models were predicting a large event but this was perhaps less of a challenge because in June the warming was already well under way, with large anomalies already present.

Overall, the forecast may be considered a successful one for forecasts at such lead time. It allowed NCEP forecasters, along with other forecast groups, to issue warnings of the event before there was any sign of warming in surface indices, and a good part of a year before the maximum warming. Quantitatively though, Figure 4.16 shows a more modest view of the skill of such forecasts. It directly compares forecasts issued regularly over several years with the actual SST anomalies for the Niño-3 region (index region shown in Figure 1.5). In each case, the verifying observations are only available after the forecast is made. In the format shown, each actual forecast produces one point on each plot. For instance, a forecast issued in March of 1997, for the average SST of April–June, is termed 3-month lead, since it is based on the average of 3-month forecasts initialized in January, February and March. The forecasts from March for July–September and October–December are termed 6-month and 9-month lead forecasts. Thus the series of forecasts shown in each plot are each independent from the last in that they began from a different time. If random initial data were used, or the forecast model were random, they would not be related to each other.

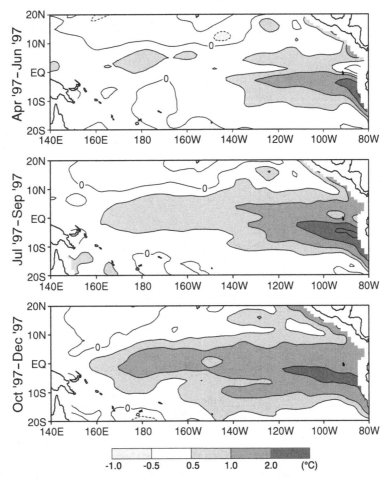

Forecast of SST anomalies (as 3-month averages) made from March 1997 for subsequent months, during the onset of the 1997–98 El Niño from the NCEP ocean–atmosphere model. Courtesy of the National Center for Environmental Prediction. Contour interval 1°C with additional contours at ±0.5°C.[9]

As one expects, the shorter lead time forecasts compare better with what was subsequently observed. While the 9-month forecast predicted a moderate warming, what actually happened was a large El Niño, with a very rapid onset. Later forecasts (at shorter lead time) did somewhat better at the amplitude, but by then the El Niño was under way. This gives a fair picture of the state of ENSO forecasts. They can be very useful, but they are not perfect, especially at longer leads.

4.7.1 Limits to skill in ENSO forecasts

The loss of skill in forecasts can be divided into two categories: (i) imperfections in the forecast system, and (ii) fundamental limits to predictability. The imperfections in the forecast system are often something we can change if willing to spend enough time and money.

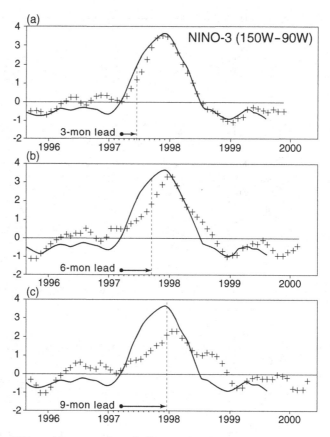

Series of forecasts of SST anomalies averaged over the Niño-3 region of the equatorial Pacific. The solid line gives the observed, while each cross represents a forecast, made from several months previous. To the extent that the crosses agree with the observed, the forecasts were accurate. Note that at the end of the time series, the data are not displayed because they were not available at the time that the last forecast was issued. (a) Forecasts at 3-month lead time. (b) Forecasts at 6-month lead time. (c) Forecasts at 9-month lead time. Courtesy of the National Center for Environmental Prediction.

They include: error in model parameterization of sub-grid-scale motions (like atmospheric convection and ocean mixing), the relative scarcity of points with data, errors in measurement of the data, and so on. However, even if we were to achieve almost perfect models and tremendous data coverage, we would not be able to predict ENSO (or other climate phenomena) indefinitely into the future. The fundamental limits to predictability come from the complex, chaotic nature of the system.

Weather forecasts are limited to less than two weeks of useful skill. This is because of chaotic behavior in weather systems. In fact, the primary innovation in the invention of chaos theory came from Edward Lorenz of the Massachusetts Institute of Technology (MIT), precisely from work to explain why weather prediction ran into problems at long leads. Suppose you take two copies of the same weather model (based on the equations of Chapter 3, or even simplified versions) and try to predict the weather in one (the control) using the other. There are thus no model errors. The data at the initial time of the forecast can be made almost perfect, except that there are tiny errors due to round-off error in the computer. These errors grow exponentially. After a characteristic time for this error growth, the weather in the forecast model is completely unrelated to the weather in the control. For different chaotic systems, the characteristic time differs, but the principle remains the same. For weather, two weeks is about the time scale for such errors to grow, even at the large scales (for small-scale motions, such as individual thunderstorms, it is shorter).

For climate prediction, such as ENSO forecasts, the forecast quantities are averaged over many weather systems. Thus to some extent the problem of unpredictable behavior in individual weather systems is reduced. But the problem is not entirely removed. From the point of view of El Niño, the fluctuation of weather about the slowly changing average looks like a random forcing. Since chaos causes weather maps to be essentially unrelated after two weeks, we could approximate weather by randomly choosing a new weather map every week or two (except we have to be careful to retain the more slowly evolving effects of SST on the atmosphere). Such random processes are known as *noise* (by analogy with, for instance, the static that competes with the signal on a radio). The term *weather noise* is thus used for the effectively random (on long time scales) weather variations that can impact ENSO.

Consider the development of an El Niño event following the Bjerknes hypothesis. The weakening of the SST gradient leads to a weakening of the trade winds. But now suppose the weather that month happened to have storms, or other transient phenomena, that caused the trades to weaken less than they would on average for this SST gradient. Then the onset of the El Niño might be delayed, or the warming might be weaker than expected. Likewise consider a slowly ending La Niña. If a weather event producing a westerly wind anomaly appeared, it might hasten the onset of the next El Niño. Thus not only can El Niño affect the weather, but the weather can, to some extent, have effects on El Niño.

To illustrate how the cumulative effect of weather noise can gradually degrade the predictability of ENSO, Figure 4.17 gives an idealized view in which weather events "kick" the ocean once in a while, causing jumps in SST anomaly. In this simplified example, a

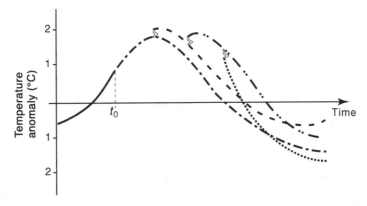

Fig. 4.17 The effects of weather noise on the ENSO cycle, schematized as causing random perturbations to the Niño-3 SST anomaly. The SST evolution between noise events is idealized as being smooth and predictable. A forecast from time t_0 is given by the dot-dashed line. The dashed line gives the SST evolution after the first noise event, as it would occur if there were no subsequent noise events. The dash-double-dot line gives the evolution after a second noise event, and so on.

forecast from time t_0 is perfect up until the first noise event, and then the evolution follows a slightly different path. The first forecast still is of some use, but it has incurred some error. After a second noise event, there is even more error. Occasionally noise events can have partially cancelling effects, but on the whole, the more noise events, the more the error tends to grow. The longer the lead time of the forecast, the more noise events will have occurred, so the less accurate one expects the forecast to be. In the example above, suppose a second forecast were made just after the first noise event. Its initial data would include the effects of the first noise event so this forecast would have no error until the second noise event, but then the forecast would begin to accrue error. Thus the growth of the error depends on the length of the forecast, i.e. the amount of time between the last data given to the forecast system and the time period for which the prediction is given. In a more realistic setting, the weather noise would be perturbing the system all the time, but most of the kicks would be small, and larger-amplitude weather effects would occur only occasionally, as shown. The error would tend to grow steadily with the length of the forecast.

The above description makes the point that loss of predictability in the atmosphere can eventually lead to fundamental limits to predictability in the coupled system. In the coupled system, errors in the ocean initial conditions can be equally important. A small error in the sea surface temperature (either from the initial conditions, or from subsurface error affecting the surface) can grow by feedbacks with the atmosphere, as in the Bjerknes mechanism. Furthermore, once the atmosphere starts to change, the development of weather patterns is perturbed and the weather noise effects discussed above affect the predictability. Overall, small errors in the initial conditions in either ocean or atmosphere lead to a growth of error. The time scale of a few seasons for this error to become large limits the predictability.

Figure 4.18 shows an example of such effects in an ocean–atmosphere forecast model from the European Centre for Medium-range Weather Forecasting. A set of forecasts, known as an *ensemble*, has been made over the same period. Each forecast in the ensemble is an integration of the ocean–atmosphere system forward in time. They have slightly different ocean conditions at the initial time. This then changes both the atmospheric response to sea surface temperature and perturbs the sequence of weather patterns. The atmospheric differences feedback on the evolving La Niña, so each forecast behaves differently. One can easily pick out the average behavior of the ensemble, but the variations among the different cases increase with time. We can imagine that one of these runs is the actual evolution of the real Pacific climate system, and the rest are forecasts with a model that was perfect except for small errors in the initial conditions. The error growth slowly makes the forecasts less reliable as predictors of the real system. Note that after 6 months, the ensemble spread (the noise) is still not as large as typical SST variations (the signal) so forecast information is still useful, even if imperfect.

These real-time forecasts contain a combination of errors that can be reduced by better observations and model improvements and fundamental error growth due to weather noise. Recent research suggests that weather noise may limit the range of useful forecasts to less than half an ENSO cycle, and may be a dominant contributor to error by the end of a one-year forecast. Exactly how fast the error due to weather noise grows is an important question, since weather noise sets fundamental limits to predictability. No matter how much work is done to improve the models, they cannot predict at longer leads than the weather noise

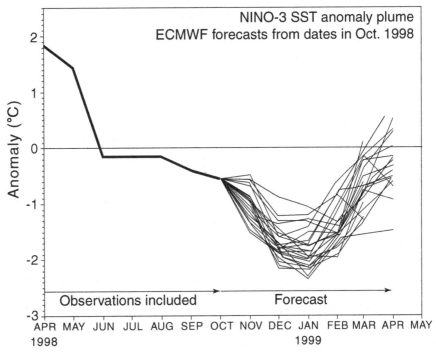

NINO-3 SST anomaly plume
ECMWF forecasts from dates in Oct. 1998

Fig. 4.18 An ensemble of forecasts during the onset of the 1998–99 La Niña. The SST anomaly over the Niño-3 region is plotted as a function of time. The thick solid curve gives the SST anomalies estimated from observational data included up to the time the forecast begins (October 1998). The series of thinner curves gives forecasts that differ in random initial differences in the ocean conditions. Forecast data courtesy of the European Centre for Medium-range Weather Forecasting.

permits, probably less than two years. One can also ask the question, what if we lived in a world with no weather? In some climate models, this can be arranged. In that case, chaotic behavior may also occur in the ENSO cycle itself. It appears that this probably has quite long error growth times compared with the weather noise effect. It occurs mostly because of the complex interaction of ENSO with the seasonal cycle of the climate state in the Pacific.[10]

4.8 El Niño remote impacts: teleconnections

Although the origin of the El Niño phenomenon lies in the Pacific Ocean, its effects are felt over a large portion of the Earth. These remote effects are called teleconnections. The changes in atmospheric patterns occur over broad scales, although they have preferred paths of influence. They have reliable effects on quantities that humans care about only in certain regions. In order to assess if a region has an ENSO connection, it is necessary to take as many years of data as possible and perform statistical analysis on the relation of the weather to ENSO indices. Since there are not that many El Niño events in the time for which reliable data are available, and since weather has large chaotic (effectively random) fluctuations of its own, it is necessary to be skeptical of claimed El Niño relations unless they are shown

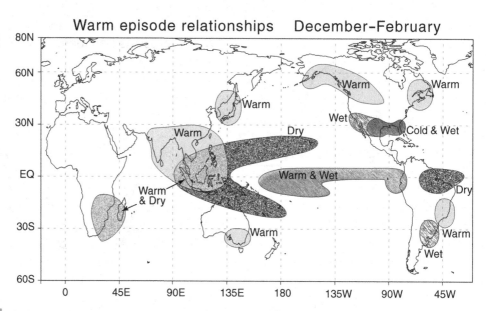

Fig. 4.19 Regions with statistically reliable relation of precipitation and surface air temperature to El Niño. After Ropelewski and Halpert (1987, 1989), Halpert and Ropelewski (1992), and updated with data from the International Research Institute as described in Mason *et al.* (1999).[11]

to be statistically significant. On the other hand, it is also possible that among hypothesized relations, there are some that will be proven with time. Figure 4.19 shows regions that have been shown with some degree of reliability to be affected by warm ENSO phases. The effects for cold phases are in approximately the same regions, but with the opposite sign, although La Niña episodes tend to be smaller in magnitude and thus have weaker effects.

The regions that have significant impacts change with season because the ENSO influences interact with other features of the climate system. In the tropics, the position of the climatological convection zones is important, while in midlatitudes the positions of the jet streams and preferred regions for storm development (*storm tracks*) are important. In Figure 4.19, the northern hemisphere winter map is shown. In the northern summer season, midlatitude effects in the northern hemisphere are reduced and more effects are seen in the southern hemisphere. Tropical effects tend to continue all year long.

The basic mechanism by which remote influences are produced in the atmosphere is via large-scale atmospheric waves as discussed in section 3.7. Atmospheric Kelvin and Rossby wave dynamics are akin to those seen in the ocean for El Niño in that they involve transfer of atmospheric mass by anomalous winds, which in turn affects pressure, while the winds are influenced by the Coriolis force. They differ in that the atmospheric mean winds strongly affect the Rossby wave propagation (section 3.7.3). The regions along the equator over South America that are dry during El Niño and wet during La Niña are believed to be due to atmospheric versions of the Kelvin wave. The regions of surface effects seen in Figure 4.19 over western Canada and the northwestern USA, the east coast and Florida are in part related to the stationary Rossby wave pattern of Figure 3.15, although other atmospheric processes tend to complicate matters.

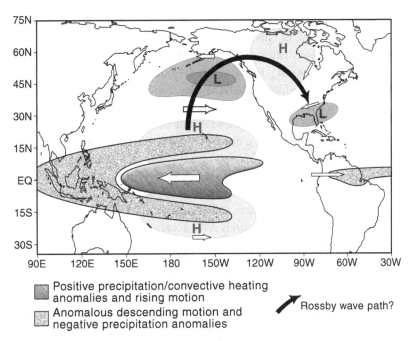

Legend:
■ Positive precipitation/convective heating anomalies and rising motion
▨ Anomalous descending motion and negative precipitation anomalies
↗ Rossby wave path?

Fig. 4.20 Patterns of typical response to El Niño observed for northern hemisphere winter. The regions denoted H or L show areas of anomalously high or low pressure (or geopotential) in the upper troposphere. Upper-level wind anomalies are indicated with open arrows. The solid arrow indicates an originally hypothesized path of Rossby wave influence connecting these highs and lows to the tropics.[12]

Figure 4.20 shows observed effects of El Niño in the upper troposphere, as well as some of the steps in understanding dynamical connections from the tropical regions to the midlatitudes. The upper tropospheric anomalies are shown because they tend to be good indicators of the large-scale atmospheric processes that communicate El Niño effects over long distances. The regions marked H and L are regions of anomalously high or low pressure at a height of 8 or 10 kilometers. Equivalently, they are also regions of high or low geopotential height on a given pressure surface (say, 200 mb). The latter is commonly used since atmospheric dynamics is usually done in pressure coordinates (section 3.1) but the interpretations are the same in height coordinates. The pattern consisting of a low in the Pacific off the west coast of North America, a high over the Canadian Provinces and a low over Florida is known as the Pacific–North-American (PNA) pattern. The highs and lows in this pattern tend to be correlated with each other and with El Niño. In the tropics, a typical El Niño pattern of increased precipitation in the central Pacific is shown (with decreased precipitation to the west). The increased precipitation corresponds to increased convective heating in the region of anomalous warming during El Niño. The thermal circulation argument (Chapter 3, Figure 3.5) implies that there will tend to be high pressure at upper levels in the warmed regions. The maxima of these high pressure anomalies are slightly off the equator because the winds responding to the anomalous pressure gradient also have to come into balance with the Coriolis force.

Consider now the anomalous high in the northern hemisphere subtropics that is directly related to the El Niño warming. To its north, there tend to be anomalous westerly winds, approximately in geostrophic balance with the pressure gradient. These produce an increase of the subtropical jet stream in that region. This increase in the subtropical jet is itself important to El Niño impacts but it also has consequences for the PNA pattern. In general any perturbation to the circulation in one region must be balanced by compensating flow elsewhere. Because the Coriolis force varies in latitude, a compensating return circulation around a single high or low at a different latitude cannot simultaneously be in perfect geostrophic balance. Instead the flow sets up a Rossby wave pattern with a sequence of highs and lows. Rossby waves occurring in a region of westerly climatological winds have a characteristic shape (Figure 3.15) very much like the pattern of highs and lows seen in the figure.

A Rossby wave pattern is fairly simple for a climate model to capture, so if this original hypothesis had been entirely right, forecasting impacts over North America would be easier than it has turned out to be. It is now known that while the Rossby wave argument was a step in the right direction, other effects also enter along the way, including changes to the storm tracks. Storms in turn transport momentum, as outlined in Chapter 2. Thus changes in the region with a large amount of transient variance actually feed back to enhance the changes in the jet stream. Furthermore, the upper-level high and low pressure anomalies interact with similar patterns associated with ocean–continent contrast in the climatology. The details of these processes are a subject of active research.[13]

Figure 4.21 shows the changes to the storm tracks associated with ENSO, superimposed on the pattern of the upper tropospheric highs and lows from Figure 4.20. The thermally forced high in the subtropics tends to increase the subtropical jet in the eastern Pacific during El Niño. The low to the north (in the PNA pattern) helps reinforce this. Under normal conditions the most intense part of the jet occurs in the western subtropical Pacific, so this intense jet is extended during El Niño. This affects the storm tracks in two different ways during El Niño. First, the extended jet carries the storms that develop in the western Pacific further eastward (on average). Second, because the tropics is warmer there is a larger meridional temperature gradient that may slightly enhance the number of storms. The extension of the Pacific storm track during El Niño in the region of the enhanced jet stream may be seen in Figure 4.21.

During La Niña, the jet stream track not only changes but becomes more variable from month to month within the winter. Overall the storms tend to end up deviating further north in the Pacific. Qualitatively the impacts are fairly common sense, once one knows how the jet stream changes. For instance, Californian El Niño impacts are, roughly speaking, due to shifts in the jet stream dumping a few of the storms that would normally hit Seattle onto Los Angeles.

The impacts tend to be greater for larger SST anomalies. For a small amplitude El Niño or La Niña event, the effect may not be easily detectable compared to the essentially random variations of the weather in a particular region. Furthermore, it is not accurate to say that a given storm is due to El Niño. It is more realistic to say, for instance, that El Niño shifts the probability distribution for total precipitation over a month or a winter. An example is schematized in Figure 4.22. By chance, one could still get a winter with lower than normal precipitation during El Niño. But if you bet on higher than normal precipitation during

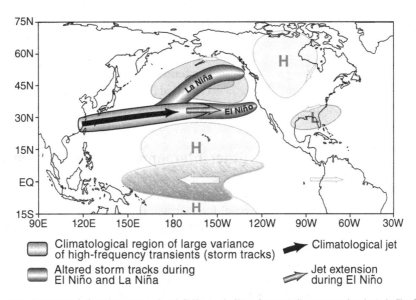

Climatological region of large variance
of high-frequency transients (storm tracks)   Climatological jet

Altered storm tracks during
El Niño and La Niña   Jet extension
during El Niño

Fig. 4.21 Jet stream and storm track changes associated with El Niño or La Niña. Arrows indicate upper-level winds. The filled arrow shows the position of the jet in the climatology, and the shaded arrow shows the anomalous extension during El Niño (see key). The jet position is more variable during La Niña. The dark shaded region indicates the storm track (see key), as measured by large variance of high-frequency transients. Extensions during El Niño and La Niña are indicated. Patterns of ENSO influence in the upper troposphere and the central Pacific precipitation anomaly are repeated from Figure 4.20 for reference as they occur for El Niño. For La Niña similar patterns occur but with reversed sign, i.e. high pressure instead of low, etc.

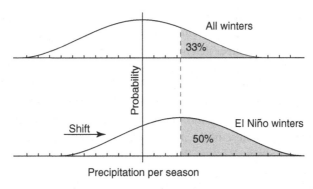

Fig. 4.22 Schematic indicating an idealized shift in the probability distribution for total precipitation in a given winter. The upper curve illustrates the probability calculated for all winters. The center is the average precipitation, but there is considerable probability of getting higher or lower rates just through random weather fluctuations. The shaded area shows the top tercile, i.e. the top 1/3 of all the winters. The lower curve illustrates the probability distribution given that there is an El Niño. For the case shown, the probability of getting precipitation with a value in the top tercile of all the winters increases to 1/2. This is roughly what holds for rainfall averaged over California. For Southern California, the shift in the distribution would be larger, so that about 70% of El Niño winters qualify as being in the top tercile.[14]

several El Niño winters, you would tend to come out a winner in the long run. This highlights the fact that forecast skill does not have to be perfect to be useful. The required level of skill depends on the user. The insurance industry, for instance, can benefit from very low levels of skill, as long as there is some information about shifts in the probability distribution. An individual farmer, trying to decide what type of crop to plant, would be unwise to base his decision purely on an El Niño forecast if the cost of being wrong were high.

4.9 Other interannual climate phenomena

Success at producing useful forecast skill for some El Niño-related phenomena opens the question: what other aspects of the climate system can be predicted? This is an area of current research.[15] Examples of possible targets include rainfall in the northeastern part of Brazil, known as the *Nordeste* region, and in the Sahel region of Africa, a semi-arid region that lies between the Sahara desert and the tropical convergence zone. In the case of the Nordeste, Atlantic SST anomalies appear to influence the rainfall. To the extent that these SST anomalies can be predicted, advance warning of a good or bad rainy season might be invaluable to this region, which gets rain only in a few months of the year. Other possible targets include the Southeast Asian and Indian monsoon. It was the failures of the Indian monsoon that first motivated Sir Gilbert Walker to begin studies in this area. Success so far has not been easy, but perhaps the coming decades will see some progress.

4.9.1 Hurricane season forecasts

Forecasting the development and paths of individual hurricanes has long been one aspect of weather forecasting. By turning one's attention to questions of the statistics of hurricanes, such as the number that occur in a given year for a region, one enters the realm of climate and climate variability. For a number of years, experimental forecasts have been issued for certain statistics of the Atlantic hurricane season, and this is a form of climate forecast. The Atlantic hurricane season is often defined as June through November, the months in which hurricanes form in the northern tropical Atlantic. This has received particular attention because of the impacts on the Caribbean, Central America and the southeastern United States. Western Pacific and Indian Ocean hurricane statistics are also of interest. Because hurricanes tend to move westward initially, carried by tropical easterly winds, eastern Pacific hurricanes are usually of less concern as they normally do not make landfall, although occasionally they can reach Hawaii or Baja California.

Meteorologists have conventions for naming hurricanes and for stages of hurricane development. When a region of enhanced convection develops its own circulation with an identifiable low center and sustained winds over 10 m s^{-1}, it is termed a tropical depression; when winds pass 18 m s^{-1} it becomes a tropical storm and receives a name from a predetermined list (alphabetical lists of male and female names in alternating years). Statistics are often compiled for "named storms," that is, for a combination of tropical storms and hurricanes. At 33 m s^{-1} (74 miles per hour), a storm is designated a hurricane. As it develops

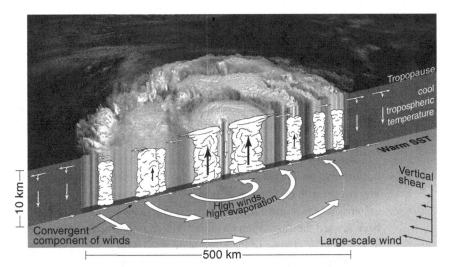

Fig. 4.23 Schematic of a hurricane, with some of the large-scale factors that affect development. A vertical cross-section through the storm is overlaid on a three-dimensionalized satellite view of cloud top, with surface winds (large white arrows) in the foreground, spiralling about the center of the storm. The wind component that converges toward the center at the surface and diverges outward at upper levels near the tropopuase is important for the strong convection (schematized as clouds in the cross-section) and associated upward motion (dark arrows). Compensating downward motion occurs away from the storm. Smaller regions of downward motion occur in the "eye" region at the center, and sometimes in spiral bands within the storm. From a climate perspective, the questions of interest are how changes in climate variables such as SST, tropospheric temperature and winds at scales larger than the storm affect the statistics of hurricane development and movement.[16]

there are gradations of intensity, designated as categories 1 through 5, with category 5 having winds in excess of $69 \, \mathrm{m \, s^{-1}}$ (155 miles per hour), known as the Saffir–Simpson scale.

Figure 4.23 shows a schematic of a hurricane and some of the large-scale factors that can influence development. The typical diameter of the region with strong convection and winds is on the order of 500 km. The component of the wind rotating about the center of the storm is by far the fastest, but the smaller component converging at low levels toward the center is important to the storm development. Air converging at low levels and rising in the strong deep convective regions at the center of the storm imports moisture, which feeds the convective heating and precipitation. The converging air also helps to spin up a vortex, a rapid circular motion of the winds about the center of the storm. As the air converges, the Coriolis force causes the winds to rotate about the low pressure at the center. Conservation of angular momentum implies that as the rotating wind is carried inward, it must speed up. The high winds at the ocean surface lead to high evaporation and this supplies moisture to feed the convective heating. If the feedback between convection, winds and evaporation is strong enough, the storm intensifies.

High sea surface temperatures, which tend to heat and moisten the boundary layer, favor the development of hurricanes. However, just as for other motions involving convection, this also depends on the temperatures in the troposphere. In general, a rising parcel of air must be warmer than the surrounding air to be buoyant, so warmer temperatures at upper levels tend to disfavor convection. For hurricanes, there is a theory due to Kerry Emanuel of the Massachusetts Institute of Technology that treats the storm as a heat engine.[17] The amount of energy available to be transformed into the kinetic energy of the hurricane winds depends on the difference in temperature between the warm sea surface and cold upper tropospheric temperatures typical of the air in the upper-level outflow. Thus warm SST with normal or cooler than normal troposphere favors hurricanes. If both the SST and troposphere are warmer than normal, as is expected to occur under global warming, then the question is more subtle, and depends on the relative warming.

The pattern of winds at spatial scales larger than the hurricane itself also influences the evolution of the hurricane. These winds exert a steering effect on the hurricane movement. The typical path of a hurricane is thus toward the west in the deep tropics where winds are easterly. As the storm drifts poleward it enters the region of westerly winds and is carried back toward the east, often changing to look more like a conventional midlatitude storm as it goes. The change of the large-scale wind with height, known as the *vertical shear* or wind shear, appears to have a disfavorable influence on hurricane formation, presumably by interfering with the formation of the vortex.[18]

The number of Atlantic tropical storms and hurricanes varies considerably from year to year, from 3–4 in low years to 14–18 in high years. An understanding of why some years differ from others, or better still, a prediction system for whether the coming hurricane system would be strong or weak would have obvious potential benefits. Since weather is known to be chaotic, the default hypothesis is always that variations occur naturally for reasons that are essentially random. Any claim to explain the variations must be subject to statistical significance tests. Given the relatively few years of data, being sure of a significant relation can prove challenging. However, physical arguments for how features of the large-scale climate can influence the development of hurricanes, together with statistical indications from the data, suggest that links to potentially predictable climate variables are plausible. One such link is to the sea surface temperature in the region of the north tropical Atlantic where hurricanes typically develop. Years with anomalously warm sea surface temperature have a slight tendency to have higher numbers of hurricanes.

Another link is to ENSO. Figure 4.24 shows the number of named storms (tropical storms and hurricanes) for July–October of each hurricane season, plotted against Pacific SST anomalies (using an equatorial box average known as Nino3–4) for April–September. The number of storms has a tendency to decrease during El Niño and to increase during La Niña. A simple linear regression model of the number of storms you expect for a given Nino3–4 anomaly is shown as a dashed line. This model would suggest that for a large El Niño you might expect 6 storms, while during a large La Niña you might expect 10 or 11 as opposed to an average between 8 and 9. The actual number of storms varies considerably from the number predicted by this line, as may be seen from the figure. One way of summarizing this is by the correlation, which is roughly -0.4, suggesting that the association is far from perfect (the negative sign is because storms go down as Pacific SST goes up). Correlations to Atlantic SST are of slightly lower magnitude.

There also is a further catch, since Figure 4.24 shows 6-month averages of each variable, with the Pacific SST anomaly beginning 2 months earlier than the hurricane season average. The 2-month lead is chosen to indicate that the information associated with SST has some

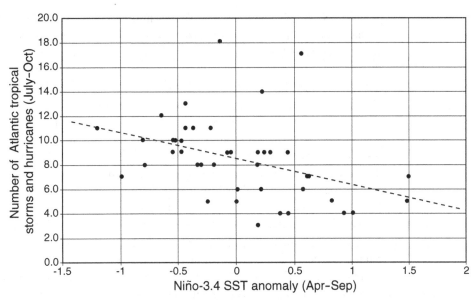

Fig. 4.24 The effect of ENSO on the frequency of Atlantic named storms (tropical storms and hurricanes). The number of named storms that occur during July through October of each year in the northern tropical Atlantic is plotted against the sea surface temperature anomaly along the equator in the Pacific "Niño3–4" region (box averages over 5° S to 5° N, 170° W to 120° W). During El Niño, when equatorial Pacific SST is anomalously warm, there tend to be fewer storms and hurricanes in the Atlantic. During La Niña, there tend to be more. A linear regression fit is shown as a dashed line. The scatter about this line indicates that ENSO is far from the only factor.[19]

degree of lead in time relative to the hurricane season. However, there is time overlap with the hurricane season, so the linear regression shown is not really a prediction system in the sense of having the predictor variable in advance of the time to be predicted. At longer lead time, or shorter averaging time, the correlation tends to degrade. Similar considerations tend to apply to the Atlantic SST. Thus one further requires a prediction of ENSO or Atlantic SST anomalies during the hurricane season, made in advance of the hurricane season.

This being said, one can make predictions with climate models or statistical models of these climate variables that have relations to hurricane numbers, of the sort illustrated in Figure 4.24. Experimental prediction schemes have been created and used for number of storms and for other hurricane variables, such as the average intensity of storms in a season. These schemes are currently based entirely on statistical models, rather than physically based climate models.[20] It should also be noted that there is still considerable debate in the climate and meteorological communities over such predictions. The expected outcomes of such predictions have to be understood rather carefully because the correlations to climate variables are typically of a similar nature to the relationship to ENSO noted above. In a given year, the actual number of storms may very well deviate substantially from the number suggested by predictor climate variables. Nonetheless, if one were to make a series of risk avoidance decisions based on such a relation averaged over enough years, one could benefit in the long term.

Why does El Niño tend to suppress Atlantic hurricanes? ENSO has teleconnections into the Atlantic that include changes in the winds, tropospheric temperature and even a tendency to influence Atlantic SST in some seasons. The precise connection is not fully determined at this time, but the roles of wind shear changes in disrupting storms and of upper-level warming in reduced convective instability are both plausible contributors.

Before fully accepting statistical relations based on short time series, climate modelers prefer to have corroborating evidence from the climate models themselves. When a relationship hinted at by the data can be simulated in a model, the model can be run many times with different initial weather to create greater statistical significance for the relation, and often cause and effect can be better sorted out through numerical experiments. In the case of hurricanes, the spatial scale of 500 km lies uncomfortably between the large scales that are typically simulated explicitly in a climate model and the small scales that must be parameterized. Weather forecast models and regional models have grid sizes that permit hurricanes to be simulated. Unfortunately, such models currently do better at predicting evolution of a developed hurricane a short period forward in time from observational data that included the hurricane than they do at predicting the development from a time before the hurricane was initiated. Simulating statistics of hurricane development for a large number of years with differing SST and large-scale flow conditions is now possible. Many climate models have storms with hurricane-like features. It is likely that the near future will see progress on tropical storm statistics both from global climate models with grid sizes that permit hurricanes, and from regional models.

4.9.2 Sahel drought

The Sahel is a region in Africa, south of the Sahara desert, that lies at the margin between arid regions to the north and the regions of strong convection and plentiful rainfall to the south, closer to the equator. The Sahel receives essentially all of its rainfall during June to September when the convection zones move northward in northern summer. This movement is part of the African monsoon that was seen in Figure 2.13. As with certain other regions that lie between arid and moist regions, the Sahel rainfall is vulnerable to variation on both year-to-year time scales and decadal time scales. Essentially, if the monsoon rains fail to extend quite as far northward as normal during the rainy season, it can yield a substantial deficit in the Sahel annual rainfall. Figure 4.25 shows the average annual rainfall as a departure from its long-term mean for an average over a large region roughly corresponding to the West African Sahel. The region has experienced decades of drought from 1970s until very recently, compared with 1950s and 1960s, with great negative impact on agriculture and society.[21] Lake Chad, formerly the sixth largest lake in the world (by surface area; it is always shallow), shrank to roughly one-tenth of its normal size.

An early hypothesis for this drought was that land surface changes, such as reduction in green plant cover, in part by grazing and other human activities, might have increased the albedo of the region. In that case, a larger fraction of incident sunlight reflected back to space would leave less energy to be transferred from the surface to the atmosphere to drive convection, reducing rainfall. While such feedbacks between albedo vegetation and rainfall are likely to have been important on paleoclimate time scales (for instance, 6000 years ago

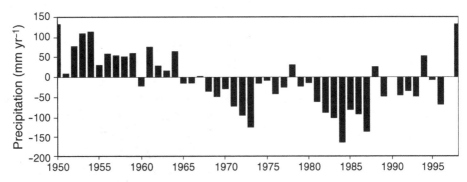

Fig. 4.25 Annual rainfall anomalies (vertical bars) over the West African Sahel (13° –20° N, 15° W–20° E) from 1950 to 1998. Data from Hulme (1994).

when the Sahara was vegetated), this hypothesis has given way to a different view of the Sahel drought. A number of studies have since suggested that sea surface temperature anomalies in the surrounding oceans (Atlantic and Indian) create changes in atmospheric circulation and that the Sahel rainfall reductions are due to these teleconnections. The default explanation for the sea surface temperature variations is a natural climate variability, although it has also been suggested that there may be an anthropogenic contribution. So far, the ability of climate models to predict Sahel drought appears limited, although some models can approximately reproduce it when given observed sea surface temperatures.

4.9.3 North Atlantic oscillation and annular modes

The *northern annular mode* and *southern annular mode* are leading modes of atmospheric variability on monthly and longer time scales. They produce a large-scale surface pressure pattern with a pressure anomaly of one sign covering a large area near the pole, and a pressure anomaly of opposite sign at midlatitudes. They are referred to as annular modes because the winds tend to blow around the globe in a ring (annulus) at the latitude between the high and low pressure region. A similar pressure pattern and associated winds extends through the troposphere and lower stratosphere. The northern and southern modes (surrounding the North and South Pole, respectively) are independent, but have roughly similar characteristics. The northern annular mode tends to be asymmetric, and in the Atlantic sector is essentially the same as a pattern of variability that had been previously named the *North Atlantic Oscillation* (NAO). The term North Atlantic Oscillation is more widely known because it has a long history (dating back to Gilbert Walker), and because of its association with weather impacts on Europe.[22]

Figure 4.26 shows patterns associated with the NAO for the *positive phase* of the oscillation, in which anomalously low pressure near the pole and anomalously high pressure at midlatitudes is associated with westerly wind anomalies that tend to increase the climatological jet and shift it northward as it approaches Europe. This tends to increase the number of storms that hit northern Europe, relative to southern Europe, creating a tendency to increase precipitation in the north and decreasing it in the south. In the negative phase of

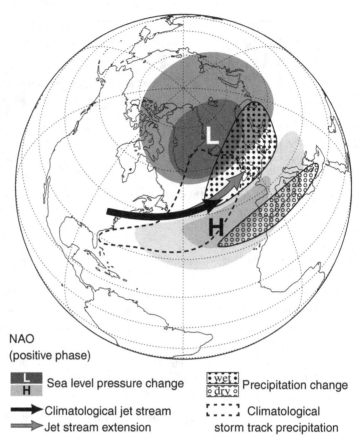

NAO
(positive phase)

| L H | Sea level pressure change | wet dry | Precipitation change |

➡ Climatological jet stream

➡ Jet stream extension

⌐⌐⌐ Climatological
storm track precipitation

Fig. 4.26 Patterns associated with the positive phase of the North Atlantic Oscillation, schematically showing patterns of anomalously high and low surface pressure, and anomalous winds that tend to extend the strong region of the climatological jet stream across the Atlantic. The resulting precipitation anomalies that can be associated with this are shown as stippled regions.[23]

the oscillation, the anomalies are approximately reversed, with easterly anomalies tending to weaken the climatological westerlies, resulting in precipitation anomalies of opposite sign in approximately the same locations.

Notes

1 Secondary effects that also help warm SST are that upwelling is not quite as strong as under normal conditions, and westward currents carrying cold water westward are a bit weaker than normal.

2 For a more rigorous treatment of adjustment of the surface height to thermocline depth, it is also necessary to consider the speed of waves associated with displacement of surface height and currents through the entire depth of the ocean (known as the barotropic mode). The adjustment

time of the barotropic mode contribution to surface height is very fast compared with the evolution of the thermocline. For treatment at the graduate level see Philander (1990).

3 The Tropical Ocean–Global Atmosphere (TOGA) program ran for exactly one decade from 1985 through 1994. Funding for TOGA science came through federal agencies led by the National Oceanic and Atmospheric Administration and the National Science Foundation. It provided support both for the large observational programs and for climate model development.

4 The TOGA measurement system is described in review articles by Halpern (1996) and McPhaden *et al.* (1998). These give some history of tautly moored surface buoys in equatorial current systems (Halpern 1987; Hayes *et al.* 1991). McPhaden *et al.* (1998) also describe other parts of the tropical ocean observing system. In addition to the TOGA–TAO array, other *in situ* measurement systems include drifting buoys for surface velocity (Niiler *et al.* 1995), a volunteer observing ship program in which expendable bathythermographs are dropped from ships along their regular commercial tracks, and an island tide gauge program (Wyrtki 1975). A moored array similar to TAO has subsequently been expanded into the Atlantic (Bourles *et al.* 2008) and Indian oceans.

5 The typical annual budget for the TAO array is on the order of several million dollars per year (note this is a rough figure that depends on how the accounting is done). The TOPEX-POSEIDON satellite observation system cost roughly 800 million dollars, but this included much development of new instrumentation. A satellite for SST measurements would be less than 100 million dollars and a satellite produces several years of data.

6 Equatorial subsurface temperature anomalies are from the TAO array; surface temperature anomalies from satellite retrievals.

7 The westward expansion in Figure 4.13 occurs as an equatorial Rossby wave, obeying balances very similar to the simpler Rossby wave in Figure 3.15 and propagating at a specific speed. More precisely, it is a Rossby wave front since the eastern side of the low pressure region is set by the wind stress so the pattern expands as the western edge moves west with the Rossby wave speed while the rest comes into steady balance with the wind.

8 For review of the delayed oscillator conceptual models of ENSO (Schopf and Suarez 1988, Battisti and Hirst 1989) versus other models, see Neelin *et al.* (1998).

9 The NCEP SST anomaly forecast was published in the *Climate Diagnostics Bulletin*, March 1997 (published quarterly by the Climate Prediction Center, Camp Springs, Maryland; senior editor V. G. Kousky). The forecasts are averages of three sets of forecasts initiated on approximately 15 January, 15 February and 15 March 1997. The NCEP coupled model is described in Ji *et al.* (1996).

10 For a review of ENSO chaos see Neelin *et al.* (1998).

11 Motivated by schematics following Ropelewski and Halpert (1987), the areas in this schematic are based on International Research Institute for Climate and Society's seasonal-to-interannual prediction data for below-normal, normal and above normal terciles.

12 For review of observed teleconnections see Wallace *et al.* (1998).

13 For review of teleconnection dynamics see Trenberth *et al.* (1998). For El Niño effects on storm tracks see, e.g. Straus and Shukla (1997), or Chen and van den Dool (1997).

14 The schematic of the probability distribution shift used to illustrate the impact of El Niño on California precipitation (Figure 4.22) is simplified in several ways. The distribution of precipitation can differ substantially from the familiar Gaussian distribution used in the schematic, and the distribution shape can change in more complex ways than the simple shift shown. The schematic shows half of El Niño winters in the upper tercile to emphasize the point that one can have a significant shift and yet still sometimes have dry winters; this is roughly consistent with California averages discussed in Schonher and Nicholson (1989), although they use a more complex methodology. For Southern California, they find approximately 70% of El Niño winters in the upper precipitation tercile, as do more recent unpublished analyses by the US Climate Prediction Center. For further discussion of El Niño impacts on the US hydrological cycle in streamflow in other qualities, see for instance Cayan and Webb (1992), Piechota and Dracup (1996), McCabe and Dettinger (1999) and references therein.

15 Examples of experimental seasonal-to-interannual forecasts for various regions may be found in the Experimental Long-Lead Forecast Bulletin and the Climate Diagnostics Bulletin. The CLIVAR program, organized under the auspices of the World Meteorological Organization, is an umbrella for various national research projects undertaking work on seasonal-to-interannual climate prediction among other topics.

16 NOAA GOES-9 satellite photo of hurricane Linda, September 12, 1997, 1800 UT, from Laboratory for Atmospheres, NASA Goddard Space Flight Center. Initial rendering by Marit Jentoft-Nilsen. Cross-section follows Emanuel (1988a).

17 See, for instance, Emanuel (1988b).

18 For work on hypothesized hurricane wind shear relations, see for instance Goldenberg and Shapiro (1996) and references therein.

19 Data analysis courtesy B. Tang following Tang and Neelin (2004).

20 For examples of statistical hurricane prediction systems, see, e.g. Gray (1984), Bell (2008) and Wang *et al.* (2009).

21 Interaction of the Sahel drought with societal impacts and policies is discussed in Glantz (1994). Change in Lake Chad area is from the US Geological Survey. Discussion of factors affecting the drought may be found, for instance, in Xue and Shukla (1993), Nicholson *et al.* (1998), for land surface effects; Rowell *et al.* (1995), Zeng *et al.* (1999), Giannini *et al.* (2003) for SST forcing and possible interaction of SST forced teleconnections with land surface processes; and Held *et al.* (2005) and Biasutti and Giannini (2006) for anthropogenic contributions including aerosol effects.

22 Walker and Bliss (1932), Hurrell (1995, 1996) and Thompson and Wallace (1998) give defining descriptions of the NAO and annular modes respectively. Visbeck *et al.* (2001) provide a brief summary.

23 The spatial patterns in the NAO schematic are based on correlation maps of NOAA Climate Prediction Center NAO index over 1949 to 2006 to seasonal (December–March averages) of sea-level pressure and 200 mb zonal wind from NCAR/NCEP reanalysis and CMAP precipitation data, respectively. The overlay of several patterns in the schematic follows a concept by Martin Visbeck, personal communication.

Climate models

5.1 Constructing a climate model

Construction of a climate model is a task whose principles are easily understood but which in practice involves mastery of a multitude of technical details. In this chapter we thus present the generalities in this first section, while an introduction to the more quantitative aspects is given in subsequent sections. A reader who is interested primarily in the output of climate models or in an overview of their construction can thus restrict their attention to section 5.1, skipping the more detailed material of sections 5.2 through 5.4. Section 5.5, which introduces issues of model climate equilibration and errors, is useful background for Chapter 7, and section 5.6 provides a sense of how well models simulate current climate.

The climate models discussed in section 5.1 are the general circulation models (GCMs). Other types of climate models are summarized in section 5.4, along with models that are relevant to climate studies such as weather prediction models, atmospheric regional models and cloud resolving models. Many of the principles used are the same as for the GCMs, each type of model having a particular aim that leads to variations within the same overall modeling approach.

For computational representation, the continuous fields of temperature, pressure, velocity, etc. in the atmosphere and ocean must be approximated by a finite number of discrete values. The most intuitive approach to this *discretization* is to divide the fluid up into a number of *grid cells* and approximate the continuous field by the average value across the grid cell or the value at the center of the grid cell. This can approximately capture the behavior of motions at space scales much larger than the grid cell but obviously omits the infinite number of values that a continuous temperature field has at different points within the grid cell. There are a number of different techniques of discretizing the equations of motion of a continuous fluid and other approaches (such as spectral methods) are outlined in section 5.2 along with more details of the discretization discussed in this section. Likewise, a brief outline of the parameterization of small-scale processes is given in section 5.1.2, with additional details in section 5.3.

5.1.1 An atmospheric model

Figure 5.1 shows the representation of the atmosphere as it occurs in many climate models, focusing on a particular region out of a global grid. Figure 5.1a shows a three-dimensional

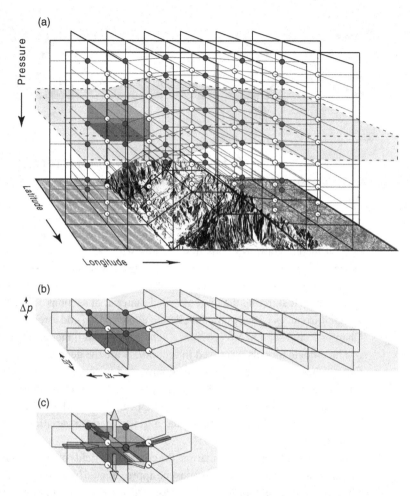

Fig. 5.1 A section of the grid for a typical atmospheric general circulation model. The grid would continue globally.
(a) The atmosphere is divided into a finite number of grid cells in the vertical and horizontal. Dots indicate vertices of
grid cells. The levels in the vertical are of variable spacing and are usually in a modified pressure coordinate that
follows large-scale topography. (b) A particular level, broken out from (a). The horizontal grid size corresponding to
longitude and latitude is indicated as Δx and Δy, while the vertical grid increment for a region without topography
is indicated as Δp. (c) Transports or fluxes into and out of a particular grid cell in both horizontal and
vertical.

view of the grid, including a region over elevated topography. Views of a particular level
are blown up in the subsequent panels.

Features to note include:

• For each of the discrete grid cells, there would be a single value of each variable (e.g.
  temperature). This would represent, for instance, the average value across the grid cell.
  Any feature of smaller scale than the grid cells cannot be explicitly represented in the
  model and must be parameterized.

- The vertical coordinate is essentially a pressure coordinate, but modified so it follows the large-scale topography. (This is usually done by defining $\sigma = p/p_s$, where p_s is the surface pressure, and using sigma as the vertical coordinate.)
- Grid spacing is not constant in the vertical. Typically the boundary layer has finer vertical resolution.
- The horizontal grid is in latitude and longitude. The zonal length of the grid cell (in kilometers), denoted Δx, thus varies with latitude. The meridional grid cell length Δy may also vary. In discussion below, when a grid size is given for a model, it should be understood as a typical grid size, since variations occur within the model.
- The partial differential equations of motion for atmosphere and ocean are replaced by a finite number of *difference equations* involving differences between values in neighboring grid cells.
- Each grid cell communicates with its neighbors, as schematized in Figure 5.1c. The arrows indicate transports (or fluxes) of mass, energy, and moisture into a particular grid cell. The fluxes are proportional to differences between the grid cells for each variable; the budgets of these fluxes are associated with the respective equation. For the velocity equations, neighboring grid cells affect each other via the pressure gradient force, since this depends on the difference of the pressure between a grid cell and its neighbors.
- By considering the balance of forces, fluxes etc. on a given box, the time rate of change of wind, temperature, etc. for a given grid cell can be computed. This is then used to calculate a new value for the wind, temperature, etc. one short *time step* (typically half an hour or less) later. This is then repeated for all the other boxes so the solution is available everywhere for the next time step.
- The time integration proceeds one time step at a time until the desired length of simulation (e.g. 100 years) is reached (about 2 million half-hour time steps).
- The equations are local in the horizontal in the sense that they only involve a grid cell and its immediate neighbors. However, since effects are passed from neighbor to neighbor, the solutions eventually involve the entire grid. Thus as the simulation is carried forward in time, the solution involves the whole globe.

5.1.2 Treatment of sub-grid-scale processes

In addition to treating the effects at the large scale that are explicitly treated by the transports and forces among grid cells, it is essential to model the physics of processes on scales smaller than the grid cell. For the atmosphere these are particularly complex. In GCM jargon, the part of the computer code that deals with these processes is called the "physics package," while the processes involving interaction among grid boxes due to large-scale forces and transports are referred to as the "dynamics." Figure 5.2 shows schematic examples of the treatment of moist convection and radiative transfer. These processes are not local in the vertical, that is, grid cells are affected not only by the nearest neighbors but by grid cells anywhere in the column. However, the convective and radiative processes depend only on the column, not on horizontally adjacent columns. This is because of the ratio of the vertical dimension to horizontal grid size: the entire troposphere is only about 10 km deep, while the horizontal grid size is, say, 200 km.

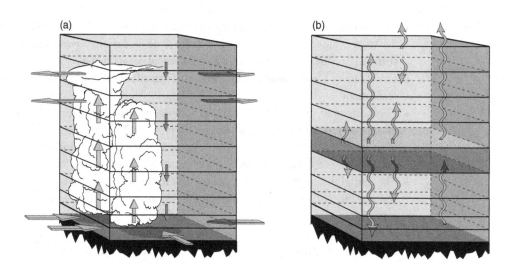

Fig. 5.2 A vertical column showing two examples of the parameterized physics of small-scale processes within a single column in a GCM. (a) Moist convection. Clouds are shown schematically with convective updrafts (not to horizontal scale; the fraction of the horizontal area of the grid cell that each cloud would occupy is about 0.0001). The individual clouds are not calculated but the average of all the updrafts in the domain is parameterized as a function of grid-scale averaged temperature and moisture. (b) Radiative transfer. The infrared emission (wiggly arrows) from a single layer (dark shading) is shown affecting other layers and the surface, where absorption occurs. Emission from the surface that is absorbed in the layer and emission from the layer and the uppermost layer to space are also shown. Emission and absorption at every layer interact with all other layers.

The convective parameterization illustrated in Figure 5.2a emphasizes the role of convection in vertical motions. Small-scale convective motions within the grid box have updrafts in the deep convective clouds (there are also downdrafts, not shown, but updrafts dominate). One approach to convective heating is to calculate the net cumulus mass flux that would occur as an average over the grid box for given temperature and moisture profiles over the column. Subsiding motion also occurs within the grid box which tends to cause warming but does not balance the cumulus mass flux. The average effect of clouds within the grid box therefore affects the large-scale dynamical equations, with inflow at low levels and outflow at upper levels.

The infrared component of the radiative transfer scheme is partially schematized in Figure 5.2b. Infrared radiation is emitted both downward and upward from every layer according to the temperature of that layer (for radiative transfer it is easier to think in terms of layers of air, so layer is used rather than level). This radiation is partially absorbed in other layers according to the thickness of the layers and the concentration of greenhouse gases. Part of the downward emission reaches the surface and part of the upward emission is lost to space. The layer absorbs infrared radiation from each of the other layers and from the surface. The intensity of each flux depends on the temperature of the layer doing the emitting. The interaction of every layer with every other layer and with the surface must be calculated. Furthermore, the calculation is done for a large number of wavelengths of infrared radiation, taking into account that absorption by each type of greenhouse gas depends on wavelength. This is thus quite a substantial calculation which is repeated for every column, at every time step.

5.1.3 Resolution and computational cost

The term *resolution* is used to describe the degree of refinement of a climate model grid. A model with a small grid size is termed fine resolution or high resolution, while one with a large grid size is coarse resolution. This is consistent with terminology, for instance, for microscopes or photography where high resolution images can distinguish smaller-scale features.

Figure 5.3 uses the topographic height for the North American region to illustrate the effects of resolution on a familiar example. The topography is contrasted for a larger grid size (lower resolution) typical of a climate model and a smaller grid size (higher resolution) typical of a weather prediction model. The high resolution case is shown as a blowup of the topography over a subregion, first, so that it can be seen more clearly, and second, because this resolution is also typical of a regional climate model, discussed further in

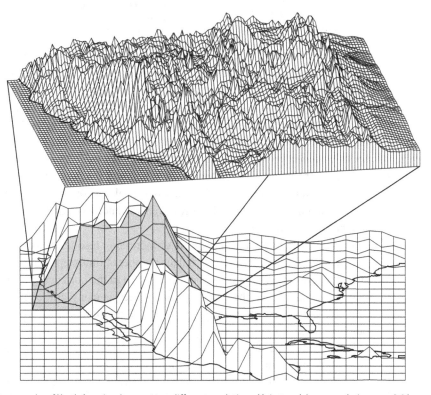

Fig. 5.3 The topography of North America shown at two different resolutions. Main panel: lower resolution, on a 2.0 by 2.0 degree grid typical of climate models. Blowup of the Western US region: high resolution, on a 0.2 by 0.2 degree latitude–longitude grid, typical of global weather forecast models or regional climate models.

section 5.4. Much more detail can be seen in the high resolution case while the lower resolution case captures the broad outline of the main mountain ranges. The grid size of 2 degrees in latitude and longitude used to illustrate the lower resolution case is typical of the most common resolution for the climate models used in the IPCC (2007) report. Some coarse resolution models of roughly 5 by 4 degree latitude–longitude grid were used, while the highest resolution models had grids between 1 and 1.5 degrees in each direction. Two degrees of latitude corresponds to slightly over 200 km; the high resolution case at 0.2 degrees corresponds to slightly over 20 km in latitudinal grid size. Typical grid sizes get smaller every few years, as outlined below.

Given that the topography looks so much more impressive at high resolution, why do climate models not run at even higher resolution? The main reason is computational cost which increases dramatically for high resolution. Additional reasons are outlined at the end of the section.

Computational time for a simulation is in principle given by:

$$\textit{Computational time} = (\textit{computer time per operation})$$
$$\times\ (\textit{operations per equation})$$
$$\times\ (\textit{number of equations per grid box})$$
$$\times\ (\textit{number of grid boxes})$$
$$\times\ (\textit{number of time steps per simulation})$$

where an operation is, for instance, a single multiplication or addition, several of which would occur in calculating each of the terms of each of the equations for each variable in each grid box. In practice, there are additional effects associated with factors such as the type of operation, the time taken for reading input and outputting the calculated climate variables to storage, and the extent to which operations can be performed in parallel. The latter aspect is further discussed in section 5.2. The computational cost may be taken to be roughly proportional to computational time, since faster computers tend to cost more and the longer the computational time each simulation requires, the more computers must be purchased and maintained to achieve a given goal. Of course, if the computational time for a given simulation is so long that a researcher cannot wait for it to be completed, the dollar cost becomes irrelevant.

There will be two primary impacts of increasing resolution on computational cost: the number of grid cells will increase and the number of time steps per simulation will increase. If one doubles the resolution of a climate model in the horizontal, i.e. halving the grid size in each direction, then the model has four times as many grid cells. When the horizontal grid size is reduced, it is necessary to reduce the time step proportionately, for reasons outlined below and in section 5.2. Thus the time increment must be halved as well, and twice as many time steps will be required to complete the same number of years of model simulation. It is normal to increase the number of grid boxes in the vertical as well (although this is not strictly necessary) since smaller horizontal scale features often introduce effects with smaller vertical scale and it would be odd to have high resolution in the horizontal and coarse resolution in the vertical. For the sake of simplicity, let us assume the vertical resolution is also doubled. Thus the number of grid cells, including the effects of vertical resolution, increase by a factor of $2^3 = 8$ and with the doubled number of time steps, the computational time of the run increases by a factor of $2^4 = 16$.

Comparing the two grids in Figure 5.3, there is a factor of 10 in grid size in each horizontal direction. The corresponding models would also have a factor of 10 difference in time step which implies a factor of 10^3 in computational time even before taking vertical resolution into account. Let us suppose the fine resolution model has only double the vertical resolution, leading to a factor of 2000 in cost if both are run on the same domain. Thus for the same computational cost, the coarse resolution model can run a simulation 2000 times longer. For instance, the cost of running the coarse resolution model for 40 years is about the same as running the fine resolution model for 7 days.

The reason that time step must be reduced when the grid size is smaller involves computational accuracy and *computational instability*. The time step must be small enough to accurately capture time evolution of the resolved phenomena, and for smaller grid size, smaller time scales tend to enter. One important time scale is the time it takes a given wind or wave speed to traverse the length of one grid box. For instance, if a wind of $50\,\mathrm{m\,s^{-1}}$ is the fastest velocity in the domain of a model with $200\,\mathrm{km}$ grid size, the time it takes to cross one grid box is about an hour. If the time step is longer than this, more than one grid box is traversed in a single time step, so the model will not properly resolve motion of small-scale features. Small-scale noise can be artificially amplified (further details on why this happens are given in section 5.2). The model solution then becomes useless or, more often, the model "blows up," i.e. generates enormous numbers that cause an error message in the program.

As computers become faster, climate model resolutions increase every few years. Short runs of global models are now being done with grid sizes more typical of the high resolution case in Figure 5.3. It is tempting to imagine that with more powerful computers, model grid size can become so small as to resolve all features. Parameterization of such things as moist convection would then appear to be no longer necessary. First, let us consider how much increase in resolution would be needed. To model clouds accurately, a grid size of $50\,\mathrm{m}$ would be desirable, i.e. 10 000 times smaller than a current coarse resolution model. Taking this factor to the fourth power, for corresponding increases in resolution in the vertical and horizontal and in time gives 10^{16} times more computational time for the same length of simulation. Furthermore, higher resolution does not immediately imply a better simulation. The model will definitely produce more small-scale features, but unless it simulates them accurately, it will not necessarily help. A model with a $50\,\mathrm{m}$ grid could resolve large updrafts in convective systems, but it would still not resolve the process of raindrops coalescing, ice crystals forming, etc. These would then have to be parameterized appropriately. Finally, a model with very high resolution would not necessarily lead to better understanding of the climate system than a model with an accurate set of parameterizations of small-scale processes. Thus faster computers and higher resolution are very helpful, but are not the only important factor in climate modeling. At any given time, a variety of resolutions are used for different climate models, owing to the trade-off between the usefulness of higher resolution versus being able to run many cases or carry out longer simulations. Climate models of different degrees of complexity are also used, as further discussed in section 5.4.

5.1.4 An ocean model and ocean–atmosphere coupling

Figure 5.4 shows a cross-section through a typical ocean model grid. The levels are more closely spaced (as little as 10 m thick) near the surface to resolve surface current effects. They remain quite closely spaced through the upper ocean to resolve the thermocline, but gradually increase to a vertical grid size of several hundred meters. In this example, there are 27 levels, similar to what has been used in many El Niño models. The ocean grid size shown here is 100 km. Much of the work on global change has been done with models of much coarser resolution in both horizontal and vertical. However, a 100 km grid size still cannot be considered high resolution because the typical ocean eddy at midlatitudes, such as those seen in Figure 2.4, is about 50 km. Thus a model must have a grid size substantially smaller than this to be "eddy resolving."

The approximation to the topography appears crude in the ocean model. Sloping grid cells are used in some models but the approximation shown should be viewed in perspective. If the entire Pacific basin were shown, for example, the domain in the figure would be only about one-twentieth of the width. And the total depth of the figure is only about one-twentieth of the width of a single grid box. The island shown is about the size of Hawaii. Many smaller islands would be omitted from the ocean model, or would be approximated as sea mounts, since the base would be resolved but the narrower top portion would be smaller than the grid size.

The lower levels of an atmospheric model are also shown in Figure 5.4, for an atmospheric model grid size of 2 degrees of longitude. It is common for the ocean model

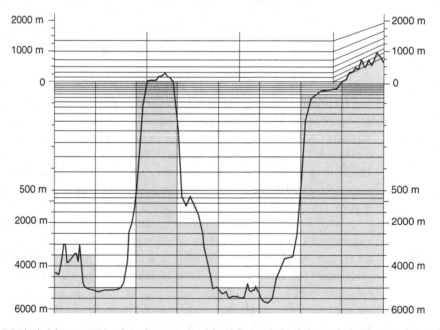

Fig. 5.4 A longitude–height cross-section through an ocean model grid. The vertical scale is stretched in the upper 500 m of the ocean so that the closely spaced levels near the surface can be seen. The solid curve gives the true bottom topography and the dark shading shows the model approximation to the topography.

to have smaller grid size than the atmosphere for two reasons. First, some climatically important features in the ocean have smaller scale (such as equatorial upwelling and equatorial jets, western boundary currents like the Gulf stream and ocean eddies). Second, ocean models are slightly less computationally expensive per grid cell because atmospheric radiation schemes, cumulus convection parameterizations, etc. add considerably to the computation.

Because of the difference in grid size between the atmosphere model and ocean model, some care must be taken in interpolating between the grids at the surface, apportioning the fluxes properly among the grid cells. Notice also that the atmospheric model, with its coarser grid, treats the island in Figure 5.4 as essentially flat, and has a slight offset in where the topography begins to affect the atmospheric grid at the coast. These are details as far as large-scale climate simulation is concerned, but worth being aware of if trying to apply climate model results to a regional problem.

Figure 5.5 shows a single atmospheric column, one grid cell wide in the horizontal, and the ocean grid cells below it, expanded to illustrate the coupling effects across the ocean–atmosphere interface. The various energy fluxes are illustrated for one of the ocean grid cells, and the wind stress for the other. The amount of solar radiation reaching the ocean surface depends on atmospheric variables such as clouds and thus is calculated in the atmosphere model. The solar radiation penetrates into the upper few ocean levels. Exchange of sensible heat and latent heat (evaporation) occur between the uppermost ocean level and lowest atmospheric level. Infrared radiation is exchanged between the uppermost ocean level and all levels of the atmosphere. Near-surface heating is carried down by mixing in the ocean, which is parameterized to depend on atmospheric wind stress, ocean surface currents, etc. Because sea surface temperature (SST) remains relatively constant over a few atmospheric model time steps, whereas atmospheric variables tend to have more

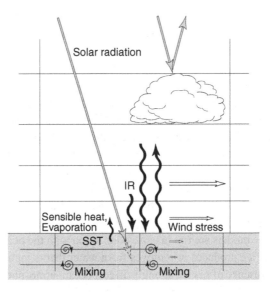

Fig. 5.5   Atmosphere–ocean coupling in a GCM via energy fluxes and wind stress.

variation, SST is typically passed to the atmospheric model where all the heat fluxes are then calculated. Wind stress is calculated in the atmosphere and passed down to the ocean model, acting as a force on the uppermost level. Mixing effects then redistribute this force downward to accelerate the levels below.

5.1.5 Land surface, snow, ice and vegetation

In addition to the atmosphere and ocean model equations presented in Chapter 3 a number of other processes must be included in climate models. One important component is a land surface model. Land surface models are often run as a subcomponent of an atmospheric model since land quantities interact with atmospheric variables on short time scales. The variables typically include: land surface temperature; soil moisture, which depends on precipitation and evapotranspiration; and snow cover. The land surface temperature equation is essentially governed by energy balance at the surface, since land stores little heat. For both this and the soil moisture equation, evapotranspiration must be calculated, which depends on the properties of the vegetation in each grid box.

It is common to define a number of *land surface types*. For each surface type several properties are specified, including quantities related to vegetation type. Depending on the complexity of the model these might include rooting depth of the vegetation, the *leaf area index*, which gives the typical vertical thickness of the leaves in the vegetation cover, and parameters specifying how much soil moisture the soil can absorb before it saturates and runoff becomes strong. Because such land surface models depend so strongly on vegetation, they are sometimes termed *biophysical land surface models*. An example of the land surface types used in one land surface model, the Biosphere–Atmosphere Transfer Scheme (BATS), is given in Figure 5.6 with their geographical distribution on a 0.5 by 0.5 degree grid.[1] The land surface types also affect the *drag coefficient* felt by near-surface winds (rough surfaces like forests tend to produce more drag). Albedo depends on surface type, but is also affected by such things as snow cover. Some models now include interactive vegetation models in which properties of the vegetation, such as leaf area index, are calculated as a function of time, depending on soil moisture, temperature, insolation, etc.

A snow cover model might have one to several layers of variable depth. Snow coverage and depth depends on atmospheric inputs such as snowfall, temperature, surface radiation and low-level atmospheric temperature and moisture. Energy balances within the snow layers determine melting and sublimation. For models with state-of-the-art treatment of water at the land surface, the soil moisture and snow cover equations are linked to land surface hydrology model which includes calculation of river runoff by catchment basin.

A sea ice model would typically include the fraction of areas of open water (known as "leads") within the sea ice region, ice thickness, ice surface temperature and ice velocity. Sea ice models are run in close connection with the ocean model. The melting and freezing of sea ice produces or removes fresh water that is input to the ocean salinity equation, and the presence of sea ice strongly affects ocean–atmosphere heat transfer. Atmospheric inputs strongly affect energy balance in the sea ice layers, and winds drive much of the ice motion.

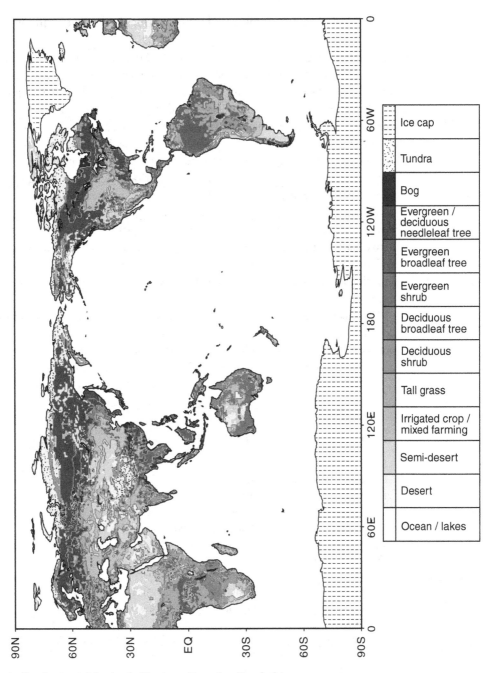

Fig. 5.6   Land surface types, following the Biosphere–Atmosphere Transfer Scheme.

5.1.6 Summary of principal climate model equations

In Chapter 3, the equations of motion used in atmosphere and ocean models were examined, and additional processes for land surface and sea ice were summarized in section 5.1.5. Table 5.1 lists the main equations of a climate model, with cross-references to the equations in Chapter 3. The exact form of the equation used in a particular implementation of an atmosphere or ocean model might vary somewhat from the form given in Chapter 3, but the principle would be the same. The equations with time derivatives are known as *prognostic equations* because they give the change in variables one time step into the future. They thus have particular significance and are noted in Table 5.1. The other equations (diagnostic equations) give relations among variables at a given moment. They can thus sometimes be combined, eliminating one variable (e.g. hydrostatic equation combined with the ideal gas law to eliminate density). The sea surface height equation in the ocean and the surface pressure equation in the atmosphere are special cases of the continuity equation (see Chapter 3 endnotes 9 and 10) but require separate treatment numerically. Sea-level height was treated in a simpler approximation in Chapter 4 in a one-active-layer upper-ocean model; ocean GCMs use a more complete equation.

There can be a number of different processes that must be computed for each of the equations listed in Table 5.1. For instance, one time step of the atmospheric temperature equation requires computation of the complex radiative transfer calculations and all of the calculations for the parameterization of moist convective heating seen in Figure 5.2. The

Table 5.1 Summary of basic equations for atmosphere, ocean and land surface models.

Equation name	Model	Comments	Corresponding equation no.
Horizontal velocity eqns.	Atm/Ocean	Prognostic (u, v)	3.4, 3.5
Hydrostatic equation	Atm/Ocean		3.8
Equation of state	Atm	Ideal gas law	3.10
	Ocean		3.11
Temperature equation	Atm	Prognostic (T)	3.20
	Ocean		3.18
Continuity equation	Atm		3.30
	Ocean		3.28
Moisture equation	Atm	Prognostic (q)	3.34
Salinity equation	Ocean	Prognostic (s)	3.35
Surface pressure eq.	Atm	1 level	Chapter 3, endnote 10
Surface height eq.	Ocean	1 level	Chapter 3, endnote 9
Surface temperature eq.	Land	1 or a few levels	
Soil moisture equation	Land	A few levels	
Snow cover equations	Land	1 or a few levels	
Sea ice equations	Ocean	Ice fraction, thickness	

Note: The component to which the equation applies is given in the second column. Equations with time derivatives are denoted prognostic. The variables with time derivatives are shown in brackets: respectively, u zonal velocity, v meridional velocity, T temperature, q specific humidity and s salinity. Equations apply at multiple levels in the atmosphere or ocean unless otherwise noted.

equations shown provide the overall structure which would be common to most climate models. Differences among models might include the degree of complexity of the parameterized processes that go into these equations. For some applications, not all of these equations would be used. For instance, sea ice might be specified from observations instead of calculated interactively. This works fine for studying tropical climate variability such as ENSO, but would not be appropriate for global warming simulation.

As climate models develop, it is common to include additional equations beyond the standard set given in Table 5.1. For instance, in the atmosphere, equations for aerosol transport, cloud water, cloud ice or concentration of particular gases; in the ocean, equations for concentrations of *tracers* (compounds whose concentration can help verify the model simulation against observations) or biogeochemistry equations. When the additional processes involve entire additional model components, one speaks of the model as a "climate system model" as described in section 5.1.7.

5.1.7 Climate system modeling

The basic climate model equations listed in section 5.1.6 are largely restricted to the physical aspects of the climate system, aside from the fact that the land surface model needs information about vegetation in calculating evapotranspiration and albedo. Current research in climate involves the chemical and biological aspects of the climate system as well. Climate modeling that seeks to incorporate the broader array of processes in the climate system is challenging enough that it has come to merit its own term: *climate system modeling*. While any model that is applied to climate questions might be called a climate model, the term climate system model explicitly points to the attempt to incorporate as many parts of the climate system as feasible. Naturally, there is not full agreement on the terminology: *Earth system modeling* is an alternate term with essentially identical usage. These models of course do not include models of plate tectonics or other aspects of the solid Earth, so "Earth system" might exaggerate slightly. On the other hand, as ecosystem representations in such models become more complex, it might be argued that the term climate system model sells the biologists short.

Additional components on the list for climate system models include the following. Representation of the *carbon cycle* is typically not a single model unit but rather implies a coordinated set of changes in the atmospheric model, the land surface model and the ocean model. The land surface model must be modified to include the growth and decay of biomass, carrying a budget of the carbon contained in it. The atmosphere must carry carbon dioxide concentration as a predicted field, interacting with the land and ocean via surface fluxes. The ocean must carry a set of equations for dissolved carbon compounds, both organic and inorganic, and the biological and geochemical sources, sinks and exchanges among these. The land surface *vegetation model* may include other aspects of the evolution of vegetation with time, both natural and anthropogenic. The *ocean biogeochemistry model* may include several groupings of organisms, concentrations of key nutrients, and the exchanges among these.

An *atmospheric chemistry model* would include equations for the concentrations of a large number of chemical species and the interactions among these. Typically even if a single compound is of interest, many others must be included because they are involved in reaction pathways that affect the compound of interest. An *aerosol model* includes chemical compounds that tend to form aerosols, as well as atmospheric transports, fall rates, and interactions with water vapor and water droplets. A model of dust sources at the land surface, dust transport and interaction with other aerosols might be included. Atmospheric chemistry models are typically so expensive computationally that a leading argument for making them interactive with the atmospheric model is not that they affect the winds, but rather that the atmospheric model is comparatively cheap, so one might as well include it.

5.2 Numerical representation of atmospheric and oceanic equations

The numerical representation of the equations in Table 5.1 is key to climate modeling and there is a vast lore of competing methods, each with its particular advantages and limitations. From a practical point of view, however, many climate model users in the current state of climate modeling will not need to undertake such discretization and coding of the resulting equations themselves. The dynamical core of a climate model already exists, and while it is occasionally revised by the research team responsible for the model, a user running the model or working with the output can treat it like any other piece of software. Some basic properties need to be known, along with some of the associated terminology and the essential strengths or limitations. The presentation here is thus oriented to the climate model user, rather than a climate model developer.

5.2.1 Finite-difference versus spectral models

The two dominant methods of discretizing the continuous fluid equations of atmosphere and ocean are *finite-difference* methods and *spectral* methods. A third method, the finite-element method, is less common. Spectral atmospheric models are widespread since this method offers some computational advantages for atmospheric applications (although this may change with changes in computer architecture). Models using finite differences are often referred to as grid-point models and this is what is presented in Figure 5.1. Ocean models are usually grid-point models because the complex shapes of realistic ocean basins present some difficulties to spectral methods. Despite the name, a large fraction of the computations in a spectral model are actually done on a grid just like that in a finite-difference model. Computations done at grid points include radiation and all the parameterized processes, such as convection, as illustrated in Figure 5.2. Part of the computation of advection terms is also done on grid points. The model transforms the principal variables from the grid-point representation to a spectral representation for part of the computation. A spectral representation gives the spatial pattern of the variables in terms of a sum of sinusoidal functions (and related shapes), as discussed below. The model output is on grid points, so to the user the spectral aspects are almost invisible except for two things: the latitude grid has somewhat varying grid spacing; and the model resolution is described by numbers like R15 or T105, which refer to the spectral truncation, defined below. The spectral methods

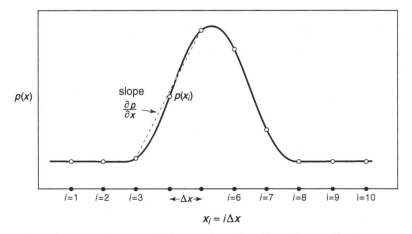

$x_i = i\Delta x$

Fig. 5.7 Example of finite differencing of a pressure field as a function of distance x and of the finite-difference pressure gradient at one of the points. The pressure gradient $\partial p/\partial x$ at point x_i is approximated by the slope of the line between the two adjacent points. Note that the errors relative to the true slope at that point depend on the grid size.

underlie a great deal of theory in atmospheric and oceanic dynamics, particularly where wave dynamics is important, so a brief investment in understanding them is worthwhile.

An example of a finite-difference method is illustrated in Figure 5.7. For a continuous field of pressure as a function of longitude, x, for example, the finite-difference method retains the value of pressure p_i at a finite set of grid points x_i, separated by a grid size Δx, where $i = 1,...,N$ is the grid-point index. The smaller Δx, the higher the resolution and the finer the detail that can be retained. The gradients are approximated using the difference between the values at neighboring grid points. For instance

$$\frac{\partial p}{\partial x} \approx \frac{p_{i+1} - p_{i-1}}{2\Delta x} \tag{5.1}$$

approximates the pressure gradient at point i. This is referred to as a centered difference. This approximation is not unique and many variants are used, including "higher order" schemes that use additional neighboring points to improve the accuracy (although this adds the cost of additional computations).

An example of a spectral representation is given in Figure 5.8. The same pressure field is shown, along with three versions of a spectral approximation to it. The pressure field as a function of x is approximated by a finite sum of sinusoidal functions

$$p(x) = \frac{a_0}{2} + \sum_{k=1}^{K} \left[a_k \cos\left(\frac{2\pi k x}{L}\right) + b_k \sin\left(\frac{2\pi k x}{L}\right) \right] \tag{5.2}$$

This is also known as a *Fourier series* representation. The wavenumber k gives the number of wavelengths in the longitudinal domain of length L, i.e. $k = 1$ has one maximum, the $k = 2$ contribution has two maxima, and so on. The more wavenumbers that are included in the sum, the more exact the approximation. If an infinite number were included, any continuous function could be fit exactly. The highest wavenumber K is known as the truncation value, since the contribution of all higher wavenumbers is truncated.

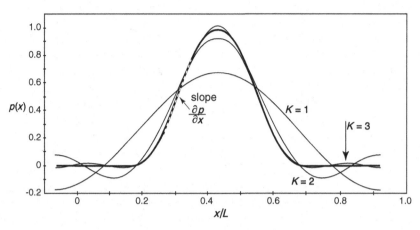

Fig. 5.8 Example of a spectral representation of a pressure field as a function of distance x and of the spectral pressure gradient at one of the points. The spectral representation is shown for three different truncations, $K = 1$, $K = 2$ and $K = 3$. The pressure gradient $\partial p / \partial x$ is shown only for $K = 3$. Adapted from chapter 9 of Trenberth (1992).

For relatively smooth fields, fortunately, a modest number of wavenumbers can give quite a reasonable approximation. In the case shown, which has a single smooth, high pressure region, using only a single sinusoid plus a constant leaves much to be desired, but adding a second sinusoid, i.e. with $K = 2$ does quite well. The case with $K = 3$ fits very well except in a few places. One disadvantage of a spectral representation can be seen in the overshoot on either side of the high pressure region. The true field goes to zero, but the spectral approximation has negative values as a consequence of fitting the high value nearby. If the field is not pressure but topographic height, with zero as sea level, this can be quite a nuisance. On the other hand, spectral methods work splendidly for problems involving wave dynamics. One reason is that gradients are exact for each wavenumber. Since the derivative of a sine is just a cosine, the pressure gradient $\partial p / \partial x$ derived from Eq. (5.2) is

$$\frac{\partial p}{\partial x} = \frac{2\pi}{L} \sum_{k=1}^{K} k \left[b_k \cos\left(\frac{2\pi k x}{L}\right) - a_k \sin\left(\frac{2\pi k x}{L}\right) \right] \qquad (5.3)$$

This is just another Fourier series that uses the coefficients of the original series multiplied by a factor involving the wavenumber. Needless to say, this is easy and efficient to implement computationally. It also makes certain types of theoretical derivations straightforward, which is why properties of atmospheric and oceanic waves are almost always derived using a spectral approach. For small amplitude waves in a spatially constant basic state flow, the solution can be obtained one wavenumber at a time, i.e. for a sequence of purely sinusoidal waves.

In order to transform from grid point representation to spectral representation, the spectral coefficients a_k and b_k must be computed from the values p_i at the grid points. These values are given by a sum of sinusoids similar to Eq. (5.2) but summed over the grid index i. A clever numerical trick called the Fast Fourier transform (FFT) greatly speeds this summation if the number of points N in the grid is given by some power of 2; for instance, $2^7 = 128$ points. This is why many atmospheric models have rather particular longitude spacing; for instance, 128 points implies a grid spacing of 2.8125 degrees in longitude.

The sinusoidal functions for a sequence of wavenumbers k in the above are referred to as basis functions. Spectral methods can be done with other basis functions as well. In fact, because the Coriolis parameter varies with latitude, it turns out to be advantageous to use a different set of functions in the latitude direction (known as Legendre functions) that oscillate like sinusoids but have an amplitude that varies with latitude. The evaluation of the transform from grid points to the latitudinal spectral representation turns out to be most efficient if a slightly uneven grid in latitude is used. The grid spacing is always included with the model output and routines to take into account this grid spacing are typically made available to the users of the output.

The truncations in latitude and longitude should usually be chosen to be consistent with each other, with a similar number of basis functions in each direction. The most common approach is known as a triangular truncation because the included wavenumbers make a triangular pattern on a wavenumber–wavenumber diagram. A model with 63 sinusoidal basis functions in longitude would be referred to as a "T63" model (T for triangular). Some classic climate model studies used a different pattern on the wavenumber diagram, known as rhomboidal truncation, for example 15 basis functions in longitude and 15 in latitude for each one in longitude, abbreviated R15 (coarse resolution by today's standards). A current high resolution model for weather prediction might use a resolution of T799, while truncations used in models in the IPCC (2007) report included T42, T63, T85 and T106.

The output of a spectral model is on grid points, so it is natural for a user to associate the spatial resolution with the grid. For instance, a T63 spectral model has a grid size of approximately 1.9 degrees in longitude (and roughly similar in latitude). The highest wavenumber that is included in the spectral representation, however, has a wavelength of $360/63 = 5.7$ degrees longitude. If you think of a localized feature, such as a region of rising motion associated with convection, the smallest size that can be represented even approximately is the half wavelength, 2.9 degrees. The grid point computations are done at slightly higher resolution because they include the computation of nonlinear terms, like velocity times moisture or temperature gradients, and these must be done at higher resolution than the spectral representation to avoid a type of numerical error known as aliasing. The parameterized processes, including precipitation, are done at the grid points, but the inputs, such as moisture and temperature, have a half wavelength about 1.5 times the grid size. In other words, the grid for a T63 model is chosen to be $2.9/1.5 = 1.9$ degrees (or $360/(106 \times 2 \times 1.5) = 1.1$ degrees for a T106 model) so that some of the information can be thrown away for numerical reasons when converting from grid points to the spectral representation. Thus in interpreting spectral model output, it is wise to bear in mind that the effective resolution is slightly poorer than the grid size would suggest.

5.2.2 Time-stepping and numerical stability

Finite differencing is used on the time derivatives in all of the equations of motion as well, choosing a *time step* Δt. Consider a very simplified temperature equation which has only a negative feedback term for perturbations of temperature T':

$$\frac{\partial T'}{\partial t} = -\frac{1}{\tau}T' \tag{5.4}$$

The temperature perturbations are relative to a solution where other terms in the temperature equation are in balance. This equation arose in section 3.3.4 as an approximation for decay of SST anomalies in an ocean surface layer, but it arises for similar reasons in other processes that have negative feedbacks, some of which have much faster decay times, i.e. smaller τ. The small values of τ are important here because they are the ones that limit the longest time step that can be used in the model. Suppose we do a simple finite differencing of this equation

$$\frac{T'_{n+1} - T'_n}{\Delta t} = -\frac{1}{\tau}T'_n \tag{5.5}$$

where T'_n denotes the value at time step n. The time derivative has been replaced by the difference between the values at subsequent time steps divided by the time step Δt. Rearranging Eq. (5.5), the value at time step $n+1$ can be obtained from the current values at time step n by

$$T'_{n+1} = \left(1 - \frac{\Delta t}{\tau}\right)T'_n \tag{5.6}$$

The procedure is then repeated to obtain values at step $n+2$ from values at step $n+1$ and so on. The solution is shown for three choices of Δt in Figure 5.9. These choices are given relative to the time scale τ. If τ is 2 hours, as might happen if the negative feedback involves a convective process, for instance, then $\Delta t/\tau = 0.1$ would imply a time step of 12 minutes. When Δt is small relative to the time scale of the process, then the numerical solution gives a good approximation to the true solution (in this case the true solution is a simple exponential decay of the form of Eq. (3.26). When Δt is less small, the solution is qualitatively reasonable although substantial error occurs, as illustrated by the $\Delta t/\tau = 0.5$ case in Figure 5.9. When Δt becomes comparable to the time scale of the process we are trying to model, the solution becomes highly inaccurate (in this example, changing sign when the true solution does not) and can even become unstable. When this *numerical instability* occurs, the amplitude of the perturbation in the numerical solution grows exponentially in time until it reaches such large values that the computer code encounters fatal errors and stops (the model "blows up"). For the time-stepping procedure in Eq. (5.6), the model is unstable for $(\Delta t/\tau) > 2$. The temperature perturbation obtained for step $n+1$ has a negative value of larger amplitude than the temperature at step n. At each subsequent step, the sign changes, the amplitude increases and the error grows, as seen in Figure 5.9. Why this happens can be seen in the very first step. The slope of the true solution is being used to estimate how T' will change a small time later. Projecting this linear estimate of the change too far forward in time results in an incorrect estimate of a perturbation whose magnitude is actually larger than the original. The model then duly returns to use the right hand side of Eq. (5.5) to estimate the slope by which the solution should be changing, and then projects this slope too far again, resulting in an even greater error.

The finite-difference approximation for the time derivative in Eq. (5.5) is very simple, involving only two time levels ($n + 1$ and n). It works fine for the simple decay equation,

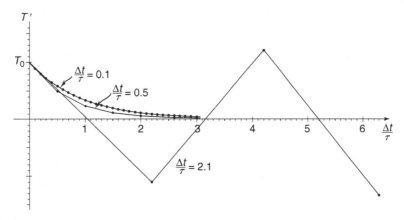

Numerical solutions to a simple exponential decay equation using a simple time-stepping scheme and three choices of the time step. Dots indicate times for which the solution is computed. The case for the smallest time step is sufficiently accurate that it appears to overlie the true solution. The largest time step is slightly larger than the numerical stability criterion so the solution erroneously increases with time.

Eq. (5.4), and thus is useful for illustration, but the errors at each time step are proportional to Δt. Certain schemes that involve an additional time level ($n + 1$, n and $n - 1$), or that estimate the time derivative and then do a correction step, have smaller errors, proportional to Δt^2, and have better stability properties for wave propagation. Nonetheless, the principle holds that both numerical stability and accuracy limit the time step to be sufficiently small relative to the fastest time scale of the resolved physical processes.

Similar considerations apply for advection terms. Consider temperature advection by the zonal wind $u(\partial T/\partial x)$. The fastest time scale that arises is when the temperature changes at the smallest scale, i.e. across one grid point. The time scale for this change to be carried across the grid box is $u \Delta x^{-1}$. The smaller Δx, the smaller the time scale, and thus the time step must be reduced when the grid size is reduced. Wave propagation speed places similar limitations on time step, with the wind speed u replaced by the wave speed. Since both gravity wave speeds and wind speeds in strong regions of the jet stream might reach $60 \, \mathrm{m \, s^{-1}}$, the time step for a model with a 200 km grid cell will have to be less than roughly an hour to be numerically stable. And as illustrated in Figure 5.9, accuracy requires that the time step be rather smaller than the stability criterion, implying time steps of, say, 15 minutes. A model with a 400 km grid spacing could take double this time step (if wave or advection speeds are the most severely limiting process).

5.2.3 Staggered grids and other grids

Grid-point models often use a technique known as a staggered grid in which different variables are offset by half a grid point from each other. An example is shown in Figure 5.10. Staggering the locations where different variables are computed is useful because of the way that gradients of different variables occur in the various equations. For instance, vertical velocity depends on horizontal divergence. In the example shown in Figure 5.10, vertical velocity occurs on the points marked T, and is located in between u points in the

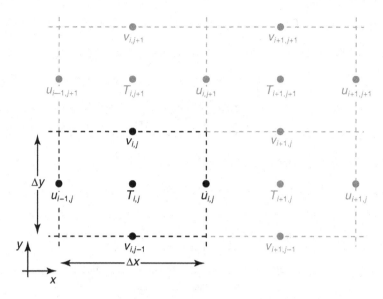

An example of a staggered grid. The configuration shown here is known as a C grid and is used in many grid-point atmospheric models. The points marked with T are the locations of temperature points. Pressure and vertical velocity are also computed on T points. The points marked u are the locations where eastward velocity is computed, while the points marked v are the locations where northward velocity is computed.

x (eastward) direction and v points in the y (northward) direction, following the convention introduced in Chapter 3 that u is eastward velocity and v is northward velocity. The difference $(u_{i,j} - u_{i-1,j})/\Delta x$ thus gives the contribution to divergence by u at the central T gridpoint. Likewise $(v_{i,j} - v_{i-1,j})/\Delta y$ gives the contribution by v. The vertical velocity at the T gridpoint is thus easily computed from the directly neighboring points (using the relation of vertical velocity to divergence from Eq. (3.29)). The vertical velocity is a major contributor to temperature change, so this location of vertical velocity is also useful for the temperature equation.

Considering the momentum equation, the distance between pressure surfaces is computed from temperature, so the eastward pressure gradient force at the point $u_{i,j}$ is computed from differences between pressures at the neighboring temperature points. Letting p have the same grid-point indices as marked for T, the eastward pressure gradient is $(p_{i+1,j} - P_{i,j})/\Delta x$. This is an improvement over the calculation of pressure gradient on the unstaggered grid shown in Figure 5.7 in which the pressure gradient was calculated over a distance $2\Delta x$.

Because different terms in the equations are more important in ocean applications, ocean models tend to use a slightly different configuration of staggered grid. From the user's point of view, the main thing is to be aware of the location of the variables. For many user applications of an atmospheric model, variables of interest are precipitation and temperature, and these are located at the T grid points. If wind is required at the same location, it must be interpolated from the neighboring u and v grid points. Often routines are supplied by the modeling center to perform such interpolations.

The latitude–longitude grids discussed here have been the most commonly used in climate models. They are also the simplest to present in discussing these models. One issue with latitude–longitude grids is that the resolution effectively increases with latitude. Longitude circles converge toward the poles, so a longitudinal grid spacing of 2 degrees is about 222 km at the equator, 157 km at a latitude of 45°, 39 km at a latitude of 80°, and goes to zero at the poles. This requires special treatment of polar regions to avoid numerical instability where the grid spacing is small. Current efforts are under way to use geodesic grids that cover the sphere with hexagonal cells (a few pentagonal cells are also required), or other methods that keep the area of the grid cells closer to equal over the whole sphere.

5.2.4 Parallel computer architecture

One means of reducing the time a computation takes to complete is to break the computation into pieces and perform simultaneous calculations on many computer processors. Such parallel architecture is a major tool in computing. The advantages gained, however, depend on the problem being computed. If the computation breaks naturally into a large number of small computations that require little communication of results between the various parts, then the speed at which the calculation is completed goes up roughly proportional to the number of computer processors one uses. In climate applications, the typical approach is to divide the atmosphere or ocean into different regions, and compute each region on a different processor. However, neighboring regions affect each other, so there must be constant communication between the different processors. The more processors, the more communication of information is required, and so the gain in speed tends to drop off as more processors are added. Thus, if one considers the formula for computational cost in section 5.1.3, one might think that using 10 000 processors would allow a climate model to be run in the same amount of wall-clock time (i.e. real time, as opposed to the amount of time spent by the sum of the computer processors) with a factor of 10 increase in resolution in each spatial direction (along with reducing the time step by a factor of 10). However, even when the coding is done cleverly, the communication required among the processors means that only a fraction of this increase can be realized. Computer systems with tens of thousands of processors and climate model codes that run efficiently on them are indeed being developed, but the costs are considerable – just the energy costs for running such enormous systems can be substantial. Thus parallel computation is very helpful, but is not an instant solution to having to parameterize climate processes that occur at scales too small to explicitly resolve.

One aspect of climate model runs falls into the category of "embarrassingly parallel," i.e. requires no extra effort to run on multiple processors: ensembles of simulations. When runs with different initial conditions are used, for instance, to evaluate impacts of weather noise on ENSO, as seen in Chapter 4, or natural climate variability in global warming simulations, as in Chapters 6 and 7, these can be run on separate groups of processors, or even separate computers. An experiment known as climate.net, earlier in this decade, used personal computers of many volunteers to run thousands of climate model runs with slightly different parameter settings.

5.3 Parameterization of small-scale processes

With the grid size determined largely by feasible computational costs, all processes that occur at sizes smaller than a climate model grid box are representeted as parameterizations, as introduced in sections 2.1.2 and 5.1. The effects of parameterizations are important both to what climate models simulate correctly and to some of the uncertainties in the simulations. This section aims to give a sense of what occurs in some of the main parameterized processes, often using the simplest example of a parameterization, since the details of complex parameterizations can be rather technical.

5.3.1 Mixing and surface fluxes

Turbulent motions at scales smaller than the grid scale cause small parcels of fluid to be exchanged between adjacent volumes at the grid scale and then to mix into their new surroundings. This causes mixing of heat, moisture or salt, and momentum. While this mixing is often a small effect over a time step at a given location, the overall effects on the flow over the long term are important. Figure 5.11 shows an example for the case of vertical

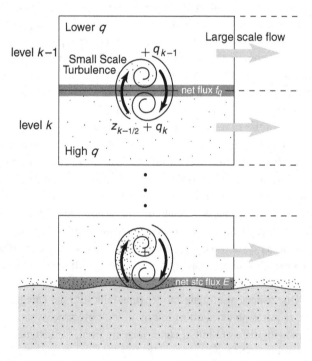

Fig. 5.11 Schematic example of vertical mixing for specific humidity, q, and of the surface flux of moisture by evaporation, showing the processes to be parameterized on a grid with k levels. Moisture at each level is denoted q_k. The net flux across the face of the grid box at level $z_{k-1/2}$ that in reality would occur by smaller-scale motions acting on the gradient in q must be parameterized in numerical models.

mixing of moisture in the atmosphere and its connection to the surface flux of moisture by evaporation at the ocean surface. When there is a gradient of moisture, in this case a vertical gradient with moisture increasing toward the surface, mixing between two grid boxes acts to increase moisture in the box with the lower value and vice versa, bringing their values closer together. For the budget of moisture in each box, the average flux F_q due to mixing across the interface between the two boxes is needed. It is approximated as proportional to the gradient in moisture between the levels. Using height z as the vertical coordinate, this is

$$F_{q(z_{k-1/2})} = K_q \frac{(q_k - q_{k-1})}{\Delta z} \tag{5.7}$$

The vertical mixing coefficient K_q typically depends on factors that affect the strength of the turbulent motions, such as the wind shear. Similar expressions are used for horizontal mixing, proportional to horizontal gradients. The source term for the moisture equation then depends on the net flux due to mixing across all faces of the box.

At the ocean surface, evaporation acts to bring air within an extremely thin layer near the surface toward saturation, while mixing continually moves the moisture up through the depth of the first layer represented in a model. The surface flux of moisture is still termed evaporation, E (bearing in mind that turbulent mixing is essential to its magnitude), and it is approximated as proportional to the difference between the saturation value of moisture at the temperature of the surface $q_{sat}(T_s)$ and the moisture at a reference level q_a, for instance at 10 m, which is computed from the moisture at the lowest model level. A simple representation is

$$E = \rho C_D V (q_{sat}(T_s) - q_a) \tag{5.8}$$

where ρ is density of air, C_D is an empirically estimated drag coefficient, and V is a wind speed that typically includes both the speed of the large-scale wind from the model's lowest level, and an estimate of small-scale gusts. Some models attempt to include such effects as the increase in gustiness when there strong moist convection. The contribution by these effects to the rate of change of moisture in the lowest model layer is then given by the difference between the surface flux by evaporation and the flux by mixing at the top.

Because the effects of mixing (as well as of large-scale transports) conserve moisture within the atmosphere (up to where other processes cause it to rain out), the surface flux has considerable importance in diagnostics of models. Other important surface fluxes associated with turbulent mixing processes are sensible heat, and the surface wind stress (in units of force per unit area) which gives the surface momentum flux.

Fluxes of sensible heat and momentum, both at the surface and due to mixing in the interior, are represented by expressions very similar to Eq. (5.7) and Eq. (5.8), respectively, but with temperature and velocity components[2] in place of q. In the atmosphere, some special treatment is required for vertical mixing by moist and dry convection, outlined below. In the ocean the mixing of salinity, temperature and momentum is rather similar. Surface fluxes for momentum and sensible heat are as given for the atmospheric side of the interface. The freshwater flux out of the ocean that affects the salinity of the layer nearest the surface is given by the evaporation in Eq. (5.8) minus input precipitation given by the atmospheric model. If the model computes air–sea exchange of other gases such as carbon dioxide, similar principles are applied, with a parameterized surface flux depending on values in the ocean topmost layer and atmospheric lowest layer, and mixing depending on the gradients. Surface evapotranspiration is discussed in section 5.1.5.

Under some circumstances, models use larger mixing values than might be ideal because it tends to aid numerical stability. Depending on the details of the spatial discretization scheme, some models may also have effects that act very much like mixing – termed numerical diffusion – owing to the approximations in advection by the large-scale flow. This can be important, for instance, in the ocean when there should be strong mixing between water masses of the same density, but little mixing that crosses density surfaces, known as *isopycnals*. To reduce impacts of numerical diffusion, some ocean models – known as isopycnal models – use density as a vertical coordinate. Mixing is then computed across and along surfaces of constant density rather than vertically and horizontally.

5.3.2 Dry convection

A special case of mixing occurs when the turbulence is driven by heating from below, as in the atmospheric boundary layer. When the rising parcels do not condense, this is termed dry convection. Figure 5.12 shows this process in terms of the temperature profile of the air on a large-scale average, termed the environmental profile, and the temperature of a rising parcel. The initial environmental lapse rate shown is quite warm at the surface since the surface is heated by the sun. A parcel moving up at the adiabatic lapse rate is warmer than the environment, so it is buoyant and keeps going up until it reaches a level where the environment is warmer. Thus the initial environmental lapse rate is *unstable*. Other parcels have to move down to balance, so on the whole the temperature in the boundary layer will get mixed along the dry adiabat. If the convection were completely successful, the new

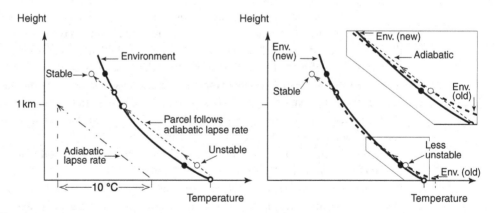

Fig. 5.12 Changing the environmental lapse rate toward the dry adiabatic lapse rate by dry convection. Open circles denote parcel temperatures, as parcels rise at the adiabatic lapse rate. Where parcel temperature is greater than environmental temperature at the same pressure, the parcel is buoyant and continues to rise, so the environment is unstable. In the right panel, the original environmental profile has been replaced by one that has been mixed by convection until it is closer to the adiabatic lapse rate, and is less unstable.

environmental lapse rate would be a dry adiabat with temperature slightly warmer at the top and slightly colder at the bottom than the original sounding. The heating from below keeps the environment slightly unstable, so convection continues. But the convection is quite a fast process, so the environmental lapse rate ends up rather close to $10\,°\mathrm{C\,km^{-1}}$. We will see below that similar arguments can give an approximate explanation of the effects of moist convection in clouds.

A useful way of keeping track of what is happening to a parcel when it moves vertically without condensation is to define a *potential temperature* θ. This is defined as the temperature a parcel from any level would have if it were brought down to the surface (to a reference pressure p_0) without any effects of heat exchange or mixing with surrounding air. The solution is

$$\theta = T \left(\frac{p_0}{p}\right)^{R/c_p} \tag{5.9}$$

where R is the ideal gas constant and c_p the heat capacity. This has the property that the temperature equation, Eq. (3.20), in the adiabatic case where there is no heating, becomes simply

$$c_p \frac{d\theta}{dt} = 0 \tag{5.10}$$

To show this, substitute the expression for θ into Eq. (5.10) and compare to Eq. (3.20). This implies that θ is conserved for motions without heating. For parcels moving up and down rapidly in the boundary layer, θ is approximately conserved. When the parcels mix with surrounding air at a given level, the effect is to mix the boundary layer toward a constant value of θ. Thus atmospheric mixing is represented in terms of gradients of θ, rather than temperature, whenever a change in pressure is involved. Atmospheric models typically include some version of the parcel stability arguments above, often more sophisticated versions, in determining the layer of strong mixing near the surface.

5.3.3 Moist convection

Moist convection has some features in common with turbulent mixing, but a number of differences that require special treatment. There is latent heat release by the condensation of water vapor; the small-scale cloud motions penetrate through a deep unstable layer that covers many model levels in the vertical; and the effects of the interactions among cloud water droplets and ice particles – known as cloud microphysics – must be taken into account. The simplest considerations for representing moist convection are outlined here, but in general representation of moist convection and the resulting rainfall and cloud effects on radiation is an ongoing challenge in climate models. A few current directions are mentioned in the last paragraph.

The moist adiabat determines whether an air column is unstable to a rising condensing air parcel. It can further be used to understand the bulk effects of convection on the large scale. Both aspects are illustrated in Figure 5.13. In each panel, the same parcel temperature as a function of pressure is shown, for a parcel with a given initial temperature and moisture rising adiabatically from the surface. It first follows the dry adiabat, and then, once condensation

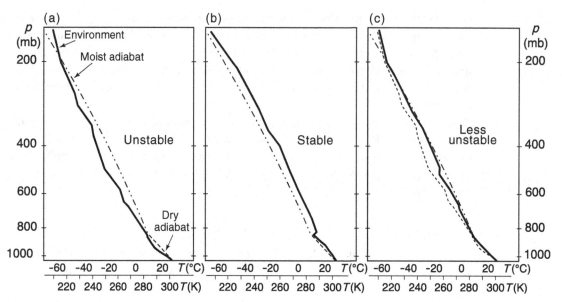

Fig. 5.13 Schematic of a parcel raised adiabatically from the boundary layer along the moist adiabat (dot-dashed curve), comparing its temperature with that of the large-scale environment (solid line) for three cases. (a) A conditionally unstable case that would most likely give rise to deep convection. (b) A stable case. (c) The dashed curve repeats the temperature of the environment from (a). The solid curve indicates a likely environmental temperature after convection has warmed the upper levels and cooled the boundary layer, so the column is less unstable than in (a).

begins, it follows the moist adiabat. This parcel path is compared with three different temperature profiles of the large-scale atmosphere (referred to as the environment).

To determine the stability of the large-scale column, the temperature of the moist, adiabatically rising parcel is compared with the large-scale temperature. If the parcel is warmer, as in Figure 5.13a, it will be buoyant and will keep rising. The sounding is said to be *conditionally unstable*. The term "conditionally" is used because it depends on the parcel being raised enough to become saturated in the first place, but typically there is enough turbulence in the boundary layer that convection will most likely occur. Convection will tend to keep occurring until the upper levels have been warmed and/or the boundary layer cooled enough and dried sufficiently that future parcels are stable and convection ceases. Figure 5.13b shows a case of an environmental temperature profile that is stable for the same value of surface temperature and humidity. The upper levels are warmer than in Figure 5.13a, so if the parcel were somehow raised, it would be denser than the environment and would sink back down.

Figure 5.13c shows a case typical of how convection modifies the environment. The large-scale temperature has been warmed by the effects of convection so that the column is less unstable than in Figure 5.13a. Convection will occur less frequently in the less unstable column, but input of heat at the surface acts to destabilize the sounding again if it is in a region that is favorable for convection. A balance will thus be established in which the large-scale temperature profile is drawn toward that of convective parcels raised from the surface, i.e. approximately toward the moist adiabat.

Thus the simplest representation of moist convection, known as *convective adjustment*, is that whenever a column is unstable as in Figure 5.13a, it is adjusted toward a moist adiabat on a time scale typical of cloud processes, e.g. 2 hours, and the moisture sink is determined by the required heating. More sophisticated parameterizations of deep convection, known as *mass flux schemes*, use a simple model of an ensemble of cloud plumes, which entrain surrounding air to varying degrees as they rise. The total amount of air rising in these plumes (the mass flux, schematized as upward arrows in Figure 5.2a) is commonly determined by principles very similar to those outlined above: the plumes alter the environment in a way that tends to reduce the buoyancy. The effect on large-scale temperature is thus fairly similar to moist convective adjustment, but these schemes can keep track, for instance, of how much cloudy air is *detrained* from cloud plumes into the surrounding air at each level, and the properties of the cloud air at the cloud top level where a plume loses buoyancy and ceases to rise. Some also keep track of the mass flux in strong downdrafts that can occur, for instance, when detrained air is cooled by evaporation.

The representation of cloud microphysics varies considerably among climate models, and this likely contributes to the uncertainties in cloud feedbacks we will encounter in Chapters 6 and 7. One approach is to include budgets of cloud liquid water and cloud ice in the model equations. The range of sizes (the size spectrum) of these droplets or crystals is important to rainfall processes, but one cannot afford to compute these in great detail. For example, a budget for cloud water (defined as droplets small enough to have negligible fall speed) and rain water (droplets large enough to fall) as well as one for cloud ice might be used. Alternately, some models parameterize cloud properties directly as empirical relations to the moisture and temperature.

5.3.4 Land surface processes and soil moisture

Biophysical land surface schemes were introduced in section 5.1.5, with Figure 5.6 showing land surface types characterized by different vegetation. Figure 5.14 illustrates some of the land surface processes, particularly those involved in the soil hydrology and evapotranspiration.

Over land surfaces with vegetation, the surface flux of moisture occurs primarily by evapotranspiration. Because plants actively regulate their moisture loss, this must be represented differently than evaporation from a water surface. The rate of the transfer of moisture from the soil to the atmosphere is limited first by the transfer within the plants, and then by the turbulent transfer in air around and above the vegetation. In models, each of these processes has a coefficient – termed a *resistance* – controlling the rate of moisture transfer. A example of a representation of evapotranspiration from a forest canopy is

$$E_c \propto (r_a + r_c)^{-1}(q_{sat}(T_c) - q_a) \qquad (5.11)$$

where r_a is a bulk aerodynamic resistance depending on wind speed, and r_c is the canopy resistance summarizing the active control of moisture transfer by vegetation. The saturation value of moisture is evaluated at a temperature T_c characterizing the forest canopy. If r_c were zero, this equation would be the same as for surface evaporation in Eq. (5.9), with r_a given by the drag coefficient, air density and wind speed.

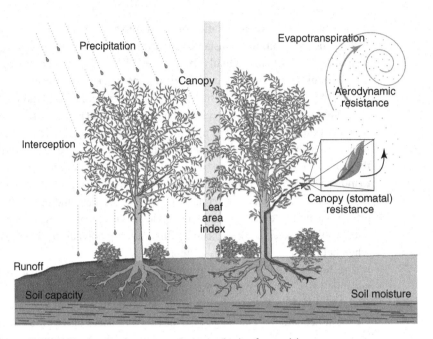

Fig. 5.14   Processes affecting soil moisture and evapotranspiration in a land surface model.

The canopy resistance depends on the water content of the soil, often termed *soil moisture* or *soil wetness*. As the soil dries out, the plants reduce the rate at which moisture is transferred, at some point reaching a wilting level where the transfer all but shuts down. It thus becomes important to include an equation for the soil wetness in the land surface model. A conservation equation is set up for several soil layers, in which the soil moisture increases or decreases as a function of input by precipitation, loss by evapotranspiration, and transfer among the layers. Another important loss term is runoff, which becomes especially large when the soil gets saturated by a heavy precipitation event. When soil moisture approaches the *soil capacity*, which must be defined for each land surface type, runoff increases. Interception of precipitation by the canopy must also be represented, since this water evaporates directly into the atmosphere. The runoff is then directed into a hydrological model for stream and river flow.

One common measure of the density of vegetation is the *leaf area index*, essentially the number of leaves that a vertically traveling ray of sunlight would on average encounter on its way to the surface. In some models the leaf area index changes as a function of season, temperature, soil moisture, etc. This affects the albedo of the surface as well as the evapotranspiration.

5.3.5 Sea ice and snow

Snowfall is determined by the atmospheric model, but once it is on the ground, it is necessary to keep track of it via a snow model until it melts from a given grid box. These models often have several layers and include representations of the changes in snow as it ages, but the essential budgets are a surface energy budget and a mass budget for snow. Since the snow surface temperature cannot rise significantly above freezing, thus tending to limit the upward energy transfer from the snow, a large downward radiative flux, for instance, in spring, will be compensated by latent heat of melting. There is also *sublimation* of snow from crystals directly into vapor (although melting, once it begins, is usually much larger). One then keeps track of the resulting loss of snow mass. Other factors to consider are the appropriate surface albedo for a snow covered grid box, which can depend on the age of the snow, and even on the vegetation type, as protruding vegetation has much lower albedo than the surrounding snow. While these may sound like annoying details, they add up to considerable scatter among different climate models in regional snow cover and its change under global warming. The snow cover can also be sensitive to the combined effects of other parameterizations in the model that affect the surface energy budget.[3]

Sea ice thermodynamics includes similar considerations to land snow cover, and a snow layer may be included at the top of the sea ice model, but many additional effects must be treated, as shown schematically in Figure 5.15.[4] Net surface heat flux exchange with the atmosphere is calculated as a sum of radiative fluxes, sublimation and sensible heat. The latter two and the heat flux exchange with the ocean below the ice are parameterized similarly to other turbulent surface fluxes in section 5.3.1. The ice model often has a few layers, and the flux between them is represented in terms of an ice conductivity giving fluxes proportional to temperature gradients. The heat transfer through ice is far slower than exchange directly between atmosphere and ocean. In cold seasons, the ocean is relatively warm at around $-2\,°C$, the freezing point of salt water. Ice formation tends to insulate the ocean, reducing the heat transfer to the atmosphere, and allowing the surface temperature to become far colder than freezing. It is important to keep track of heat transfer in leads between blocks of sea ice, since relatively large heat exchange occurs there.

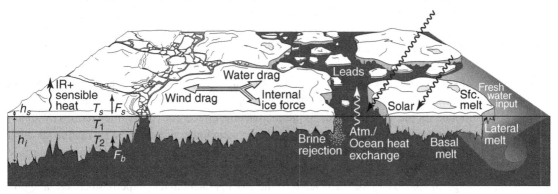

Fig. 5.15   Schematic of some of the processes in a sea ice model. Net heat fluxes at bottom and surface, F_b and F_s, are exchanged with the ocean and atmosphere above, respectively. In this example there is a snow layer of thickness h_s and temperature T_s (at the surface), and two ice layers with temperatures T_1, T_2 and an overall ice thickness h_i. The ocean also exchanges heat directly with the atmosphere through leads. Ice movement depends on a balance of several forces, the largest of which are indicated by arrows. During freezing, rejection of salty brine from the ice tends to increase ocean salinity. During melt, as increased surface heating causes surface ponding, and warming waters cause lateral and basal melt, fresh water is returned to the ocean.

For a simple case of sea ice growth, consider just the ice thickness. The contact with sea water at the bottom of the ice is held near freezing, while the ice surface is being cooled by infrared and sensible heat loss exceeding the cold-season solar radiative input. When the ice is relatively thin, the ice conductivity prevents the surface temperature from being much colder than the fixed ice bottom temperature. Thus the heat lost from this relatively warm surface is substantial. The ocean surface water cannot get colder than freezing, so the heat loss is balanced by the latent heat of fusion as freezing creates more ice. The energy budget thus dictates the added ice in the ice mass budget, giving the increase of the ice thickness.

While these simple considerations give a first approximation to ice creation (the most basic elements of thickening and thinning can be approximated in a single layer), additional effects must be considered. The addition or loss of ice can decrease or increase not only the ice thickness, but also the fraction of a given area that is covered with open water (leads) versus the fraction covered by ice. Partitioning the ice mass change into these contributions must be treated with approximations to available data. Different models use slightly different variables to keep track of the fraction of a given area covered by ice, often called the *ice fraction* or *ice concentration*. The principle is to maintain an overall mass budget for ice and the exchange with the ocean and among layers, as both the thickness and ice fraction change.

As sea water freezes, salt is rejected to form relatively pure ice, and pockets of highly saline water (brine) can remain in the ice. These brine pockets affect some of the properties of ice, including thermal conductivity. The rate at which brine is rejected into the ocean surface layer affects the salinity of that layer, increasing ocean surface density when ice forms. Conversely, when ice melts, fresh water is released into the ocean, tending to decrease salinity. The temperature profile during ice melt can differ from the coldest-at-the-top profile during the cold season. Snow falling on the ice and then aging to ice can contribute to the mass balance, has lower thermal conductivity, and has higher albedo. This motivates use of a few ice layers and a snow layer. During melting, surface ponds can play a role as they alter the albedo of the ice, and lateral melt can be important as the solar radiation warms the water in the leads (which have lower albedo than the ice). An ice model that include these effects via heat and ice/freshwater budgets is termed a *thermodynamic sea ice model*.

Ice motion, driven by winds and currents, is a significant factor in the extent and thickness of sea ice, especially in some regions. For instance, there is substantial export of ice from the Arctic Ocean to the Greenland and Norwegian seas. Ice models that include equations of motion for the ice, in addition to thermodynamic effects above, are often termed *dynamic-thermodynamic sea ice models*. Typically the main forces affecting the ice motion are wind drag, water drag and internal ice forces. There are also smaller forces due to the ocean surface slope and the Coriolis force. The ice internal forces are represented as an average over areas large enough to include many leads. If there is a gradient in ice properties between a region with few leads and thicker ice and a region with less ice, the average flow of ice would tend to be toward the lower ice region in absence of other forces. In the presence of strong wind, tending to pile up the ice (which also creates ridging effects that affect ice surface properties), the internal ice forces tend to resist further pile-up. Thus the ice internal forces are parameterized as functions of properties such as lead fraction and average ice thickness. The internal ice forces are strongly nonlinear functions of these properties, and details of simulated ice distributions can be rather sensitive to them. Models aimed at climate change have moved from thermodynamic sea ice models to models that include dynamical effects, and typically to inclusion of some of the more detailed effects in the thermodynamics. Despite work on ice models, substantial errors can persist in simulation, for instance, of the seasonal excursions of the ice edge. Some of these errors may have nothing to do with the ice model itself, since atmospheric model clouds (shielding solar radiation) and surface wind can have a substantial impact on the ice. In evaluating climate change simulations for regional effects involving sea ice, one should factor in any such known errors in the climatology.

5.4 The hierarchy of climate models

Climate modeling has benefited from a variety of approaches to constructing climate models. In particular, much benefit has come from models that aim to simulate or understand particular aspects of the climate system, without attempting to include the full complexity. This is now known as a *hierarchical modeling* approach. The hierarchy of climate models is listed in rough order of increasing complexity in Table 5.2 with the most complex models at the end.

We have already seen examples of some of these categories of model. For instance, in Chapter 3, a simple energy balance model for globally average temperature was discussed for the case of no atmosphere (and in Chapter 6 this type of globally average energy balance model is applied to global warming). In Chapter 1, the Cane–Zebiak model that was first used to simulate ENSO and conduct experimental climate predictions is classed as an intermediate complexity model. The ocean component of this model is essentially the one-layer model for the layer above the thermocline that was introduced in Chapter 4. This is considerably less complex than an ocean GCM representation of the ocean, but captures

Table 5.2 The hierarchy of complexity of climate models.	
Model type	Comments
Simple models	e.g. energy balance models
Intermediate complexity models	e.g. Cane–Zebiak model for ENSO, EMIC
Hybrid coupled models	Ocean GCM with a simple atmosphere
Atmospheric GCM with a mixed-layer ocean	
Regional climate models	Boundary conditions from global climate models
Global atmospheric GCM with a regional ocean GCM	e.g. tropical Pacific
Global ocean–atmosphere GCM	
Earth system model with interactive vegetation and chemistry	Includes interactive carbon cycle

Note: Various classes of models are listed in order of increasing complexity.

much of the essential dynamics needed to simulate the phenomenon of interest. Intermediate complexity models are defined as those that make approximations relative to full GCMs, but can still be quantitatively and extensively compared to observed climate variables. Ideally, their derivation from the full equations should involve systematic, well-stated, testable approximations (although in practice this is a high standard). Another class of intermediate complexity models that we have not covered is known as "Earth system models of intermediate complexity" (EMIC). These are models that attempt to represent most aspects of the climate system, but each in a simplified way. Typically, they extend the concept of energy balance models to include many latitude bands or a substantial number of regions of roughly continental width, together with a simple ocean, and parameterizations of many processes, including sometimes vegetation growth and decay. However, they greatly simplify the representation of atmospheric and oceanic dynamics so the system is computationally much less demanding. This permits very long computations useful for paleoclimate studies.

Hybrid coupled models use an ocean GCM but a simplified representation of the atmosphere. They were originally invented for use in ENSO studies, where the most important aspect of the atmosphere is the wind anomalies along the equator that arise from anomalies in SST. They remain competitive for ENSO forecasts with full coupled GCMs. In the long run, they will remain useful for understanding feedbacks in the coupled system.

Many of the early studies of global warming were done with an atmospheric model coupled to a *mixed-layer ocean*. This includes the effects of the ocean heat capacity in the surface layer of the ocean and a simple estimate of ocean heat transport that does not change in time. The ocean temperature is taken to be constant through a layer about 50 to 75 m deep, which gave Eq. (3.19) for SST. The surface fluxes are computed in the atmospheric GCM, while the transport terms within the layer and the flux through the bottom of the layer are approximated by a value known as the Q-flux. This is a function of space (and usually of season) but does not change from year to year. Ideally it is estimated from observations, but more often it is estimated by running the atmospheric model with observed SST and then estimating the implied value that the ocean would need to transport to maintain the observed SST. If the atmosphere model is very accurate, this is a good estimate of the ocean transports. If the atmospheric model has errors, this has the effect of compensating for those errors. This has some advantages in that it permits one to examine changes relative to a reasonably good climatology even with an imperfect model – but of course, this can be a two-edged sword if some of the errors that led to an imperfect climatology also affect the changes being modeled. Atmospheric GCMs coupled to mixed-layer oceans are very reasonable tools for a first approximation to global warming because many of the most complex aspects occur in the atmosphere. These models do not have to be run for long periods of time to bring the deep ocean into equilibration before starting global warming experiments and they get around errors that arise from feedbacks between the ocean–atmosphere system. Now that coupled ocean–atmosphere GCMs are available, they remain an additional diagnostic tool for understanding system behavior.

For simulation of phenomena that involve the ocean but have a primary region of origin, such as ENSO, one can use an ocean model that is coupled to the atmosphere only over the ocean basin of interest, such as the tropical Pacific. Observed SST is specified elsewhere (with an overlap region at the edges if they occur in an ocean region where the observed

SST and the SST from the ocean model are smoothly blended). One advantage is obviously that the ocean model has fewer grid points than if it had to cover the globe at the same resolution. A hidden advantage is that because the SST is specified over much of the globe, the model is less prone to climate drift. Furthermore the inflow in the deep ocean comes from regions where ocean vertical structure is specified. Thus the model does not have to produce an accurate simulation of the thermohaline circulation to have reasonable deep ocean temperatures.

There are currently a number of groups working with *regional climate models* – regional domain atmospheric models adapted to climate problems. These can afford to have a smaller grid size than a global climate model because they cover only a region, say the size of a continent or part of a continent, similar to the Western US region shown at high resolution in Figure 5.3. They also are typically only run in equilibrium conditions, since evolution of the large-scale ocean conditions is not included. Regional climate models face a number of challenges. The boundary conditions at the edges of the region must be specified from a global model, and there are often challenges in terms of maintaining compatibility of the simulation within the region and the larger-scale model solution. Such models are often useful for examining how global-scale climate change might affect the region, providing more detail of, for instance, effects of high resolution topogaphy on rainfall patterns. Another challenge in evaluating regional climate model simulations is that changes in climate features under global warming at fairly large scales, such as a shift in a storm track across the Pacific, affect the changes at small-scales, such as detailed precipitation patterns on the US West Coast. A regional climate model creates its small-scale precipitation change in response to the large-scale change coming in from the global climate model. When a set of different global climate models is used to drive the regional climate model, the results at the regional scale can differ greatly if the global climate model simulations differ.

Coupled ocean–atmosphere GCMs are the models that have been presented in section 5.1.4. At the even more complex end of the model hierarchy are the climate system models or Earth system models discussed in section 5.1.7. In addition to having a full ocean–atmosphere GCM, they include representations of *biogeochemistry*, that is, of aspects of the biosphere that affect the chemical composition of the atmosphere.

5.5 Climate simulations and climate drift

Once one has a climate model constructed, there are a number of things that must be done even before the most interesting experiments can begin. The time integration of the model equations by the time-stepping procedures described in section 5.2 are called *integrations, runs, simulations* or *experiments* (all four terms are approximately equivalent though with slightly different emphasis). Each run can proceed from any reasonable set of initial conditions, that is, a set of initial values of all the main variables (the ones that occur in the equations labelled "prognostic" in Table 5.1). However, when the model is first run, it takes a considerable period of model simulated time before it arrives at a state similar to the long-term mean. This is referred to as the model *spin up*. Two examples of model runs are

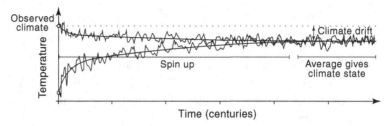

Fig. 5.16 Schematic example of model integrations during spin up to a simulated climate state. The temperature shown could be either a global or regional average. The value of temperature for the observed climate is indicated by a dotted line. Two runs are shown, one from arbitrary initial conditions and one from observed initial conditions. The climate model runs have much natural climate variability and so smooth curves are added that show the main features of the spin up. Time is in centuries, as would be appropriate for a full climate model. Similar curves but with a faster time scale would apply if only the upper ocean were included in the model or if a regional climate model were used.

shown in Figure 5.16. One run starts from some arbitrarily chosen initial conditions (that happen to be colder than observed). When a model is first run, the atmosphere might be started from constant temperature and no wind, while the ocean is started from a simple vertical profile that has no spatial variations and no currents. The model then responds to the solar forcing, warming at the equator, cooling at the poles and spinning up winds and currents. Eventually it reaches a simulated climate that resembles the observed. The initial part of the spin up, involving the atmosphere and upper ocean, tends to go fairly quickly. The atmosphere alone would spin up within a year, the global-scale features of the upper ocean in decades. To get the deep ocean reasonably close to equilibrium requires many centuries or a millennium of simulated time. The model then reaches a statistical equilibrium climate which can be defined by a long-term average. Natural variations about the long-term mean occur (including El Niño fluctuations caused by weather noise, decadal variations, etc.), but if the model does not have any change in the externally specified conditions (such as the solar input and the trace gases), then the long-term climate will be fairly stable. Some climate variability occurs on centennial or longer time scales as well, so even averages between different centuries will have some differences, but these tend to be relatively small compared with the signals we will be examining in global warming runs.

The second run shown in Figure 5.16 begins from an estimate of the observed climate. A climate model requires values of every variable at every grid point in its initial state, so to achieve the estimate of the observed, special procedures are often used to fill gaps in the data. The spin-up process in this case does not involve large changes, but it still occurs because the model adjusts toward its own equilibrium climate which is not exactly the same as the observed. This *systematic model error* is often referred to as climate drift, because when started from observations, the model "drifts" slowly toward its equilibrium state. The term *climate drift* also refers to the steady state result of this adjustment (much as a "snow drift" is the end result of drifting snow). Reducing this model error is one of the main activities of climate model research centers and it is a painstaking process. One problem is that an error in one part of the climate system tends to create errors in other parts of the climate system and these can feed back on each other. It is very common to find that the cause of an error in one variable is related to some subtle aspect of a process in some other part of the system, or even in another region.

Can a model that has some error in its simulation of current climate be used to model changes in climate? To some approximation, yes. One argument is that if a model gets the current climate correct to within a small percent error, when a change in forcing due to greenhouse gases is applied, one might expect to simulate the change to within a small percent error. There is some plausibility to this, but caution must be used as well. When applying the results of a climate model, one should carefully examine the simulation of the current climatology for the relevant variables and region, as errors in the climate definitely can impact estimates of climate change.

5.6   Evaluation of climate model simulations for present-day climate

5.6.1   Atmospheric model climatology from specified SST

A common way of testing an atmospheric model is to specify observed sea surface temperature (commonly, observed sea ice and/or observed surface albedo are specified as well). This tests some aspects of the atmospheric model in isolation from the ocean model and other components of the climate system. Distribution of precipitation, winds, energy fluxes at the surface and top of the atmosphere, and other climate variables can then be examined. This represents a non-trivial test of the atmospheric model, although there are definitely limitations to this test, as described in the coupled model section below.

While experiments specifying climatological SST can be used, a more common experiment is to specify the time series of observed SST, for instance, from 1960 to present. The long-term average of each climate variable over this period gives the model simulated climatology. Furthermore, the atmospheric model response to observed ENSO SST anomalies can be evaluated, as can other climate variability. Because they were popularized by the Atmospheric Model Intercomparison Project (AMIP), such runs are sometimes referred to as "AMIP runs."

Figure 5.17 shows a climatology of the precipitation from such a run for one model, the Community Climate System Model version 3 or CCSM3, from the US National Center for Atmospheric Research (NCAR). This may be compared with observations seen in Figure 2.13. To give a quick visual comparison, the observed 4 mm day^{-1} contour is overlaid. Now that we are getting down to detailed comparison, it is necessary to be a bit more specific about which dates are used in the climatology, and which estimate of the observed field is used. Here the CPC Merged Analysis of Precipitation (CMAP) standard product is used, as obtained from five kinds of satellite estimates by an algorithm developed in the US Climate Prediction Center (CPC). Other satellite estimates of precipitation can differ noticeably in some regions. The climatology starts in 1979 because satellite data are not available in earlier periods and station gauge data have much more limited coverage.

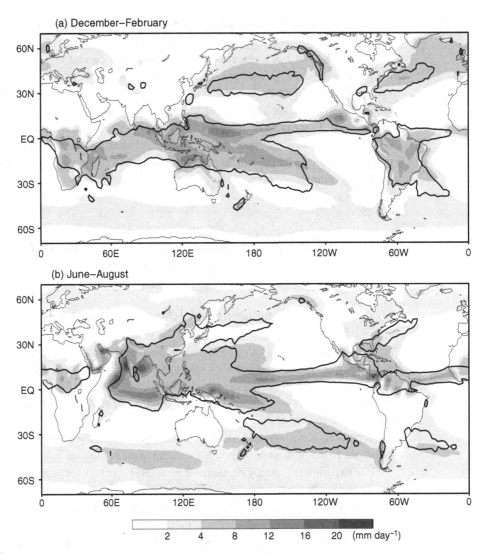

Fig. 5.17   Precipitation climatology for (a) December–February (DJF) and (b) June–August (JJA) for 1979–2000 from the
atmospheric component of the NCAR CCSM3, forced by observed SST. Contoured at 2, 4, 8, 12, 16, 20 mm per day. The
observed 4 mm per day contour is overlaid (black line) for comparison (from the CMAP precipitation product
1979–2000). Model data used to make this and subsequent figures are courtesy of the Program for Climate Model
Diagnosis and Intercomparison (PCMDI) and the modeling groups that provided the data for analysis.[5]

At the large scale, the model does rather well. The overall features of the tropical convergence zones and midlatitude storm tracks are captured, as is their movement with season.
As one pays more attention to smaller-scale detail, however, the model clearly has some
regions which are in error with respect to observed (even bearing in mind limitations of
the observational data product). For instance, in December–February, the northern hemisphere storm tracks are too strong; the 4 mm day^{-1} contour extends all the way across
both Pacific and Atlantic basins. The South Pacific Convergence Zone (SPCZ) at around

20° S in the Pacific extends too far east compared with what is observed. In the South Atlantic, the observed rainfall pattern has a "notch" around 10° S, reaching westward into northeastern Brazil between the ITCZ near the equator and the South Atlantic Convergence Zone (SACZ) near 30° S. In CCSM3, this notch is filled in – so northeast Brazil is getting spurious rain in this season – and the SPCZ is weak. In June–August, one notes that in both Atlantic and Pacific the separation between the northern midlatitude storm tracks and the tropical convection zones is missing in the model, with the summer storm tracks weak or displaced. The southern hemisphere winter storm tracks are too strong. One also notes model monsoon rain extending into parts of Saudi Arabia and too much rain in the Sahel region at the southern edge of the Sahara desert. Thus on these *regional* geographic scales that are important to human activities, a slight shift of a climate feature – which might seem small from a global perspective – can yield a large local error.

5.6.2 Climate model simulation of climatology

An even more difficult test for a climate model is obtaining a realistic simulation when the ocean, atmosphere and other climate model components such as sea ice are coupled together. As discussed in section 5.4, when ocean and atmospheric models were first coupled, the typical behavior was to drift to a simulated climate similar to observed, but with significant error. This can include alterations of the spatial patterns of climate features like the tropical convection zones, a few degrees C of error in SST and so on. This can occur even if an atmospheric model produces a reasonable simulation when tested with observed SST. If the atmospheric model has slight errors in a parameterized process, this leads to an error in the surface energy flux. For instance, an error in the parameterized cloud cover might allow tens of $\mathrm{W\,m^{-2}}$ too much solar radiation to reach the surface in a certain region. When observed SST is specified, no ocean feedback on this error is permitted, but for the coupled system, the ocean will warm, in turn affecting the cloudiness, winds, etc. Slight errors thus feed back upon each other in the coupled system.

Over the past decade or so, modeling groups have engaged in a painstaking process of evaluating sources of error and attempting to reduce each in turn. As a result, there are now a number of coupled climate models with very modest climate drift and quite reasonable simulations of observed current climatology. Climate models traditionally receive version numbers as they evolve, along with short form names useful for identifying them in papers, conferences, model intercomparison projects and funding proposals. We will examine in this section examples from the NCAR CCSM3 (defined in section 5.6.1); from a well-tested model version known as HadCM3 from the Hadley Centre, the main British climate modeling group; from the Geophysical Fluid Dynamics Laboratory in Princeton (GFDL_CM2.1); from the Max Planck Institute for Meteorology in Hamburg, Germany (MPI_ECHAM5 or ECHAM5); and from the Australian climate center (CSIRO_MK3.0).

Figure 5.18 shows a comparison between precipitation from the atmospheric component of CCSM3 (forced by observed SST, using the 4 mm day^{-1} contour from Figure 5.17 and that from the fully coupled CCSM3 in which the ocean and atmosphere have approximately equilibrated to the model simulation of the climate state over recent years, for specified estimates of greenhouse gases, etc. (known as a "twentieth-century run").[6]

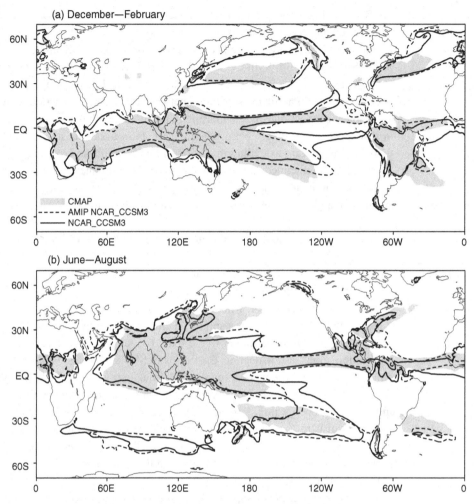

Fig. 5.18 The 4 mm per day contour for the precipitation climatology for (a) December–February and (b) June–August
(1979–2000) from each of: (i) CMAP observations (shaded over 4 mm day^{-1}); (ii) the atmospheric component of the
NCAR CCSM3, forced by observed SST (AMIP run, dashed line); (iii) the NCAR CCSM3 coupled simulation
(twentieth-century run, solid line).

The 4 mm day^{-1} contours from Figure 5.17 are repeated for the AMIP run and the observations, and the 4 mm day^{-1} contour from the coupled CCSM3 simulation is added. Again, at
the largest scale, the coupled model does well at capturing the overall climate features. As
one examines more regional details in Figure 5.18, typically any errors in the atmospheric
component alone are reflected in the coupled simulation. For instance, the winter storm
tracks are too strong. A number of regional errors noted in section 5.6.1 are exacerbated in
the coupled system. For instance, the SPCZ in December–February, which extended slightly
too far into the southeastern Pacific in the atmospheric component, extends even further east
across the basin in the coupled simulation. As will be shown below, this is associated with
changes in SST as well. The ITCZ in the Atlantic extends too far south, across the equator,

in both seasons. The Caribbean precipitation simulation is actually better in the coupled case in June–August. This may be just luck, or it may be that the SST in the coupled run has adjusted to compensate, for instance, for excessive heat flux taken out of the ocean in the uncoupled run. Analyzing such ocean–atmosphere feedbacks is one of the main challenges in improving climate model simulations.

Another interesting thing to note is that the fidelity of the precipitation simulation to observations is not necessarily related to resolution. For instance, the Hadley Centre model has one of the best coupled model simulations of tropical precipitation, seen in Figure 5.19, in several regions that tend to pose problems for climate models. This is despite having a grid size of about 3.75 degrees longitude by 2.5 degrees latitude, compared with about 1.4 by 1.4 degrees in CCSM3 (almost five times as many grid points) – and yet the CCSM3 tropical precipitation climatology has somewhat larger errors in certain regions. While noting that this is just one climate variable among many that one can examine, it illustrates that the climate models depend on the sum total of all their parameterizations interacting with each other and the large-scale flow. In comparing HadCM3 coupled model precipitation climatology with observations in Figure 2.13 (bearing in mind that the averaging periods are not identical), the large-scale patterns of the tropical convection zones, midlatitude storm tracks and their changes from winter to summer are very reasonably reproduced. Nonetheless, one can easily spot errors in particular regions. For instance, the Australian outback would be a very different place if the climate were really like that simulated in January. Summer monsoon convection that should be confined to the north of Australia spreads too far southward in the model. In winter, Italy is as rainy as the north of France – again, a slight extension of a real climate feature creates errors at the regional scale.

Figure 5.20 uses the same format as Figure 5.18 to compare precipitation simulations among several coupled models. The 4 mm day^{-1} contour from each model is used to give the shape of the precipitation features. This reinforces the conclusion that the models generally agree well with observations (and each other) at the large scale. Over a large portion of the plot, the models differ by what appear to be modest wiggles in the contours viewed from a global scale. However, these departures are often hundreds of kilometers (1 degree of latitude is slightly over 100 km) and can be very significant if examined at the scale of a state or a river drainage basin. In a few regions different models tend to have similar departures from the observed estimate. The winter storm tracks are all too strong; the SPCZ extends too far into the eastern Pacific in several models; and a number of models have trouble with a shifted boundary of the convection zone over the southern tropical Atlantic.

We now turn to two closely related measures of surface temperature, which we show for CCSM3 and HadCM3, respectively. Figure 5.21 shows the SST climatology for the CCSM3 coupled simulation. This may be compared with the observed climatology in Figure 2.16. The large-scale features, such as the strong gradient of SST through the midlatitudes, and the east–west contrast in the Pacific, are clearly simulated, although as one examines details many small shifts in position or extend of these features may be noted. Midlatitude SST contours lie close to the observed, but the equatorial Pacific cold tongue extends farther west than observed, while just south of the equator the boundary between the warm pool and the cold eastern Pacific water occurs too far east. The equatorial Atlantic is too warm in

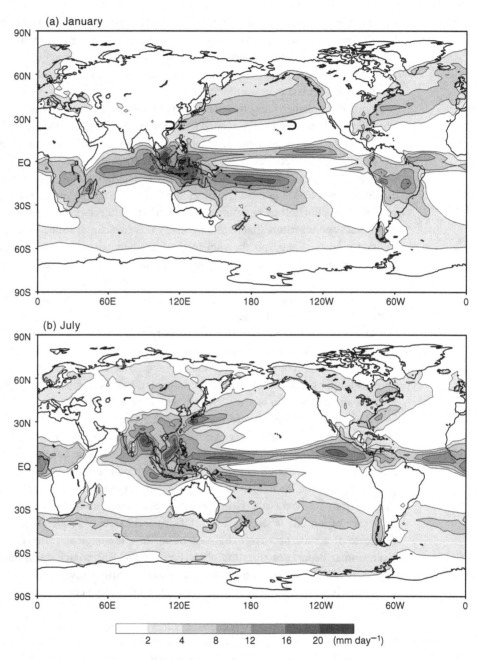

Fig. 5.19 Simulated precipitation climatology (1961–90) from the coupled climate model, HadCM3. (a) January, (b) July. Units mm per day.

December–February. The tropical precipitation features tend to follow the SST, so regions where the SST has errors tend also to be regions where the precipitation has errors. For instance, the excessive eastward extension of the SPCZ into the eastern Pacific and the southward displacement of the ITCZ in the Atlantic are both directly related to the SST error. The SST pattern in turn depends on the winds, the radiative feedbacks associated with

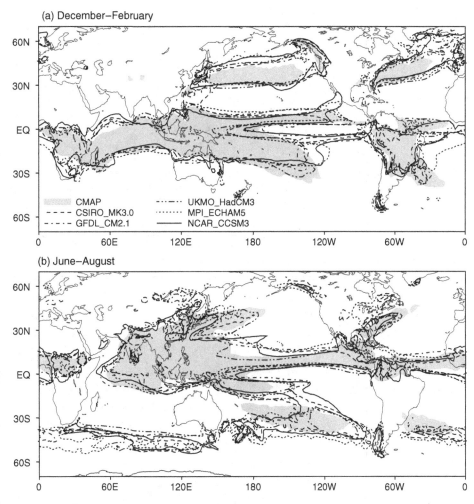

(a) December–February

(b) June–August

Fig. 5.20 The 4 mm day^{-1} contour for the precipitation climatology for (a) December–February and (b) June–August (1979–2000) as in Figure 5.18 from observations (shading) and coupled simulations from several climate models (line styles as marked).

clouds, and other feedbacks from the atmosphere. Thus these errors are often associated with ocean–atmosphere feedbacks.

Figure 5.22 shows the near-surface air temperature climatology from HadCM3. Near-surface air temperature is often used in global warming studies because it changes fairly smoothly over both land and ocean. Over the ocean, near-surface air temperature tends to be rather similar in spatial pattern to SST, but about 1 to 2 °C cooler. Over land, the temperature of the land surface itself might change rapidly over the course of a day and between different surface types (e.g. bare ground versus vegetated area). The air temperature has less severe variations and thus is easier to measure with limited samples. The general features of the climatology are reproduced well in the HadCM3 simulation (you may compare to the SST pattern in Figure 2.16, adding a degree or so for the difference between SST and surface air temperature). The movement of the regions of warm surface temperature in the tropics

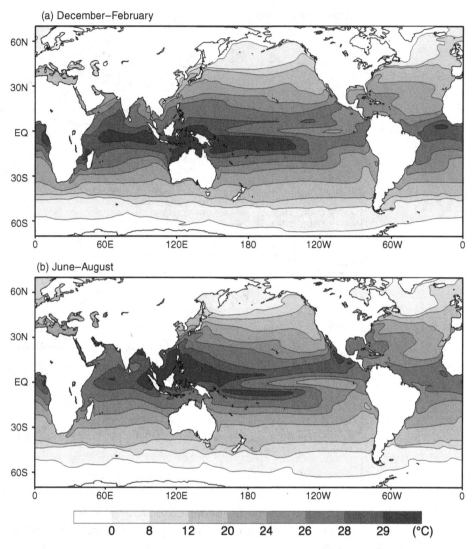

Fig. 5.21   Sea surface temperature from the coupled climatology (1979–2000) of CCSM3.

with season is well captured, as is the equatorial cold tongue. The temperature gradient from tropics to polar regions is also quite well simulated. As with the precipitation, if one subtracts an estimate of the observed fields, differences up to a couple of degrees Celsius may be found regionally. Again, any slight offset of a climatological feature can lead to errors on a region by region basis. As in CCSM3, a slight over-extension of the cold tongue to the west and insufficient temperature gradient across the Pacific south of the equator may be noted. When climate modelers find regions of common error among several models, coordinated programs of observations and modeling are often organized to address the climate processes that are not well enough represented in these regions.

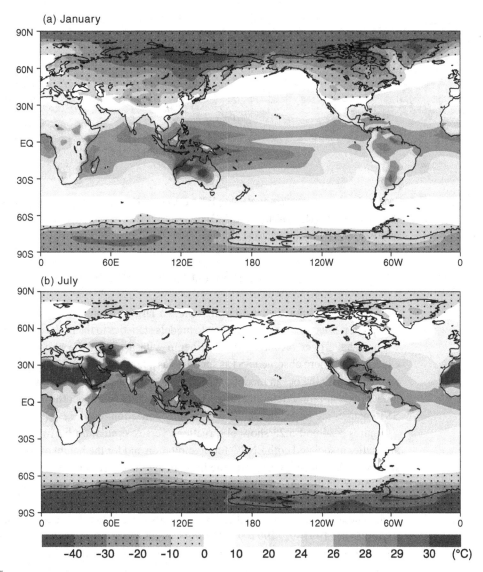

Fig. 5.22 Simulated near-surface air temperature (at 2 m height) in degrees Celsius from a coupled climate model, HadCM3. (a) January, (b) July.

Sea ice simulations in current climate are shown in Figure 5.23 for satellite observations and two ocean–atmosphere climate models. Regions that have sea ice concentration greater than 15% (i.e. more than 15% of a unit area is covered by ice) are shown for late winter and late summer, when the sea ice extent is approximately at its maximum and minimum, respectively. The models shown here do fairly well at capturing the seasonal cycle of the total area covered by sea ice, although ECHAM5 underestimates the retreat of the ice through summer, and local differences may be noted from the observations. There are some models that produce slightly larger or slightly smaller seasonal variations than observed.

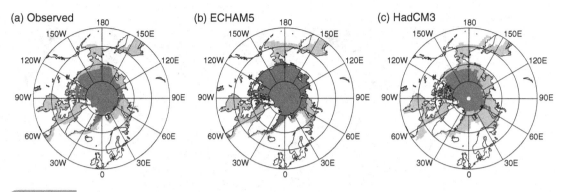

Fig. 5.23   Regions of sea ice concentrations greater than 15% are shown for March (light shading) and September (dark shading). (a) Satellite observations, (b) ECHAM5 ocean–atmosphere simulation; (c) HadCM3 ocean–atmosphere simulation. After Parkinson *et al.* (2006).

5.6.3 Simulation of ENSO response

Because the El Niño/Southern Oscillation is a large climate signal that is becoming well observed, it is useful to compare climate models statistics to the high coverage observations we now have over recent cycles. We begin with a test of the atmospheric response in experiments where observed SST is specified as a function of time. This sidesteps any errors associated with the ocean model or coupled feedbacks, and tests how well the atmospheric model does in reproducing the response, both locally in the central Pacific where ENSO SST changes are large, and in the remote teleconnection response.

Figures 5.24 and 5.25 show observations versus simulation by two climate models of anomalies associated with ENSO for precipitation and for the height of the 200 mb pressure surface in the upper troposphere (the latter being the equivalent in pressure coordinates of pressure anomalies at a given height). To better establish anomalies associated with ENSO, the quantity shown is a composite formed by taking the average over several El Niño events minus the average over several La Niña events (December–February average for 1982–83, 1986–87, 1991–92, 1994–95, 1997–98 minus those for 1983–84, 1988–89, 1995–96, 1998–99, 1999–2000). Taking the average over multiple events tends to reduce anomalies associated with random weather variations that might occur in a given month or season; and taking the El Niño anomalies minus the La Niña anomalies almost doubles the size of the anomalies we are interested in, so they are easier to distinguish from remaining effects of random weather variations. This is particularly relevant for rainfall, which has a substantial natural variability, even in averages over several seasons. Thus the anomalies shown here differ slightly from the usual definition (in terms of the difference from long climatological average), but convey similar information.

First, we should note that the "observations" shown here are estimates of the observed state, each of which has its own imperfections. The precipitation in Figure 5.24a is a satellite product based on outgoing infrared radiation, which is strongly affected by high (cold) cloud tops. An empirical relationship between daily IR and precipitation is used to provide a global estimate of precipitation. Such measurements are available since 1979. More recent satellite systems based on microwave or radar retrieval systems provide more accurate estimates,

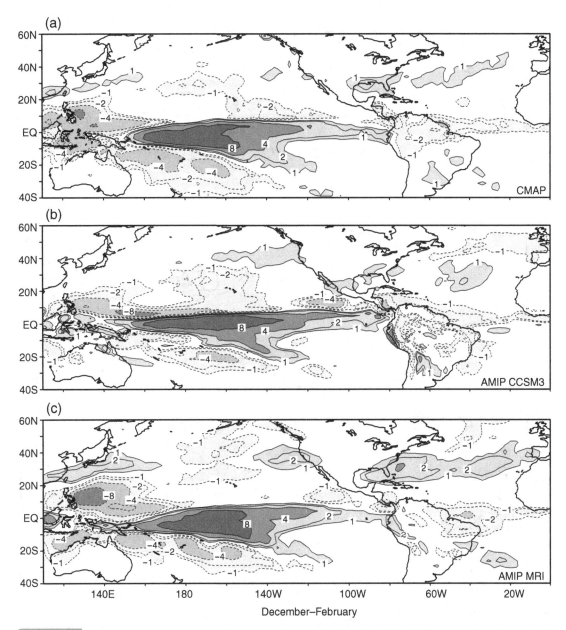

Fig. 5.24 Precipitation anomaly (mm/day) for December–February for the average of five El Niño events minus the average of five La Niña events, as observed and simulated by two atmospheric models given observed SST. (a) Observed (CMAP satellite precipitation product); (b) simulated by NCAR_CCSM3; (c) simulated by MRI_CGCM_3.2. Shaded for anomalies over 1 mm day^{-1} in amplitude.[7]

but are available over shorter periods. Observations based on rain gauges are available at particular locations, mostly over land, and thus have large gaps in spatial coverage. The 200 mb geopotential in Figure 5.25a is from what is known as a *reanalysis* product. A data assimilation system for a weather forecast model adjusts the model temperature and other

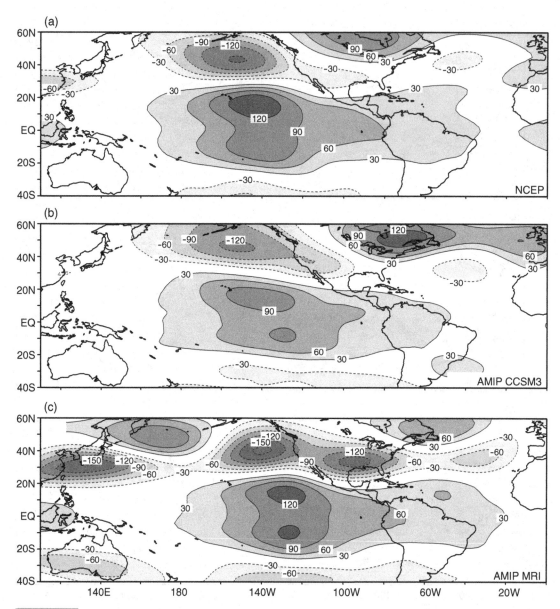

Fig. 5.25 Upper tropospheric (200 mb) geopotential height anomaly (meters) for December–February for the average of five El Niño events minus the average of five La Niña events, as in Figure 5.24. (a) Estimate of observations from the National Center for Environmental Prediction reanalysis; (b) simulated by CCSM3; (c) simulated by MRI. The simulations give the atmospheric model response to observed SST. Contour interval 30 m, shaded for anomalies greater than 30 m in amplitude.

fields to yield a best fit to the available observations every 6 hours. The model produces a full set of atmospheric variables with full spatial coverage, even where there are no observations. In the case of the geopotential shown here, this estimate is reasonably accurate because tropospheric temperature estimates with high spatial coverage are available from satellite retrievals in recent decades, and temperature is a leading effect determining upper-level pressure variations.

In tropical precipitation, the observed anomalies exhibit the basic ENSO features in the Pacific familiar from Chapters 1 and 4, with precipitation increases in the central to eastern equatorial Pacific and decreases in the western Pacific in what one might term a horseshoe pattern around the positive anomalies. The teleconnection to reduced equatorial precipitation in eastern South America and the Atlantic is also clear. The two models in Figure 5.25b,c also exhibit each of these tropical features. For the larger-scale general aspects just described, the models agree fairly well with the observations. However, as one focuses on specific regions, such as the island of New Guinea and neighboring ocean areas near the equator just north of Australia, or the exact extent and location of the rainfall reduction in equatorial South America, substantial differences may be noted, both between the two models and between each model and observations. For precipitation in midlatitudes, anomalies are generally smaller since precipitation overall is smaller than in the tropics. Some parts of the shaded regions do not meet statistical significance criteria at the 95% level, reflecting the nature of precipitation impacts in midlatitudes, in which the ENSO teleconnections are only one factor affecting the large natural variability. Other regions, notably the region extending from southern California to Florida in the observed precipitation meet this criterion, even though the amplitude is smaller than 1 mm day^{-1}. The tendency to wetter winters near southern California and around Florida seen in the observed composite is also borne out in other studies using longer local records. Both models capture something like these features, although the exact spatial location and magnitude differ.

In upper tropospheric geopotential, the observations in Figure 5.25b show an upper-level high in the Pacific corresponding to the warmer tropospheric temperatures there. The low over the northeastern Pacific, the high over Canada, and the low extending into the Atlantic form part of the wavetrain-like Pacific North American pattern, as discussed in section 4.8. In CCSM3, the midlatitude pattern matches reasonably well, despite a slightly weaker high in the tropics. In the MRI model, the tropical simulation is quite good, but there is no reduction in strength of the low as it passes across the United States.

Notes

1 The 18 BATS surface types are condensed from roughly 90 surface types compiled by Olson (1983). The BATS land surface scheme is described in Dickinson (1984) and Dickinson *et al.* (1986).

2 The surface drag representation for momentum can differ slightly, and vertical mixing typically uses potential temperature rather than temperature, but the principles are the same.

3 For comparison of snow models and their sensitivity, see Cess *et al.* (1991), Randall *et al.* (1994), Foster *et al.* (1996), Slater *et al.* (2001), and Qu and Hall (2006).

4 Figure 5.15 and associated discussion combines elements from the Semtner (1976) thermodynamic ice model, variants of which recur in many current ice models, with aspects of ice model dynamics such as occur in Hibler (1979), the force balance of which is examined in Steele *et al.* (1997). For slightly different formulations see Flato and Hibler (1992), Oberhuber (1993), Cattle and Crossley (1995), Hunke and Dukowicz (1997), Winton (2000) and Bitz *et al.* (2001). For summary of ice models from two alternate points of view at the graduate level, see Chapter 12 (by Hibler and Flato) of Trenberth (1992) or section 3.9 of Washington and Parkinson (2005). An evaluation of

the simulation of the seasonal cycle of sea ice in 11 current climate models is given in Parkinson *et al.* (2006), on which Figure 5.23 is based.

5 We thank the modeling groups, identified by their model, for providing their data for analysis via the Program for Climate Model Diagnosis and Intercomparison (PCMDI). We acknowledge the JSC/CLIVAR Working Group on Coupled Modelling (WGCM) for organizing a multi-model analysis activity that made this archived data available, with technical support from the IPCC Working Group 1 Technical Support Unit. The IPCC Data Archive at Lawrence Livermore National Laboratory is supported by the Office of Science, US Department of Energy.

6 The simulations shown here have time-varying greenhouse gases, aerosols, volcanic forcings etc. estimated from observations over the twentieth century, as further described in Chapter 7.

7 The composite ENSO response in Figure 5.24 and Figure 5.25 uses atmospheric model simulations forced by observed sea surface temperature anomalies from the National Center for Atmospheric Research Community Climate System Model version 3 (NCAR CCSM3) at 1.4 degree resolution, and the Meteorological Research Institute of Japan (MRI) atmospheric component of their coupled GCM version 3.2 at 2.8 degree resolution. Data are from the PCMDI archive acknowledged above. Observed precipitation data in Figure 5.24 are from CMAP (Xie and Arkin 1997). Geopotential data in Figure 5.25 are from NCEP reanalysis (Kalnay *et al.* 1996, Kistler *et al.* 2001). Shading in these figures is based on amplitude for clarity. Much, but not all, of the shaded region exceeds the 95% significance level for a test of the difference of means of two samples.

The greenhouse effect and climate feedbacks

6.1 The greenhouse effect in Earth's current climate

6.1.1 Global energy balance

Recall the global energy balance (Figure 2.8) from Chapter 2. Here let us review it with an eye to creating a very simple globally averaged energy balance model that may then be used to understand the results from more complex climate models.

Recall also that the upgoing energy flux leaving the surface is actually *greater* than the net energy from the Sun arriving at the surface because the atmosphere traps most of the upgoing energy and re-radiates part of it back downward. This is the greenhouse effect, as seen in Earth's current climate, discussed in Chapter 2. We begin by examining this climatological greenhouse effect in more detail, before going on to examine impacts of changes to the greenhouse effect.

In Figure 6.1, essential aspects of the energy budget diagram are repeated using the format and notation that will be used for the global-average energy balance model. The energy input by solar energy of $342\,\mathrm{W\,m^{-2}}$ is immediately reduced by 31% ($107\,\mathrm{W\,m^{-2}}$) reflected back to space, and so a net solar flux of $S = 235\,\mathrm{W\,m^{-2}}$ enters the climate system. About 20% ($67\,\mathrm{W\,m^{-2}}$) is absorbed in the atmosphere but the rest is absorbed at the surface for a net solar input at the surface of $168\,\mathrm{W\,m^{-2}}$. Considering the surface energy budget, although the solar input is the driver of the system, the largest individual terms are the infrared radiation (IR) terms. The IR emitted upward from the surface is denoted $IR_{sfc}^{\uparrow}$ and the IR coming downward from the atmosphere is denoted $IR_{atm}^{\downarrow}$. Their values from Figure 2.8 are $390\,\mathrm{W\,m^{-2}}$ and $324\,\mathrm{W\,m^{-2}}$, respectively, but they are left as symbols in Figure 6.1 because they are part of the solution that will be sought in the energy balance model. To a good approximation, all of $IR_{atm}^{\downarrow}$ is absorbed at the surface. A smaller additional upward energy flux from the surface is contributed by evaporation and sensible heat, for a total of $102\,\mathrm{W\,m^{-2}}$.

At the top of the atmosphere, a total of $235\,\mathrm{W\,m^{-2}}$ of IR are emitted upward, as must occur if the system is in equilibrium, since the IR output must balance the net solar input into the system. Of the outgoing IR to space, only $40\,\mathrm{W\,m^{-2}}$ come directly from the Earth's surface. Of the $IR_{sfc}^{\uparrow}$ emitted upward, 90% (350 out of $390\,\mathrm{W\,m^{-2}}$) is trapped in the atmosphere and only 10% exits directly to space. The rest of the upward IR toward space is emitted from the atmosphere itself, denoted $IR_{atm}^{\uparrow}$. Its value sets the characteristic emission temperature of the atmosphere.

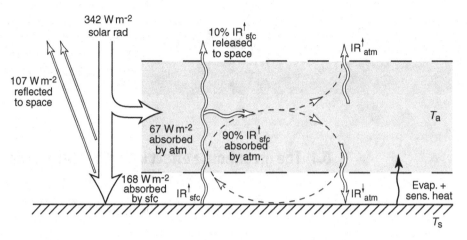

Fig. 6.1 A globally averaged energy balance model with a one-layer atmosphere. Solar radiation flux is shown on the left. Infrared fluxes are shown as wavy lines. Evaporation and sensible heat are indicated by a thin black arrow. The configuration shown approximates the global average energy budget. Simpler cases will be discussed.

6.1.2 A global-average energy balance model with a one-layer atmosphere

In the discussion of Figure 6.1 above, we have simply listed features from observations as they might be used in a globally averaged energy balance model. To make the model work, we need to consider how the fluxes depend on each other, and to include temperatures. There exist global-average energy balance models that include effects of vertical structure of temperature in the atmosphere, but here the simplest case is considered: that of a one-layer atmosphere model, which has a single atmospheric temperature, T_a. The surface has a different temperature, T_s. This model can already capture much of the basic physics of the greenhouse effect, since the atmosphere can absorb and emit differently than the surface.

Note that because the model treats only global averages, we do not have to deal with atmospheric or oceanic transports. The temperatures T_a and T_s represent globally averaged values, given the global-average solar input. Evaporation and sensible heat transfers from surface to the atmosphere can be included in this type of model, and we will include them in the discussion of processes, but for simplicity we can consider a case that neglects these. Thus IR is approximated as doing all the transfer from the surface to the atmosphere. This may seem a drastic approximation, but the IR terms are the largest individual terms in the surface budget, and evaporation and sensible heat increase with surface temperature in a manner qualitatively similar to IR. The Stefan–Boltzmann law for IR is simpler than the other dependences.

6.1.3 Infrared emissions from a layer

A fundamental property of emitted IR is essential to the greenhouse effect. For a parcel of gas of a single temperature, emissions are independent of direction. For the case of a layer, this means that upward emission is equal to downward emission. Because emissions are associated with the random bouncing of molecules in the gas (as characterized by temperature), there is no reason for them to be any different for upward or downward emission. Thus for a layer of constant temperature

$$IR_{atm}^{\uparrow} = IR_{atm}^{\downarrow} \tag{6.1}$$

6.1.4 The greenhouse effect: example with a completely IR-absorbing atmosphere

Consider a simple case where the atmosphere is "black" in the IR, i.e. it simply absorbs all of the infrared that enters it. Let us begin with a case in which the atmosphere absorbs no visible, shown in Figure 6.2a. Balances of fluxes must occur at several levels, but the balance at the top is the most important, since this is the overall input and output from

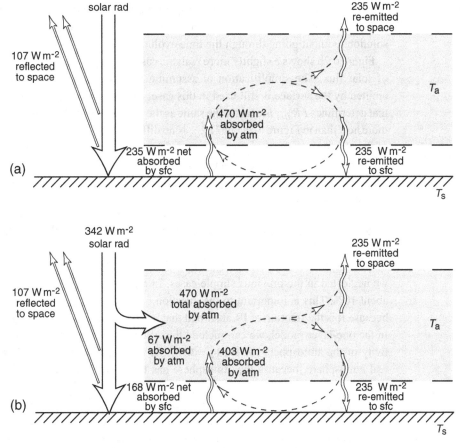

Fig. 6.2 Radiative fluxes for a simplified case where the atmosphere absorbs all infrared, as solved in the one-layer, global-average energy balance model for a given solar input. (a) The case where all solar goes through the atmosphere. (b) The case where a realistic portion of the solar radiation is absorbed in the atmosphere.

the Earth system. A net of 235 W m^{-2} of solar flux is going in, so 235 W m^{-2} of IR must leave. In this case, all of the IR emitted to space comes from the atmosphere (none from the surface) so the value of $IR^{\uparrow}_{atm}$ is 235 W m^{-2}. Using Eq. (6.1), the downward IR must also be 235 W m^{-2}. This is added to the solar flux arriving at the surface to give a total input to the surface of 470 W m^{-2}. The surface must balance this flux with $IR^{\uparrow}_{sfc}$ so this must therefore be 470 W m^{-2}. Clearly the surface must be warmer to emit 470 W m^{-2} than to emit 235 W m^{-2}, which would be the case if no atmosphere were present (if the albedo were kept the same). This is the greenhouse effect.

Note that we obtained this solution by working from the balance at the top of the atmosphere, and then considering the balance at the surface. It was obtained without actually solving for temperature, since relatively few terms enter the balances. In terms of sequence, if the solar input had been increased to the value shown from a lower value, the surface would heat up first, increasing $IR^{\uparrow}_{sfc}$ which would then heat the atmosphere. As the atmosphere began emitting to space and to the surface the surface temperature would heat up more owing to the greenhouse effect, thus heating the atmosphere more, and so on. This sequence would converge, with the temperature of the surface and atmosphere gradually approaching the balance shown. We have simply solved directly for the final equilibrium solution, without going through the time evolution.

Figure 6.2b shows a slightly more realistic case where the atmosphere absorbs 67 W m^{-2} of solar flux. The simplification of assuming that the atmosphere absorbs all of the IR emitted by the surface is still used in this case, so the balance at the top of the atmosphere that determines $IR^{\uparrow}_{atm}$ is exactly the same as the previous case. The atmosphere is absorbing more heat than in Figure 6.2a but $IR^{\uparrow}_{atm}$ is no different because it is set by the overall balance of the climate system, as seen at the top of the atmosphere. The downward IR flux is still set by Eq. (6.1). Because less solar radiation flux arrives at the ground than in the previous case (168 W m^{-2} instead of 235 W m^{-2}), the value of $IR^{\uparrow}_{sfc}$ is smaller (403 W m^{-2}), i.e. the surface is cooler.

6.1.5 The greenhouse effect in a one-layer atmosphere, global-average model

The atmosphere lets a portion of the upward IR from the surface through to space, which we neglected in the previous simple cases. In the global average energy budget, this was about 10%. This is important to the question of human changes to the greenhouse effect, because it is the fraction of IR absorbed that we are changing by adding greenhouse gases. In the one-layer model, we can include this effect, albeit crudely, by using a bulk absorptivity of the atmosphere for IR, $\epsilon_a = 0.90$. Of course, this is not exactly the same as the real atmosphere, because the atmosphere has temperature variation with height. Even if almost no IR were able to escape directly from the surface to space, adding greenhouse gases could increase surface temperature by changing the height from which upward IR primarily escapes to space, as opposed to being reabsorbed within the atmosphere. With higher greenhouse gases, the typical level within the upper troposphere from which IR is emitted to space shifts to higher heights, and the entire troposphere has to warm until these higher altitudes emit sufficient radiation to restore energy balance. The one-layer atmosphere omits these effects of vertical differences of temperature within the atmosphere, but includes the difference between the atmospheric temperature and surface temperature. This approximation is sufficient to illustrate both the climatological greenhouse effect and changes to the greenhouse effect.

We now solve the energy balance model given in Figure 6.1 for a case where we neglect evaporation and sensible heat, since $IR_{sfc}^{\uparrow}$ is the dominant cooling term at the surface. Referring to the figure we find the following flux balances (in W m^{-2}).

At the top of the atmosphere:

$$235 = (1 - \epsilon_a)IR_{sfc}^{\uparrow} + IR_{atm}^{\uparrow} \tag{6.2}$$

At the surface:

$$168 + IR_{atm}^{\downarrow} = IR_{sfc}^{\uparrow} \tag{6.3}$$

For a one-layer atmosphere:

$$IR_{atm}^{\downarrow} = IR_{atm}^{\uparrow} \tag{6.4}$$

So combining these, we find

$$IR_{sfc}^{\uparrow} = 403/(2 - \epsilon_a) \tag{6.5}$$

For $\epsilon_a = 0.90$, this gives $IR_{sfc}^{\uparrow} = 366$ W m^{-2}. This is smaller than the flux in the case of an atmosphere that was completely IR-absorbing. Changes to the greenhouse effect associated with changes in absorptivity involve only a few percent changes in the fluxes, but these can be significant to the global climate. Temperature changes implied by Eq. (6.5) when absorptivity changes will be examined in Figure 6.4, below.

Compared with the observed energy budget in Figure 2.8, the total flux up from the ground in observations (492 W m^{-2}, since evaporation and sensible heat also contribute in the real atmosphere) is greater than the upward surface flux in this simple one-layer model. This is because the downward IR flux at the bottom of the atmosphere is larger than the upward IR flux at the top, which gives an even larger greenhouse effect than in the simple model. In a more realistic model, Eq. (6.1) would apply to each layer and several layers would be needed to capture the fact that the atmosphere is warmer near the surface than high in the atmosphere. The warmer temperatures near the surface create the larger downward IR flux than in a one-layer model. Nonetheless, the one-layer model has a large climatological greenhouse effect as can be seen by comparing to the no-atmosphere case. And when studying changes to the greenhouse effect, principles can be understood in the one-layer model prior to turning to three-dimensional models for more accurate estimates.

6.1.6 Temperatures from the one-layer energy balance model

Once the fluxes have been calculated, we obtain temperatures from the energy balance model using the Stefan–Boltzmann law from Chapter 2:

$$\sigma T_s^4 = IR_{sfc}^{\uparrow} \tag{6.6}$$

$$\epsilon_a \sigma T_a^4 = IR_{atm}^{\uparrow} = IR_{atm}^{\downarrow} \tag{6.7}$$

where IR is in units of $W m^{-2}$. For the one-layer atmosphere case above, this yields $T_s = 283.5 K = 10.4 °C$. The atmospheric temperature is considerably colder, at $T_a = 249.7 K = -23.5 °C$.

We can compare this to the case with no atmosphere (but reflecting the same fraction of incoming solar radiation), where

$$\sigma T_s^4 = 235$$

yields $T_s = 254 K = -19 °C$. This is much colder than the average surface temperature of the Earth, which is about 15 °C or 288 K. It is also much colder than the temperature obtained from the energy balance model with the one-layer atmosphere (although this model does not give temperatures quite as warm as observed). The difference between the observed and the no-atmosphere case gives a measure of how much the greenhouse effect warms the planet under "normal" conditions: about 34 K. Thus the greenhouse effect is already operating very powerfully in our climate system, and is very beneficial at its current levels.

The atmospheric temperature from the one-layer atmosphere is close to the surface temperature in the no-atmosphere case, but slightly colder. In the case where the atmosphere is black in the IR, T_a is identical to the T_s from the no-atmosphere case. This is because the IR at the top of the atmosphere always has to balance the same incoming solar. If all emissions to space come from the surface, or all come from the atmosphere, the level that is doing the emitting has this temperature, which is termed the *emission temperature*. If most comes from the atmosphere, but part comes from the warmer surface, then the atmosphere is slightly colder than the emission temperature. In the observed atmosphere, this corresponds to infrared radiation escaping to space predominantly from the upper troposphere, where these temperatures typically occur.

6.2 Global warming I: example in the global-average energy balance model

6.2.1 Increases in the basic greenhouse effect

As discussed in Chapter 1, human activities are increasing the concentration of greenhouse gases – gases that significantly absorb infrared radiation – in the atmosphere. Let us first examine how this can cause warming without considering that clouds and other aspects of the climate system might also change. We will call this the "basic greenhouse effect" for ease of discussion later (terminology varies within the field). Figure 6.3 indicates schematically how these changes occur. The steps indicated in the figure are based on the one-layer atmospheric model, but hold qualitatively for more complex models. This is the same pathway as the greenhouse effect in climatology, simply increasing the effect. The atmospheric temperature in the one-layer model must be interpreted as a tropospheric temperature, since that is where the bulk of IR emissions originate. The stratospheric temperature actually cools when CO_2 is increased, as discussed in section 6.7.

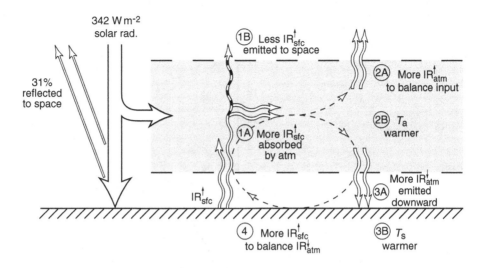

Fig. 6.3 Schematic of how increased absorption of infrared radiation by greenhouse gases leads to surface warming. Stages are shown in conceptual sequence from the point of view of energy balance requirements. (1) More upgoing IR from the surface is absorbed by the atmosphere and so does not escape directly to space. (2A) Since the input of solar energy is the same as always, while direct IR loss from the surface is decreased, more IR must be emitted from the atmosphere to achieve balance. This occurs by (2B) an increase in tropospheric temperature. A consequence of this is (3A) that more IR is emitted downward. To balance this increased input, there must be (4) increased IR upward from the surface, which occurs after (3B) surface temperature has increased sufficiently.

6.2.2 Climate feedback parameter in the one-layer global-average model

To calculate temperature increases in the one-layer global-average model when the greenhouse effect is increased, it is simple to increase ϵ_a to a new value, e.g. from 0.9 to 0.93, then repeat the calculation of total temperatures and subtract the temperatures that were calculated for the "normal" climatological value $\bar{\epsilon}_a$. This is a procedure similar to what is done in full climate models. When comparing models and effects of various processes, it is useful to define a *climate feedback parameter* α that gives the change in surface temperature per change in *radiative forcing*. This applies only approximately and for small changes. This parameter is illustrated here in the simpler model and given in a more general case in the following section with a related parameter, the climate sensitivity parameter. The term *forcing* generally refers to an external cause driving a response in a system. The radiative forcing here will be the change in infrared radiation in response to a specified change in greenhouse gas; this will then lead to a chain of feedbacks involving atmospheric and surface temperature and, as discussed in section 6.3, other parts of the climate system.

Figure 6.4 shows the dependence of surface temperature ϵ_a on in the one-layer global average model, using Eq. (6.5) and Eq. (6.6). Both equations are nonlinear – Eq. (6.5) depends inversely on $(2 - \epsilon_a)$, and Eq. (6.6) yields a fourth root dependence – but for the range of interest, this curve is fit very well by a linear approximation matching the slope of

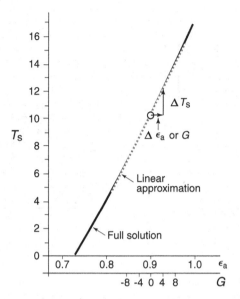

Fig. 6.4 Surface temperature (°C) as a function of absorptivity ϵ_a (unitless) in the one-layer global-average energy balance model. A linear approximation is shown as a dashed line. Using the solution at $\epsilon_a = 0.9$ as the reference climatology (open circle), the x-axis is also given as the top-of-the-atmosphere greenhouse radiative forcing G in W m^{-2}. Temperature change ΔT_s is also shown.

the nonlinear curve at the climatological value. This is an example of *linearization*, a useful and general approximation process that can also be done analytically on the equations.[1]

In a more complex model, the equivalent of ϵ_a is not a single number, since the absorptivity is calculated at many levels in the atmosphere and depends on concentrations of various greenhouse gases (GHG), among other things. In all cases, the radiative balance at the top of the atmosphere is key to the greenhouse effect, so by expressing the simple model in terms of changes in fluxes at the top of the atmosphere, we obtain an example that extends to other models. Consider adding an increase of greenhouse gas, corresponding in the simple model to an increase in absorptivity $\Delta\epsilon_a$, trapping more upgoing IR in the atmosphere. Before any temperature change occurs to compensate, there will be a deficit G in the outgoing IR causing an imbalance at the top of the atmosphere. The value of G, in W m^{-2}, is a good measure of the greenhouse radiative forcing (as a change from normal climatology). Note that G is independent of the particular gas that does the absorbing. In the case of the simple model, the total upward IR at the top of the atmosphere (from the right hand side of Eq. (6.2), using Eq. (6.3)) is

$$IR_{top}^{\uparrow} = (1 - \epsilon_a)\sigma T_s^4 + \epsilon_a(\sigma T_a^4) \qquad (6.8)$$

The first term is the IR that gets through the atmosphere. The second term is $IR_{atm}^{\uparrow}$ which also depends on ϵ_a because emissivity equals absorptivity. Keeping the temperatures (temporarily) fixed at their climatological values, denoted $\bar{T}_s$ and $\bar{T}_a$, the reduction in outgoing IR due to a change in absorptivity $\Delta\epsilon_a$ is

$$G = \Delta\epsilon_a \sigma \bar{T}_s^4 - \Delta\epsilon_a \sigma \bar{T}_a^4 \qquad (6.9)$$

The first term is the increased trapping of IR in the atmosphere, labelled 1B in Figure 6.3. This is the key term, causing the warming. The second term occurs because an increase in absorptivity implies an increase in emissivity. This term acts to reduce the effect of the first term, but is always smaller since the atmosphere is colder than the surface. Using the solution for $\epsilon_a = 0.90$ as the reference ("normal") climatology gives $G = 146.0\Delta\epsilon_a$. Computations from more complex models for doubled CO_2 give a value of about $4\,\mathrm{W\,m^{-2}}$ for G,[2] so using $\Delta\epsilon_a = 0.03$ gives a roughly comparable radiative forcing in the simple model.

After greenhouse gas increases cause the outgoing IR to decrease by G, temperature must warm to bring outgoing IR back into balance. For a linear approximation,[3] valid for sufficiently small ΔT_s as shown in Figure 6.4, ΔT_s is just proportional to G:

$$\alpha_T \Delta T_s = G \qquad (6.10)$$

The quantity α_T is the *climate feedback parameter* (in units of $\mathrm{W\,m^{-2}\,K^{-1}}$) for this model. Here we are using α_T to denote the climate feedback parameter that arises only from changes in temperature (without any changes in water vapor, snow, ice clouds, etc.), and we will use α for a more general case in the next section. Note that ΔT_a can be obtained from ΔT_s using similar linearized balances. While G gives the changes in greenhouse gases as a change in radiative forcing, α summarizes the climate model feedbacks that determine the response ΔT_s. This value of α_T is proportional to the increase in outgoing IR at the top of the atmosphere per unit increase in surface temperature. Because T_a and T_s are linked by energy balance conditions, α_T takes into account also the effects of increases in atmospheric temperature.

The greenhouse effect is thus separated into two parts: (i) the additional absorption of IR in the atmosphere due to increased greenhouse gases. This is measured by the imbalance in outward IR, G, that occurs before temperature increases; (ii) the increase in outward IR to space due to warmer temperatures. This is a negative feedback which balances the heating effects of the forcing G. If the forcing were to disappear then this negative feedback would cause temperatures to go back toward normal.

The calculation of climate feedback parameters is subject to a number of caveats, and different variants appear in the literature. For example, different vertical structures of radiative heating change may have different effects on surface temperature even if they have the same change in the radiative fluxes at the top of the atmosphere. Nonetheless, the climate feedback parameter can still be used to illustrate the relative role of different feedbacks.

6.3 Climate feedbacks

The anthropogenic change due to the basic greenhouse effect acting by itself would be cause for only modest concern, since the amount of predicted warming would be fairly small on a global average. However, there are a number of climate feedbacks that modify changes due to the basic greenhouse effect, and some of them may amplify it considerably. The main feedbacks are the *water vapor feedback*, associated with increases in water vapor, which itself is a greenhouse gas; the *snow/ice feedback*, due to decreases in snow and ice, causing global albedo to decrease (less solar radiation is reflected); and *cloud feedbacks*, due to changes in cloud cover, which affect the cloud contribution both to the greenhouse effect and to albedo.

The physics of these feedbacks will be discussed in more detail in the following subsections. First, let us use the framework from the global-average energy balance model to produce a simple notation for keeping track of the relative importance of feedbacks. Results from more complex models can then be diagnosed and discussed in this notation.

6.3.1 Climate feedback parameter

We wish to generalize the climate feedback parameter, so we can use the equation

$$\alpha \Delta T_s = G \tag{6.11}$$

to analyze the results of full climate models, where ΔT_s is the surface temperature change. If we can calculate a greenhouse forcing, G, then the ratio of the forcing to the temperature response can be defined as the climate feedback parameter α.

In the one-layer model of section 6.2, we had greenhouse gas forcing in terms of changes of the bulk absorptivity of the one-layer atmosphere, $\Delta \epsilon_a$. But for more complex models with many layers, the changes in absorption are not characterized by a single number. To calculate G from a complex model: (i) hold temperature, moisture, clouds, etc. *fixed* at their present climatological values; (ii) increase greenhouse gas concentrations; (iii) calculate the changes in outgoing IR flux at the top of the atmosphere (conventionally the top of the troposphere is used to avoid complications involving stratospheric response); (iv) take the global average. This gives a single number G (in $W\,m^{-2}$) that characterizes the change in global-average radiative forcing of the system.[4]

Note that because temperatures are held fixed in this calculation, the greater amount of IR trapped by the increased greenhouse gases can produce the imbalance in top-of-atmosphere fluxes that gives G. When temperatures are allowed to change, they re-establish balance in the top-of-atmosphere fluxes, and α simply measures the size of temperature change that is needed to produce equilibrium.

In order to measure the effects of different feedbacks, one holds different parts of the climate system constant. For instance, to measure the negative feedback associated with increased IR loss to space as temperature increases, hold water vapor, ice, snow and clouds fixed at their climatological values, but allow temperature to vary. This defines the temperature contribution α_T to the climate feedback parameter. Contributions to α are approximately additive if the changes are small enough:

$$\alpha = \alpha_T + \alpha_{H_2O} + \alpha_{ice} + \alpha_{cloud} \tag{6.12}$$

where α_{H_2O} is the contribution of the water-vapor feedback, α_{ice} the contribution of the snow/ice feedback, α_{cloud} the net contribution of the cloud feedbacks. Formally, they are defined by examining the change in the top-of-atmosphere energy balance with surface temperature

$$\alpha = \frac{\partial I R_{atm}^{\uparrow}}{\partial T_s} - \frac{\partial S}{\partial T_s} \tag{6.13}$$

where S is the net solar flux (global average) at the top of the atmosphere. Since incoming solar flux does not depend on Earth's surface temperature, the second term of α is really associated with changes in albedo, which change the amount of solar flux reflected back to space. In practice the diagnostics may be a bit more complicated.

Naturally, a simplified view like Eq. (6.11) cannot account for every aspect of climate change. In practice, the contributions of various feedbacks do not add as neatly as Eq. (6.12) would suggest, especially the cloud and snow/ice feedbacks. Other ambiguities include: different spatial distributions of flux or temperature may give different results even if the global average is the same; a change that is small in the global average may be large in some regions; and climate feedback parameter estimates by different approaches often do not agree exactly. It is simply a handy way of getting a feel for the behavior.

6.3.2 Contributions of climate feedbacks to global-average temperature response

Bearing these caveats in mind, Table 6.1 gives an estimate of the various contributions to the climate sensitivity parameter based on estimates from current climate models. The values in the table are a plausible example, taken from one study that uses a particular set of approximations. Other choices are sometimes used. For instance, this study used top-of-atmosphere fluxes for G and flux changes associated with each feedback. Other studies have used values at the tropopause, since the stratosphere actually cools, as discussed in section 6.7.1. This example illustrates how uncertainties in some feedback processes can give rise to considerable range of uncertainty in the computed warming. The overall range of uncertainty in ΔT_s in this study differs slightly from the range of equilibrium response of models in IPCC (1990) and IPCC (1992) to doubled CO_2, in which the range of 1.5 to 4.5 °C was listed. This range has remained similar in IPCC (1996) and IPCC (2001) and other recent estimates.

In Table 6.1, the first column shows the contribution of the feedback listed at the left, i.e. α_T, α_{H_2O}, α_{ice}, α_{cloud}, respectively, in Eq. (6.12). Physically, these correspond to the radiative heat loss to space per degree Celsius rise in surface temperature. The sign is thus positive for a negative feedback, i.e. an energy loss that opposes the warming, as is the case for the basic radiative feedback. The second shows the total climate feedback parameter including the feedback of that line and all feedbacks from previous lines. For instance, the value given beside ice/snow feedback includes the basic greenhouse effect (infrared cooling) and the water vapor feedback. The change in temperature in the third column is given by this value of α, with a value of the greenhouse radiative forcing of $G = 4.3 \, \mathrm{W \, m^{-2}}$.

As may be seen from the implied temperature change, the size of the global average warming increases substantially when positive feedbacks are included. These tend to reduce the climate feedback parameter, so for the same forcing there is more warming. While a

Table 6.1 Example of the contributions of various feedbacks to the climate feedback parameter α and the surface temperature increase $\Delta T_s = G/\alpha$, using $G = 4.3$ W m^{-2} for doubled CO_2.

Feedback	Radiative flux to space per degree increase in T_s (W m^{-2} K^{-1})	Cumulative climate feedback parameter, α (W m^{-2} K^{-1})	Cumulative change in equilibrium temperature ΔT_s(K)
Infrared cooling (Negative feedback)	3.7 to 4.4	3.7 to 4.4	1.0 to 1.2
Water vapor (Positive feedback)	−2.0 to −1.5	2.0 to 2.4	1.8 to 2.1
Sea ice/land snow (Positive feedback)	−0.3 to −0.1	1.7 to 2.3	1.9 to 2.5
Clouds (Positive to small)	−1.2 to −0.1	0.9 to 1.6	2.7 to 4.8

Note: While methods of estimating these feedbacks have limitations, the ranges of uncertainty typify current climate models, based on 12 models examined in Soden and Held (2006). Note that the range given for the cumulative climate feedback parameter is from the actual model values and is not the same as taking the sum of the lower and higher values of the first column. For instance, the model that has $\alpha_T = 3.7$ is not the same as the model that has $\alpha_{H_2O} = -2.0$, so the lowest value of 2.0 for $\alpha_T + \alpha_{H_2O}$ is larger than $3.7-2.0$.[5]

1 °C warming (if it occurred slowly) might be acceptable, a 4 °C warming would have substantial consequences. And unfortunately, the feedbacks are in some of the most complex and difficult to model parts of the climate system. This results in the differences among the models that give the ranges shown, which can be viewed as rough estimates of the error bars on the representation of each process.

Thus, the climate feedbacks act to (i) increase the warming due to anthropogenic greenhouse gases; and (ii) increase the uncertainty in the estimate of this warming.

6.3.3 Climate sensitivity

Another common way of expressing the sensitivity of climate change to feedbacks is by the inverse of the climate feedback parameter

$$\lambda = \alpha^{-1} \tag{6.14}$$

This *climate sensitivity parameter*, λ, gives the surface temperature change per W m^{-2} change in the radiative forcing of the system. The climate sensitivity parameter is less used in recent literature, because the climate feedback parameter has advantages for comparing contributions of various feedbacks and when time dependence is included in the system.[6] Unfortunately, both of these parameters are based on approximations. They are sufficient to get a general idea of how climate feedbacks contribute to the size of the expected warming and to the error bars on climate model projections. As one asks more refined questions, however, the approach becomes awkward.

Table 6.2 The mean, standard deviation and range of doubled-CO_2 climate sensitivity (global-averaged surface temperature response) from models included in recent IPCC reports. Range refers to highest and lowest values from among all models.[7]

Publication	No. of models	Mean	Standard deviation	Range
IPCC (1996)	17	3.8 °C	0.8 °C	1.9 to 5.2 °C
IPCC (2001)	15	3.5 °C	0.9 °C	2.0 to 5.1 °C
IPCC (2007)	18	3.2 °C	0.7 °C	2.1 to 4.4 °C

A different approach is to define a standardized experiment that is relevant to future warming but straightforward to conduct with many models. This can then be used as a benchmark for comparing models. The standard experiment is to double CO_2 in the model and then run the simulation long enough that the model comes to a new equilibrium climate state, as described in more detail in section 6.8.2. The change in the long-term average then defines the model *doubled-CO_2 response*. The global-average surface temperature response ΔT_{2x} (subscript $2x$ is often used for 2 times CO_2) is then used as a measure of climate sensitivity. It is referred to as the *doubled-CO_2 climate sensitivity* or just the *climate sensitivity*.

Rather than attempt to separate different feedbacks as is done in Table 6.1, another approach aims to summarize the overall climate response. The last line of the ΔT_s column in Table 6.1 is an example of the doubled-CO_2 climate sensitivity. The range is illustrative of the spread among different climate models but actually underestimates the full range. Table 7 gives the spread in sensitivity among recent climate models in two ways: as a mean and standard deviation among the set of models, and as the range from lowest sensitivity to highest sensitivity. Note that the uncertainty, as measured by the spread among models, has not reduced much with time although the models have considerably improved representations of many processes.

6.4 The water vapor feedback

In rough terms, the physics of the water vapor feedback is very simple. Warmer air can hold more water vapor. In a warmer climate, the troposphere will have the potential to hold larger concentrations of water vapor. Water vapor is an effective greenhouse gas (the most important greenhouse gas in the climatological greenhouse effect), and thus it can add greatly to the initial warming if it increases. Figure 6.5 shows this process.

Because this is a positive feedback, in which warming due to the water vapor feedback further increases water vapor, one might worry that this effect might compound itself indefinitely. Indeed it might, if the increase in water vapor were enough to overcome the negative feedback that is always present in the system: that due to the increased IR loss with increased temperature. On Earth this does not happen, as can be seen from Table 6.1,

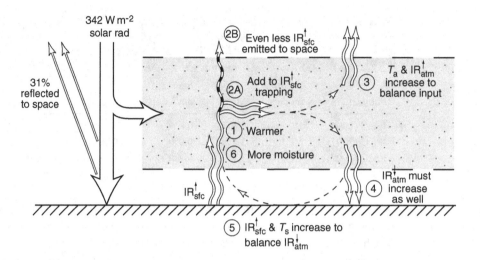

Schematic of water vapor feedback in the greenhouse effect. (1) The initial warming by the basic greenhouse effect allows the atmosphere to hold more water vapor. (2) This increases IR trapping and reduces the IR lost directly to space from the surface, just as in the basic greenhouse effect. (3) The atmospheric temperature must warm until increased upward IR from the atmosphere compensates for the reduction in IR escaping from the surface to space. (4) Since atmospheric temperature has increased, downward IR increases also, resulting in (5) increased surface temperature until increased upward IR from the surface can balance the extra input from the atmosphere. (6) The warming due to the water vapor feedback itself increases water vapor.

since the water vapor impact on outward IR is only half that of the feedback due to temperature. If the positive feedback contributed more warming per temperature rise than the blackbody effect contributes cooling, one could have a "runaway greenhouse effect" in which temperatures continue to increase until some other regime of behavior is reached. It is sometimes speculated that such an effect might have occurred on Venus sometime in that planet's history.

The water vapor feedback is not without complications. Figure 6.6 shows in more detail the step between the fact that a warmer atmosphere can hold more water and the assumption that it actually does contain more water. The arrow marked Warmer, an example of the range of typical vapor pressure values after warming, assumes that relative humidity stays in a comparable range to current climate. While the shift in the mean vapor pressure is small compared with the range of variation, this case indeed implies a mean increase in water vapor. And competition among evaporation, transport and condensation indeed tends to keep the relative humidity more constant than the humidity itself, at least in the lower troposphere where contact with the moist lower boundary, combined with vertical mixing by convection, tends to keep parcels from becoming very dry. A parcel with low relative humidity reaching the surface causes evaporation to increase, and a parcel with high relative humidity that is perturbed significantly upward tends to lose water by condensation and precipitation.

In the upper troposphere, on the other hand, much of the water vapor is supplied by lateral mixing at very large scales. This carries water from the tropics, where frequent

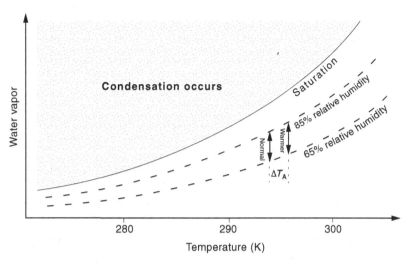

Fig. 6.6 Water vapor (as measured by vapor pressure) versus temperature, schematizing how water vapor might increase in global warming. The solid curve is the saturation value of vapor pressure as a function of temperature. Condensation would occur in an air parcel with initial vapor and temperature values to the upper left of this curve. An air parcel might have any combination of temperature and vapor corresponding to a point below this curve; the fraction of the saturation value at a given temperature is the relative humidity. Dashed curves show vapor pressure values as a function of temperature for 85% and 65% relative humidity. Atmospheric values of water vapor quite commonly fall between these lines. For a particular latitude and height, the arrow marked "Normal" indicates a typical range of water vapor pressures in the normal climate. After global warming, if the *relative* humidity remains in the same range, the arrow marked "Warmer" shows the range of vapor pressure in the warm climate.

deep convection tends to moisten the upper troposphere, to higher latitudes, where deep convection is more rare. This process largely conserves water during the trajectory from the tropics, and it is not clear that fixed relative humidity is a good approximation to its net effects. Recalling from section 6.1.6 that IR emission to space occurs more effectively from the upper troposphere, this upper tropospheric contribution to the water vapor greenhouse effect is at least as important as the lower troposphere, even though the latter has more water.

Until recently, it was difficult to get good data with global coverage for upper tropospheric water vapor. Standard weather balloon soundings often had inaccuracies, and enormous gaps in spatial coverage. Recently satellite sensors have been designed to estimate upper tropospheric water vapor. This is helping to determine how accurately climate models can simulate this quantity, and should begin to narrow the range of uncertainty on this feedback.

6.5 Snow/ice feedback

Given an initial warming by the basic greenhouse effect, snow and ice cover will tend to be reduced. Since both are highly reflective surfaces, they normally are responsible for a significant amount of reflected sunlight at high latitudes. Thus more sunlight is absorbed

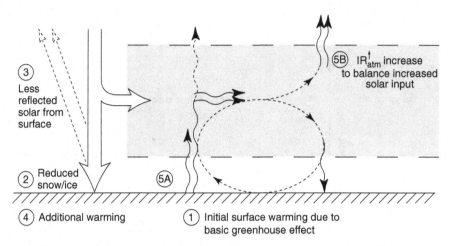

Fig. 6.7 Schematic of the snow/ice feedback in the global energy balance. Beginning with the initial warming at the surface caused by the basic greenhouse effect, the reduction (2) of snow and ice leads to (3) less reflected solar, i.e. greater net solar input into the climate system. This leads to (4) additional warming of the surface until (5) increased upward heat flux from the surface and warming of the atmosphere give enough increase in upward IR to space to balance the additional solar input.

at the Earth's surface in the warmer climate. This warms the climate still further. This is schematized in Figure 6.7, from the point of view of the global energy balance.

Of course, the snow/ice feedback, also known as the surface-albedo feedback, is felt most strongly in fairly high latitudes, and has a seasonal dependence, since it occurs at the boundary of the snow/ice cover. In Table 6.1, it appears to be only a modest effect in the global average, but this is an average of a large effect at higher latitudes with no effect in the tropics. Thus the snow/ice feedback can be quite important to the impact of global warming. Although the basic principle of this feedback is simple, it has recently been realized that there can be considerable interaction with cloud effects, since both reflect sunlight, and since snow cover can affect the local conditions under which clouds are forming.

6.6 Cloud feedbacks

Cloud feedbacks are challenging to represent in climate models for three reasons:

(i) Clouds are small-scale motions compared with the grid size of climate models. Their average effects at the grid size must be parameterized on large-scale motions.
(ii) Clouds have opposing effects in infrared and solar contributions to the energy budget.
(iii) Several types of cloud properties can affect radiative processes: these include cloud fraction, cloud top height, cloud depth, and cloud water and ice content. Different cloud types will thus have different feedbacks. Cloud amount is usually measured as *cloud fraction*, i.e. for a given area, such as a 200 km square grid box, the fraction that

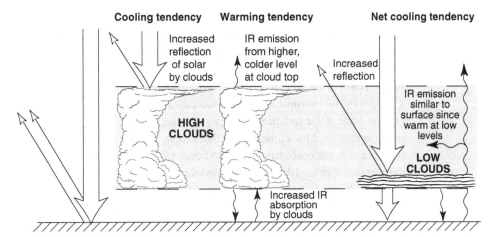

Fig. 6.8 Schematic of effects of cloud amount in the global energy balance. The feedback depends on whether the cloud fraction increases for a given cloud type. This figure shows cases of increased deep cloud fraction and low cloud fraction. Solar effects are shown on the left cloud, and infrared effects on the right hand cloud. If cloud fraction increases in a warmer climate, solar effects give a negative feedback, while IR effects give a positive feedback. For deep clouds these effects are similar in magnitude. For low clouds, the IR effects are smaller because the cloud top temperature is closer to surface temperature, so IR emitted from the cloud top is not changed as strongly.

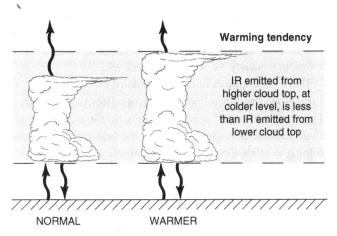

Fig. 6.9 Schematic of cloud top feedback. Cloud top tends to reach higher (and thus colder) levels because low-level moisture and temperature are increased. IR emissions from colder cloud top are thus decreased (a positive feedback).

is covered by cloud of a given type. Among several cloud feedbacks currently being studied, the schematics in Figure 6.8 and Figure 6.9 illustrate two of the better understood ones. Note that the cloud fraction feedback is really two opposing feedbacks, one involving reflection of solar and the other emission of IR from cloud top. The case illustrated for deep clouds in Figure 6.8 – in which there is considerable cancellation between these two effects – tends to apply also to other clouds with high tops, such as cirrus.

In Figure 6.8, the feedback depends on fraction of cloudy versus clear areas. Whether or not the fractional area covered by a certain type of cloud increases or decreases in a warmer climate depends on many aspects of the circulation. On a global average there will tend to be slightly more precipitation. However, increased intensity of convection in one region could actually lead to a reduction in some types of clouds in a neighboring region. Considering for example the case of increased cloud fraction, there are still two competing effects. More solar radiation will be reflected, since clouds have high albedo. This will likely tend to cool the climate, so if the warmer climate has more clouds, this is a negative feedback. On the other hand, the additional cloud region will trap more upgoing IR from the surface. It emits IR, but from cloud top, where the temperature is much colder, hence less IR will be emitted. This has a warming effect. For high clouds, the two effects tend to cancel fairly strongly (although not exactly). For clouds with tops at low levels, such as stratus clouds, the IR emission is from a temperature that is less cold (and more similar to the surface temperature), so the difference in IR emitted upward in cloudy or clear sky is smaller. Thus the solar effect dominates for low clouds.

Another feedback, the *cloud top feedback*, is illustrated in Figure 6.9. Even if the cloud fraction were to remain constant, the average height of the cloud top might change for certain cloud types in a warmer climate. If the surface layer were warmer and moister, but the temperature at upper levels did not warm enough to compensate, rising cloud parcels would tend to go slightly higher before reaching their level of stability, as discussed in Chapter 3. In climate model simulations, this tends to happen for deep convection. If the cloud top is higher, it occurs at a cooler temperature since the temperature drops with height. The IR emitted from cloud top is thus smaller than in the normal climate. This implies a warming tendency, i.e. a positive feedback.

There are a number of other postulated cloud feedbacks, including possible increases in cirrus cloud fraction, or increases in the reflectivity of clouds due to increased water content. These are subjects of current research. So far, it does not appear that new information on cloud feedbacks is greatly altering the best estimate of climate sensitivity, but rather is helping to refine the range of uncertainty.

6.7 Other feedbacks in the physical climate system

6.7.1 Stratospheric cooling

While the surface and troposphere warm when greenhouse gases are increased, the stratosphere tends to cool. This is because when greenhouse gases increase in the stratosphere, the absorbtion and emission of IR both increase. However, the balance of warming versus cooling effects differs from that in the troposphere. Consider adding a stratospheric layer of temperature T_{strat} onto the global average energy balance model. A simple stratospheric layer heat balance is

$$2\epsilon_{strat}\sigma T_{strat}^4 = Q_{ozone} + \epsilon_{strat} I R_{trop}^\uparrow \qquad (6.15)$$

The first term is cooling by IR emission from the stratosphere (both upward and downward, yielding the factor of 2). This is balanced by heating by ozone absorption of solar radiation Q_{ozone} plus absorption of IR coming up from the troposphere and surface $I R_{trop}^{\uparrow}$. After the surface and troposphere regain energy balance with the new levels of greenhouse gases, $I R_{trop}^{\uparrow}$ is approximately the same. Increased greenhouse gases in the stratosphere cause ϵ_{strat} to increase on both sides of the equation. However, $I R_{trop}^{\uparrow}$ is not twice as large as the normal climatological value of ϵT_{strat}^{4}, so the cooling on the left hand side would initially increase more until T_{strat} can adjust to compensate. This implies that T_{strat} must decrease to maintain balance.

6.7.2 Lapse rate feedback

When the surface warms, the lower troposphere warms a roughly similar amount. The warming tends to increase slightly with height, so the upper troposphere experiences a larger warming. This implies that the lapse rate is decreasing slightly, hence the name of this feedback. The slightly increased warming in the upper troposphere causes more IR emission to space from the upper troposphere than if the warming were constant with height. This is part of the negative feedback by temperature, opposing the warming, given by α_T in Table 6.1. It can be useful to split α_T into a part α_0 that assumes the warming is constant with height and a lapse rate feedback contribution α_{LR}, such that $\alpha_T = \alpha_0 + \alpha_{LR}$. The reason for this is that the lapse rate change can vary from model to model, and usually changes considerably as a function of latitude since the lapse rate is set by deep convection in most of the tropics, while at high latitudes it is determined by balances involving atmospheric transports at upper levels versus boundary layer effects at low levels. Breaking out the lapse rate feedback separately helps gauge the uncertainty in these processes. In Table 6.1, α_0 is about 3.2 W m^{-2}, with only small changes among models. The lapse rate feedback accounts for most of the uncertainty in α_T. However, there tends to be substantial compensation between uncertainly in the lapse rate feedback and the water vapor feedback. A model with slightly larger warming in the upper troposphere (larger negative feedback) tends to hold more water vapor in the upper troposphere (larger positive feedback). This is why the range in the cumulative feedback parameter $(\alpha_T + \alpha_{H_2O})$ in Table 6.1 is smaller than one might guess from the individual ranges of α_T and α_{H_2O}.

6.8 Climate response time in transient climate change

6.8.1 Transient climate change versus equilibrium response experiments

Experiments where the climate forcing is increased initially and then held fixed are termed equilibrium response experiments because the climate model is run until it reaches a statistical equilibrium state with a warmer climate. This is useful for comparing climate models and obtaining a view of the features of warmer climate in a controlled situation where the spatial pattern of the warming is not continually changing. The model can be run long

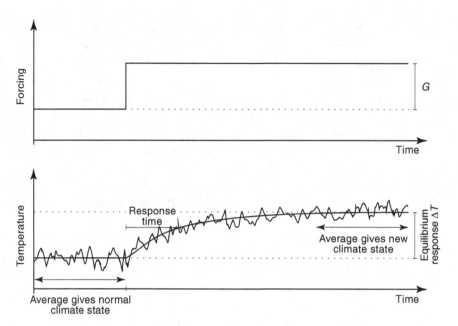

Schematic of an equilibrium response experiment. The upper panel indicates the time dependence of the climate forcing. This could be the concentration of greenhouse gases, but is more often measured as the radiative forcing G resulting from an increase in concentration. The lower panel shows global average temperature response. The smooth curve gives the response of an energy balance model with a simple ocean. The jagged curve gives a time series more like a full climate model with weather and interannual variations (for clarity, interdecadal variations are less here than in a full climate system). The time axis covers a few hundred years. To accurately assess climate change, averaging periods longer than those shown would in fact be required.

enough in the warmer climate state that reliable statistics can be produced by averaging over a longer time period. Figure 6.10 shows a schematic example of such an experiment. The climate forcing, in this case by greenhouse gases, is increased suddenly at a certain time. The initial radiative imbalance at the top of the atmosphere is a good measure of the climate forcing. This would be equivalent to the forcing G discussed in the simple energy balance model of section 6.2. A typical experiment is to double CO_2 at the initial time, since this is a forcing that we are likely to reach in the middle of the twenty-first century. The initial radiative disequilibrium for this case is about $4\,W\,m^{-2}$. Such doubled-CO_2 experiments are used as benchmark of the climate sensitivity of models. A model that has a larger equilibrium warming in such an experiment is said to have a larger climate sensitivity, as discussed in section 6.3.3.

After greenhouse forcing is increased, there is a time period during which the model global average temperature warms toward the new equilibrium. The typical time scale of this warming is termed the response time of the climate system. It tends to be on the order of decades for global average surface temperature, owing to the heat capacity of the upper ocean. The climate system equilibrates with more than one response time, so a complete picture would first have the upper ocean approach near-equilibrium on a time scale of decades, then continue to adjust by smaller amounts over longer periods as the deep ocean

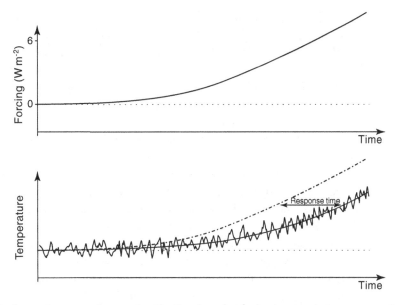

Fig. 6.11 Schematic of a greenhouse warming scenario with a time-dependent forcing, i.e. a transient response experiment. The upper panel shows the forcing as a function of time, typical of a 1% per year increase in carbon dioxide. In the lower panel, the smooth curve shows the temperature response for a simple energy balance model and the jagged curve shows a typical time series for a full climate model. The dashed curve gives the response of the system in the idealized case where the atmosphere is in equilibrium with the forcing.

equilibrates. In Figure 6.10 averaging periods to define the climate before the increase in forcing and after equilibration are indicated. Owing to natural interdecadal variability in the climate system, decades are required to establish accurate climate statistics. For quantities with high variance, such as regional precipitation, longer averaging periods would ideally be used.

In reality, greenhouse gas concentrations are increasing smoothly as a function of time, and experiments using such time-dependent forcing are termed *transient response experiments*. Because the forcing continues to change during the run, the climate response never quite catches up to the new forcing. During the time it takes the system to respond to previous increases in forcing, the forcing has increased even further. This may be seen in Figure 6.11, where for comparison an idealized case where the system is in equilibrium at every moment is included (dashed curve). This may be thought of as a case where the ocean has small heat capacity, although in practice it would be computed differently, since reducing ocean heat capacity would affect the seasonal cycle. In practice, it would have to be computed by running the model to equilibrium at each new value of the forcing, which would be very costly in a full climate model but is straightforward in an energy balance model.

In such a transient response experiment, there is no statistically steady averaging period that can be used to define a new climate. In order to separate natural variability from the forced increase, the method is to perform several climate model runs with the same forcing, but slightly different atmospheric and oceanic initial conditions. The set of such runs is known as an *ensemble*. Weather variations and interannual climate variations such as

El Niño will evolve differently in each run, whereas the response to the forcing will occur the same way. When many such runs are averaged, this *ensemble average* gives the response to the forcing. Unfortunately, it is computationally expensive, so only a few such runs are typically done in practice.

The climate model experiments we will see in Chapter 7 are affected by the heat capacity of the ocean in the manner schematized in Figure 6.11. The time-dependent response of the climate system lags behind the forcing and thus the warming is substantially less at each time than if the response had come to equilibrium with the GHG concentrations of that time. A consequence of this is that if society waits until a substantial warming has been detected before taking any action, the eventual warming will be larger. This is illustrated in Figure 6.12 using a case in which greenhouse forcing increases initially as in Figure 6.11, but then is suddenly held constant, corresponding to society suddenly realizing the problem is serious and stopping all emissions on the spot. There is still substantial warming until equilibrium with the forcing is achieved. This additional warming is termed the *additional warming commitment*, because at a given time we have committed the planet to warm this additional amount further than it has already warmed, even if we cease greenhouse gas emissions. A consequence of this additional warming commitment is that if one waits until

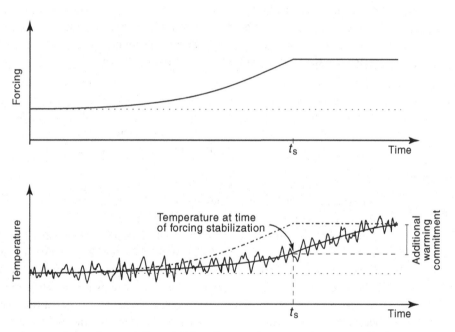

Fig. 6.12 A transient response experiment as in Figure 6.11, but where greenhouse gas emissions are suddenly stopped at time t_S, so the forcing stabilizes (upper panel). In the lower panel, the dashed curve shows the temperature response if the climate system were in equilibrium with the forcing. Since the ocean slows the response of the climate system, the actual temperature response (solid curve) lags behind the forcing. Thus the temperature continues to rise even after the emissions are stopped. The transient response eventually reaches equilibrium, but not until many decades later. The difference between the temperature at a given time (here shown for time t_S) and the equilibrium value (dashed curve) is termed the additional warming commitment.

it is clear that global warming is going to be large, there are substantial risks. This factor adds to the need to plan ahead in considering strategies to reduce emissions.

To provide a more detailed view of the spatial structure of warming in equilibrium versus the warming at a particular time, we turn to a classic climate model experiment that corresponds conceptually to the case shown in Figure 6.12. An experiment in which CO_2 increases at 1% per year (similar to observed increases over recent decades) will be examined. The warming that occurs around 70 years into the run, at the time when CO_2 has doubled, will be compared to the equilibrium response in a doubled-CO_2, experiment. This corresponds to the temperature the model would eventually reach if the CO_2 increase were stopped at $t_s = 70$ years and held constant until the model reached equilibrium.

6.8.2 A doubled-CO_2 equilibrium response experiment

As noted above, doubled-CO_2 experiments are used for understanding the general features of the response to increased gases, for comparing climate models, and as a measure of climate sensitivity. Unless special circumstances happen (such as a drastic collapse of the thermohaline circulation), the equilibrium response is independent of the history of the forcing and initial climate conditions, so such an experiment could be conducted, for instance, as shown in either Figure 6.10 or Figure 6.12 and the long-term results would be the same.

Figure 6.13a shows results for the equilibrated climate change for surface air temperature, with one of the climate models that was important in much of the early work on global warming. This model had a climate sensitivity on the high side of the range for current climate models but the main results continue to hold. A number of features of the spatial pattern of global warming may be seen in the equilibrium response in Figure 6.13a which will be discussed in more detail for more realistic, time-dependent scenarios. The surface warming is widespread over the globe in this model with a global-average equilibrium warming of about 4.5 °C. Warming occurs also throughout the troposphere. The warming is greater over midlatitudes than in the tropics and is even greater in polar regions, as will be discussed further in Chapter 7.

6.8.3 The role of the oceans in slowing warming

Turning to the time-dependent run from the same model, Figure 6.13b and c illustrate the spatial dependence of the ocean effect in slowing warming. In Figure 6.13b, the surface air temperature is shown for a 20-year average centered on the time at which CO_2 has doubled in a 1% per year CO_2 increase experiment. While the GHG radiative forcing in Figure 6.13b is the same as in Figure 6.13a, the heat capacity of the upper ocean (above the thermocline) has reduced the warming everywhere relative to the equilibrium response. Furthermore, in certain regions the warming is reduced even more effectively. This occurs in the North Atlantic and in the Southern Ocean near Antarctica where the thermohaline overturning carries heat down into the deep ocean. The warming is still largest in the high latitude regions of the northern hemisphere, but the oceanic effects have greatly delayed the

(a) Equilibrium temperature response

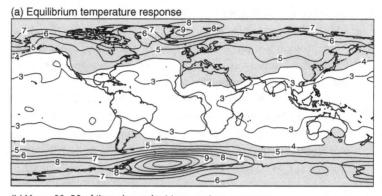

(b) Years 60–80 of time-dependent temperature response

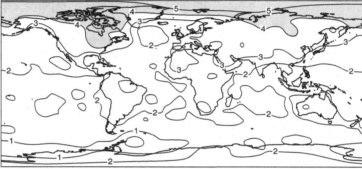

(c) Ratio of time-dependent response to equilibrium response

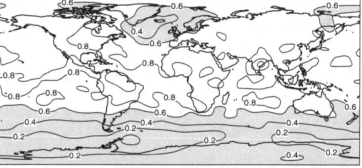

Fig. 6.13 Annual average surface air temperature response from an earlier version of the Geophysical Fluid Dynamics Laboratory climate model comparing equilibrium response to time-dependent response. (a) Equilibrium response for a doubled CO_2 experiment shown as difference from a control run. Contour interval 1 °C. (b) Coupled ocean–atmosphere model response to a 1% per year CO_2 increase shown at approximately the time that CO_2 has doubled in the model (averaged from years 60–80 of the model run) shown as a difference from a similarly averaged period from a control run. Contours as in (a). (c) Ratio of the time-dependent response in (b) to the equilibrium response in (a). Shading in (c) indicates regions where the time-dependent response is less than 60% of the equilibrium response. After Manabe *et al.* (1991).

warming in Antarctic regions. Northern hemisphere continents tend to have warmed more than neighboring oceanic regions.

The difference between Figure 6.13a and Figure 6.13b gives the global warming commitment at the time of doubled CO_2, as estimated by this model. If concentrations were held fixed at the time represented by 6.13b, the Earth would gradually warm to the situation in Figure 6.13a as the oceans approached equilibrium with the GHG forcing. Another view of the slowing of the warming by the oceans is shown in Figure 6.13c which displays the ratio of the two cases. In the time-dependent experiment at year 70, much of the globe has warmed by about 60% to 80% of the equilibrium value. The shaded area indicates the large regions that have warmed by less than 60% of what they will eventually attain, some by as little at 20%. These ratios may be taken as indicative of how the warming we might see in the middle of this century would compare to the long-term warming if GHG concentrations were capped at that time.

6.8.4 Climate sensitivity in transient climate change

For additional insight into the role of the oceans in slowing the warming, we take the simple globally averaged energy balance model of Eq. (6.11), and extend it to include the heat storage in an ocean surface layer of a specified depth, H. Heat storage in the atmosphere is negligible on these time scales, so any imbalance in the energy balance at the top of the atmosphere is associated with the ocean heat storage. The heat capacity per unit area of the layer is $C = \rho c_w H$, where ρ and c_w are the density and heat capacity per unit mass of sea water, as in the surface layer discussed in section 3.3.1. This simple climate model, including a heat storage term proportional to the rate of change of surface layer temperature, becomes

$$C \frac{\partial \Delta T_s}{\partial t} + \alpha \Delta T_s = G \qquad (6.16)$$

Figure 6.14 shows a simple example that helps to understand the effects of different climate sensitivity in different climate models in the context of transient climate change. The greenhouse forcing G is a simplified case where the radiative forcing begins to go up linearly in 1970, at a rate that roughly mimics experiments in climate models. If the heat capacity of the ocean were negligible, the first term could be dropped and the equilibrium value of the global average temperature change would be

$$\Delta T_s = \frac{G}{\alpha} \qquad (6.17)$$

and would go up linearly with G. These equilibrium curves are shown as dashed lines in Figure 6.14.

The ocean response time introduces a lag given by $\tau = C/\alpha$ (to see this, substitute the solution $\Delta T_s = g(t - \tau)/\alpha$ into Eq. (6.16), using $G = gt$). This lag gives the response time indicated in Figure 6.10 or Figure 6.11. After an initial adjustment period the temperature increases at the same rate as the equilibrium case, but the actual temperature lags the equilibrium solution by this response time. Different values of the climate feedback parameter α thus have two effects. If α is smaller (i.e. climate sensitivity is higher), then the

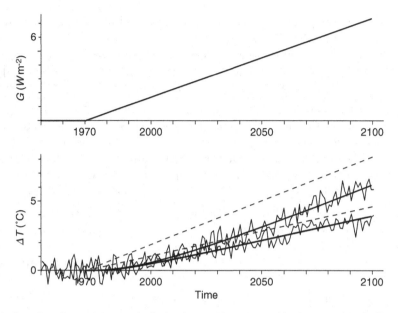

Fig. 6.14 Schematic of transient climate response to the forcing shown in the upper panel by climate models of different climate sensitivities. In the lower panel, the upper solid curve gives the global-average surface temperature response for a climate feedback parameter typical of the more sensitive climate models in IPCC (2001), while the lower curve would be typical of the less sensitive models. The dashed lines show the temperatures for each case that would occur if the ocean did not act to delay the warming. Because the response time is longer in the high sensitivity case, the curves initially look similar even though the long term response is very different. Wiggly lines indicate the natural climate variability that would occur about the simplified response curves.[8]

temperature increases more quickly in the long term. However, smaller α also implies a longer adjustment time, so it takes longer for this increase to begin. Not only does the heat capacity of the ocean slow the warming initially, it slows it more for the higher sensitivity case. Thus in the early stages after the forcing starts to increase, the curves for both the low sensitivity model and the high sensitivity model look very similar, even though the long-term outcome will be very different. Another way to see this is directly from Eq. (6.16). At early times, the change in temperature is still small, so the $\alpha \Delta T_s$, term does not contribute much to the balance. Rather, the ocean heat storage is the main balance for the radiative forcing initially. This delays the warming by an amount of time that depends on the heat capacity of the ocean (which enhances the storage term) versus α (which affects the size of the $\alpha \Delta T_s$ term).

This effect has considerable practical implications. We would like to know which model, the higher sensitivity model or the lower sensitivity model, better represents the actual climate system. We might imagine that we could wait, increasing our observational time series, until we could distinguish that one case was clearly a better fit. Because both remain similar for a long period, the disadvantage of this strategy is that a substantial warming would have already begun and an additional warming commitment would have been incurred. This is exacerbated by the presence of natural variability in the climate system which makes the two curves hard to distinguish in the early stages of the warming. We will see in Chapter 7 how these principles apply to runs with full climate models.

Notes

1 Linearization can be applied more generally, for multiple variables, in combination with Taylor series. Recall that for a smooth function F of variables X and Y $\Delta F(X, Y) \approx \frac{\partial F}{\partial X} \Delta X + \frac{\partial F}{\partial Y} \Delta Y +$ (higher order terms), where changes are defined as $\Delta X = (X - \bar{X})$, and derivatives are evaluated at X, Y. Higher order terms means quadratic and higher powers in the changes, which can be neglected if the changes are small. For the case of the one-layer energy balance model, the equations can be linearized when a small temperature increase ΔT_s is to be calculated. For instance, the change in upward surface flux is approximately

$$\Delta I R_{sfc}^{\uparrow} \cong \left(\frac{\partial \sigma T_s^4}{\partial T_s} \right) \Delta T_s = 4\sigma \bar{T}_s^3 \Delta T_s \qquad (6.18)$$

The first step applies when the changes are sufficiently small and the second step is simply differentiating Eq. (6.5). Overbars indicate climatological terms.

2 In IPCC (1996) and IPCC (1990), $G = 4.4\,\mathrm{W\,m^{-2}}$ was used. Myhre $et\ al.$ (1998) suggest that, accounting for stratospheric adjustment and some solar absorption by CO_2, a more suitable number would be $3.7\,\mathrm{W\,m^{-2}}$.

3 While Figure 6.4 shows a numerical linearization, for this simple model it can also be done analytically. When only small changes are considered, it is possible to separate the changes in the fluxes due directly to changes in absorptivity $\Delta \epsilon_a$ from those due to changes in temperature. The change in IR at the top of the atmosphere is

$$\Delta I R_{top} \approx -G + (1 - \bar{\epsilon}_a)\Delta\left(\sigma T_S^4\right) + \bar{\epsilon}_a \Delta\left(\sigma T_a^4\right).$$

The balance is approximate because we have dropped a term $\Delta \epsilon_a \Delta I R_{sfc}^{\uparrow}$ that involves changes multiplying changes ("quadratic" in the changes). This is small since the changes are only a few percent of the climatological terms. Dropping the quadratic terms is one step in the linearization. One then applies a proceedure similar to Eq. (6.18) to obtain a linear relation.

4 Some authors use the global average flux at the bottom of the atmosphere. This can give self-consistent notation as well, although some of what is treated as feedback here is treated as forcing in that case.

5 The Soden and Held (2006) study from which Table 6.1 is adapted uses a method described in Held and Soden (2000) in which changes in temperature, water vapor and surface albedo (primarily sea ice and snow) from various models have their radiative effects evaluated individually using a single radiative scheme. Cloud feedbacks are estimated as a residual from other contributions and the overall value of α_T. Climate changes are from 10-year averages shortly after CO_2 doubling, and they use a $4.3\,\mathrm{W\,m^{-2}}$ nominal forcing for all models to evaluate α from the temperature change. For comparison, IPCC (1990) gave $\alpha_T = 3.3\,\mathrm{W\,m^{-2}\,K^{-1}}$ and $\alpha_{H_2O} = -1.2\,\mathrm{W\,m^{-2}\,K^{-1}}$, owing to different choices of how to evaluate these. The range of cloud feedback in the climate models examined in Cess $et\ al.$ (1989, 1996), excluding outliers was roughly -0.8 to 0.6 (i.e. contained uncertainty even in sign), while for surface albedo feedback it was similar to Table 6.1. Overall, however, the relative contributions of the various feedbacks are fairly consistent as noted in the survey by Colman (2004).

6 Notation in the field varies. Sometimes α is used for the climate sensitivity and λ for the climate feedback parameter.

7 Based on IPCC (2001) Table 9.4 for the first two rows, and IPCC (2007) section 8.6.2.2 (see also section 10.5.2 and Box 10.2) for the last row.

8 The curves in Figure 6.14 use climate feedback parameter values of $\alpha = 0.9$ and 1.9 for the high and low sensitivity case, respectively. This would roughly correspond to the CSM1.0 and CSIRO models cited in IPCC (2001) section 9.3 and Appendix 9A. These are relatively high and low sensitivity models, although not the most extreme cases. The slope of the radiative forcing is $0.043\,\mathrm{W\,m^{-2}\,yr^{-1}}$, roughly mimicking the B2 scenario with moderate GHG emissions of Figure 7.2. The response times use an ocean mixed layer depth of 200 m, to crudely represent upper ocean adjustment for the layer above the thermocline, giving roughly 30 and 14 years for high and low sensitivity, respectively.

7 Climate model scenarios for global warming

7.1 Greenhouse gases, aerosols and other climate forcings

7.1.1 Scenarios, forcings and feedbacks

Climate model predictions for global warming differ from predictions for ENSO because the system is responding to a *forcing* that is being continuously applied, namely the radiative effects of greenhouse gases (GHG) in the atmosphere. The term *driver* of climate change is sometimes now used for communicating to the public, but forcing is the standard scientific term. In Chapter 6, we saw that the top-of-atmosphere radiative imbalance caused by GHG provided a useful measure of the radiative forcing to which the climate system responds. In a forced system, in principle, if we know how the forcing behaves as a function of time we can obtain the system response. In the climate system, there is always chaotic natural variability about the forced response but if there is an ongoing forcing that tends to warm the climate, then the variations will occur about the forced warmer state. This is why, if the forcing were known, climate model predictions could be made a century ahead and have some meaning, whereas even the most perfect ENSO prediction system would have no skill after a couple of years.

Future GHG emissions are, of course, not actually known at this time. Instead climate researchers collaborate with economists and social scientists to construct possible *scenarios* of future emissions and the resulting concentrations. These scenarios may be quite complex attempts to project energy demand and population growth, taking into account assumptions such as a possible phase-out of nuclear energy or improvements in solar technology. Or they may be assumptions that approximate what might happen but are simplified to aid implementation of experiments and understanding of the results. An example of this is a 1% per year increase of CO_2 concentration, which has been used as one standard scenario to approximate likely increases in a combination of GHGs, with CO_2 equivalent radiative effects used as a proxy for an imprecisely known combination of all GHGs. Scenarios may be even simpler, such as an instantaneous doubling of CO_2 at a certain year, with the concentration held fixed thereafter. The latter scenario is used as a standard experiment for examining climate sensitivity as discussed in Chapter 6. The IPCC Special Report on Emissions Scenarios (SRES) in 2000 outlined a specific set of assumptions for a number of scenarios, aimed at covering the range of possibilities for population growth, energy supplies, phase-out of CFCs, etc., discussed in more detail in section 7.1.3.

7.1.2 Forcing by sulfate aerosols

When trying to accurately model climate evolution in the past century and when projecting climate change into the future, estimating the role of anthropogenic effects on aerosols is a challenging additional factor. Among the most important of the human effects on aerosols is the emission of sulfur compounds (for instance by coal and oil burning and other industrial activities). Chemical alteration into sulfates produces (in addition to acid rain) suspended particles that produce polluted haze. The sulfate particles tend to attract water molecules which increase their size to the point where they can reflect sunlight, although they still are suspended in the atmosphere. The increase in reflected sunlight tends to cool the climate system. Certain other types of particles, such as soot, may also act as absorbers but the overall contribution of these is much smaller. The impact of the aerosols themselves on the radiative forcing is referred to as the *direct effect of aerosols*. Changes in the number or size of aerosol particles may also affect cloudiness by providing additional cloud condensation nuclei. Such effects are referred to as *indirect effects* of aerosols, and these effects are currently poorly estimated. We focus here on the direct effects when referring to aerosol forcing.

Aerosols have a relatively short residence time in the atmosphere (typically on the order of weeks) compared with greenhouse gases. As a result, they do not travel very far from their source, and aerosol forcing tends to be largest in industrialized continental regions. Because sulfur emissions result largely from fossil fuel burning, the emissions of GHG and sulfate aerosol production are highly correlated. However, the sulfate aerosols resulting from fossil fuel consumption fall out of the atmosphere, whereas the carbon dioxide tends to remain over long periods of time. The aerosol effect thus tends to reduce warming in the short term but not in the long term.

Figure 7.1 shows the contrasting patterns of the radiative forcing due to greenhouse gases and to sulfate aerosols estimated for the recent times relative to preindustrial climate. The large-scale forcing due to GHG produces a warming tendency everywhere, with a gradual dependence on latitude that ranges from almost $3 \, \mathrm{W \, m^{-2}}$ in tropics to $1.5 \, \mathrm{W \, m^{-2}}$ at higher latitudes. The larger values in the tropics occur simply because the tropical temperatures are warmer, and so the IR fluxes are larger. A given concentration of GHG trapping a fraction of a larger IR flux produces a larger forcing.

The aerosol forcing is localized near source regions, for instance, the east coast of the United States, Europe, southeast Asia and China. The sign is opposite the GHG forcing, i.e. it produces a cooling tendency in these regions. The maximum magnitude of the forcing is similar to the magnitude of the GHG forcing, but it occurs only in small regions. Over most of the globe, the aerosol forcing is near zero, so in the global average its impacts are smaller. Atmospheric dynamics will tend to average the cooling impact of the aerosols over a larger region than where the forcing itself occurs.

7.1.3 Commonly used scenarios

The scenarios outlined in the Special Report on Emissions Scenarios provide a convenient means of investigating the climate impacts for specific guesses about future human

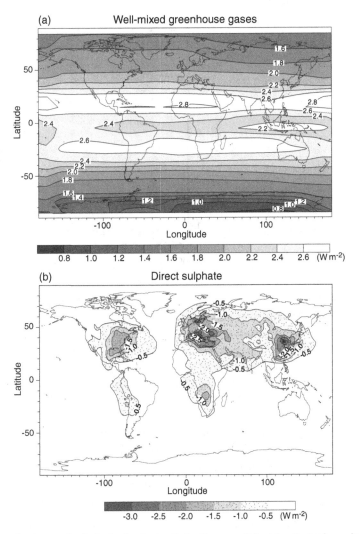

Fig. 7.1 Spatial patterns of estimates of radiative forcing due to effects of human activity: (a) well-mixed greenhouse gases from the preindustrial to present day (contour interval 0.2 W m^{-2}); (b) direct effects of sulfate aerosols (contour interval 0.5 W m^{-2}, negative sign denotes cooling tendency). After Shine and Forster (1999).[1]

activities. Estimates of natural forcings, such as aerosols due to volcanic eruptions, are also included, and estimates of observed anthropogenic and natural forcings are used for the period up to present. Various climate modeling groups then run their models specifying these past and future concentrations of greenhouse gases, sulphate aerosols and other factors that affect the radiative balance of the atmosphere. Figure 7.2 shows the radiative forcing at the top of the atmosphere associated with each of these scenarios.

Some of the climate model runs discussed in this chapter make use of inputs from the scenario known as "SRES A2." This scenario assumes continued population growth and that technological change varies among nations, i.e. a heterogeneous world with regionally

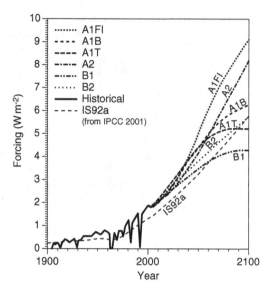

Fig. 7.2   Radiative forcing (top-of-atmosphere radiative imbalance tending to cause warming due to the net effects of greenhouse gases and other forcings) as a function of time for estimated twentieth-century observed forcing and various climate forcing scenarios for the twenty-first century. Scenarios from the Special Report on Emissions Scenarios are denoted SRES A1FI (fossil intensive), A1T (green technology), A1B (balance of these), A2 (heterogenous world), B1 (greenest future human activities) and B2 (see text). IS92a denotes a scenario used in many studies before 2005. It is shown here with the estimate of twentieth-century observed forcing from IPCC (2001). Values from Meehl *et al.* (2007).

fragmented economic growth. The B2 scenario assumes slower population growth and more environmental protection than A2, while a set of SRES A1 and B1 scenarios assume population declines after mid-century with various degrees of technological change toward resource-efficient technologies. Most of the other scenarios have smaller emissions than A2 associated with societal factors and choices. The A2 scenario has more the flavor of what was formerly called a "business-as-usual" scenario.

The six SRES scenarios were chosen by the IPCC from among a larger set, aimed at capturing the range of opposing tendencies: between economic values and environmental values on the one hand; and between globalization and regionalization on the other. In more detail, they are described as follows. The A1 scenario family: a future world of very rapid economic growth, global population that peaks in mid-century and declines thereafter, and rapid introduction of new and more efficient technologies. A1FI achieves high economic growth with fossil fuels; A1T uses advanced technology to reduce emissions in mid-century and essentially cap the radiative forcing by late twenty-first century; while A1B lies in between. The B2 scenario assumes local solutions to economic, social and environmental sustainability, with continuously increasing population (lower than A2) and intermediate economic development. The lowest of the SRES scenarios, the B1 scenario, is for a convergent world with the same global population as in A1 but with rapid changes in economic structures toward a service and information economy, with reductions in material intensity, and the introduction of clean and resource-efficient technologies, i.e. leaning most heavily toward choices of equity and environmental protection. Note that the SRES was not supposed to consider particular treaties or protocols on GHG emissions. Concentrations of GHG and aerosols from the SRES scenarios are used for the climate model runs in IPCC (2007).

The main climate model runs in IPCC (2001) were based on an older scenario "IS92a" or on a scenario of 1% CO_2 increase per year, which is roughly consistent with the IS92a scenario. These were used to typify what would be likely to happen if strong societal action is not taken. The SRES A2 scenario has somewhat larger GHG emissions and lower sulfate emissions through this century than the IS92a scenario. As a rough way of comparing these, for a 1% CO_2 increase, doubled CO_2 (relative to the concentration at the start time) occurs 70 years into the run. For the SRES A2 scenario, a CO_2 concentration of 650 ppm, double the 1970 value, occurs in 2072, while for IS92a this concentration occurs in 2086. Note that doubled CO_2 is often taken relative to the time period for which we have good measurements of climate variables. Doubled CO_2 relative to the value of about 290 ppm from the 1870s would occur in 2060 in the SRES A2 scenario. The SRES A2 and IS92a scenarios have other GHG as well as CO_2, so the radiative forcing due to GHG reaches a level equivalent to a doubled-CO_2 run somewhat earlier than the years listed above.

Along with time-dependent scenarios for future change, there is an ongoing effort to estimate as accurately as possible the historical forcings that have been operating since the beginning of the Industrial Era, i.e. since the late 1800s. These forcings include the change in greenhouse gases and the change in aerosols, airborne particles that result from industrial pollution. Also included are small variations in solar input to the climate system, and the effects of volcanic eruptions, the aerosols from which can increase reflection of sunlight producing slight variations in forcing in particular years. The solar and volcanic forcing impacts tend to be secondary corrections, so discussion here is focused on the impacts by GHG and anthropogenic aerosols. It may also be noted that more recent estimates of the observed forcing slightly exceed the earlier estimates associated with the IS92a scenario.

Model runs from the mid-1800s to 2000 using estimates of observed GHG, aerosol, solar and volcanic radiative forcings are termed "twentieth-century runs." These are runs with coupled models, so each model will have its own representation of natural climate variability, such as El Niño. One does not expect the natural climate variability to match the observed on a year-by-year basis, owing to inherent limitations to predictability. An ensemble of runs with the same climate model and the same radiative forcings but from different initial conditions will differ in the time series of natural climate variations, but have the same temperature trend due to GHG increase.

7.2 Global-average response to greenhouse warming scenarios

We begin by considering an example of the relative impacts of GHG and sulfate aerosols, from a study at the time these effects were being established. Figure 7.3a shows the time dependence of radiative forcing by GHG, sulfate aerosols and both combined, estimated from 1860 to present and projected into the future with a 1% per year CO_2 increase scenario.

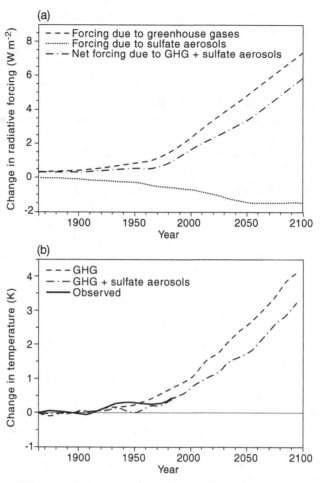

Fig. 7.3 (a) Radiative forcing by greenhouse gases, sulfate aerosols and the net forcing due to both combined. The forcing is an estimated change from 1860 to 1990, followed by a projected 1% per year CO_2 increase and a compatible sulfate aerosol scenario (see text). (b) The global-average surface temperature response of a climate model to GHG forcing alone and the combined GHG plus aerosol forcing, compared to observed global-average surface temperature. A running decadal average has been applied. After Mitchell and Johns (1997).

The combined effects of GHG are approximated by a 1% per year CO_2 increase from 1990 onward (for which a doubling of CO_2 relative to 1990 occurs in about 2060). The estimate of future aerosol forcing depends on an aerosol model driven by the IS92a industrial sulfur emissions scenario. The radiative forcing contribution by sulfate aerosols is opposite in sign to that of the GHG, i.e. it produces a cooling tendency that partially cancels the GHG warming tendency. The sulfate aerosol forcing levels off after a time because the residence time of aerosols is relatively short. The aerosols do not remain in the atmosphere, so the aerosol concentration depends only on the rate of emissions in a given year. This may be contrasted to the GHG forcing which continues to increase as the integrated emissions build up over decades and centuries. Thus while the aerosols tend to reduce the net radiative forcing considerably in past and present, in the future the GHG forcing is expected to dominate. As an extra consideration, if we were to suddenly stop burning fossil fuels today, sulfate aerosol concentrations would rapidly be reduced, but the GHG concentration would remain relatively constant over the coming century.

Figure 7.3b shows the global-average surface air temperature response of a climate model to these forcings. The climate model is a coupled ocean–atmosphere model from the Hadley Centre (this study was done with HadCM2, the version prior to HadCM3 listed below). Two experiments are shown with radiative forcing increasing with time. The temperature change is given relative to a 130-year model control run in which forcing was held constant at the value in 1860. In the first experiment, only the forcing by GHG is included in the model. The GHG effect produces an increasing warming with time in response to the ongoing GHG increase through this century. In the second experiment, both the GHG forcing and the sulfate aerosol forcing are included (this would correspond to the net forcing curve shown in Figure 7.3a). Early in the experiment, including the present time, the sulfate aerosol effects decrease the warming by a substantial fraction of the warming that would occur if GHGs acted alone. As time goes on, the aerosols still reduce the warming but by a fixed amount that becomes a smaller fraction of the total as the warming increases. In this model, the global-average surface temperature increase in 2050 due to GHG alone reaches about 2.6 °C. This is reduced to about 1.8 °C by sulfate aerosols when both effects are included.

An estimate of the observed surface temperature increase since the 1860s is also included in Figure 7.3. Considering the model warming from 1860 to present, GHG forcing alone produces a warming slightly less than 1 °C. Aerosols oppose this and reduce it to about 0.7 °C. This appears to be generally consistent with the warming in the observed temperature record, in terms of long-term trend. Some of the interdecadal variations in the observed record do not match the model simulation. This is to be expected if the observed interdecadal variations arise from natural variability – a model might be able to capture the statistics of natural variability, but is not expected to reproduce the exact time sequence of the variations. It may also be that factors have been omitted from the forcing that affected the observed system on decadal time scales.

Several caveats should be noted. The exact amount of warming for a given scenario varies somewhat among different models, as will be discussed below. Furthermore, the best estimate of the sulfate aerosol radiative forcing has been revised since these experiments were carried out, so for a level of sulfur emission that corresponds to the IS92a scenario, the sulfate aerosol impacts would probably be smaller than those shown here. However, the general features of the model simulations shown here provide a reasonable view of how future GHG emissions may be expected to produce a substantial and increasing warming, even taking effects of sulfate aerosols into account. More recent model simulations, discussed below, include both GHG and aerosols in the forcing scenario.

One means of assessing the degree of uncertainty, effectively establishing an estimate of error bars, on the projected warming, is to compare the results of many climate models. This cannot, of course, include potential sources of error due to processes that might be omitted or misrepresented by all of the modeling groups, but it does give an estimate of the effects of different parameterization and modeling approaches. Figure 7.4 shows results from several climate models from different research centers. The models are all run with the

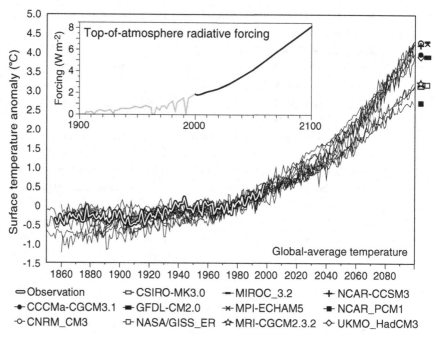

Fig. 7.4 A sample of the global-average surface temperature response of several climate models (each denoted with a different acronym – see endnotes) to estimated observed forcing through the twentieth century, followed by GHG and aerosol forcing from the SRES A2 scenario for the twenty-first century. Temperature changes are shown relative to the 1961–90 mean for each model and for observations (through 2000). Inset as in Figure 7.2. Data from the Program for Model Diagnosis and Intercomparison (PCMDI) archive. The sample is shown so that individual model curves can still be seen.[2]

same SRES A2 scenario, including GHG plus aerosol effects. The comparison thus shows the extent to which the models agree on the response to this estimate of past and scenario for future warming. Seven climate models are included, so that the behavior of individual models can still be seen, while illustrating the range spanned by an ensemble of models. Figures in later sections will show results with a larger ensemble of models, in which the range spanned by the models is shown rather than the individual model results. Each model is denoted by an acronym denoting the group that produced a model and their name for the model and a version number.

The level of natural variability about the long-term trend seen in each climate model simulation in Figure 7.4 is larger than that seen in Figure 7.3 simply because annual average rather than decadal average values of surface temperature are displayed. This variability includes contributions due to El Niño and other interannual and interdecadal phenomena. It also includes variability due to weather noise (as discussed in Chapter 4 for its effects on El Niño) which can drive effectively random variations in many parts of the climate system. On long time scales, this natural variability is unpredictable and is not expected to have the same time sequence for different models, or even in two simulations with the same model. Thus disagreement between models is significant only when the differences exceed the typical range that might occur due to natural variability. Over the historical record from 1860 to present, the warming trend in all the models agrees quite well. Comparing to observations, the long-term warming simulated by the models appears consistent with that observed, within the level of natural variability. An exception is the CCCMa model which has a larger trend than observed from the 1800s to present. An observed warm period in the 1940s is potentially unexplained, although some warm decades seen in the models due to natural climate variability are comparable in magnitude.

The projections of future warming in all the models agree fairly closely in the early part of this century. Later in the century the projections diverge to a greater extent. The range of warming later in the simulation is due to the range of values of climate sensitivity in the various models, corresponding to a range of climate feedback parameters as examined in Chapter 6. In Table 6.1 an example of the range of equilibrium response that results from uncertainties in various feedbacks was presented. The range of predicted warming is the counterpart of this in the time-dependent simulations. For the models in Figure 7.4, MIROC3.2, MPI-ECHAM5 and CNRM-CM3 lie at the more sensitive end of the range, while NCAR-PCM1, GISS-ER, CSIRO-MK3.0 and MRI-CGCM2 lie at the less sensitive end. The models shown here suggest that a range of warming of global-average surface temperature from 3 to 4.2 °C might be expected by the end of the century if the emissions of GHG continue in a manner similar to this scenario, even taking into account the effects of sulfate aerosols. The range increases slightly when more models are added, and effectively provides an estimate of error bars on the projected warming. Despite substantial error bars on the late-century warming, the agreement among the models, especially out to the middle of the century, provides basis for the substantial concern over the impacts of this climate change.

7.3 Spatial patterns of warming for time-dependent scenarios

Figure 7.5 shows the spatial patterns of surface air temperature response as a function of time for the same SRES A2 scenario for GHG and sulphate aerosol forcing for which global-average response was seen in Figure 7.4. A single model, the Hadley Centre model (HadCM3), is presented first as an example. Thirty-year averages are shown to reduce the influence of natural variability. In this section, the response is shown as a change relative to the model mean value for 1961–1990. The warming trend that was noted in the global-average temperature is clearly seen but with a characteristic spatial pattern. The most prominent feature, as in Figure 6.13, is that high latitudes exhibit substantially greater warming than the tropics. This is known as *poleward amplification* or *polar amplification*. In time-dependent simulations, the polar amplification of the warming is seen first in the northern hemisphere. As early as the 2020s, the annual average effects are substantial in Arctic regions. Continents tend to warm before the ocean regions, since the heat capacity of the upper ocean slows the warming. At each time in this simulation, there is a substantial commitment to additional warming, as the climate system has not warmed to equilibrium with the GHG concentration. The reduced warming in the Southern Ocean and northern

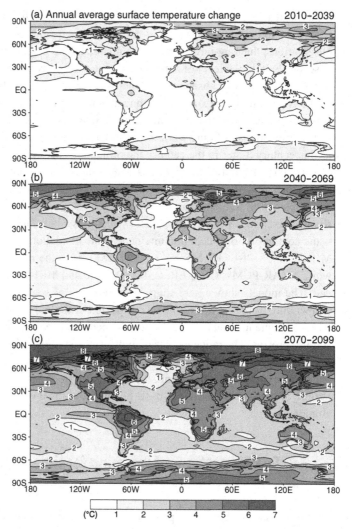

Fig. 7.5 Response to GHG and sulfate aerosol forcing from the Hadley Centre climate model (HadCM3). The change in surface air temperature, relative to the average during 1961–90, is shown as 30-year averages centered on (a) 2025, (b) 2055, (c) 2085. Contour interval 1 °C.

Atlantic is consistent with the additional slowing of the warming by involvement of the deep ocean circulation in those regions noted in Figure 6.13c (the GHG forcing in this model would be comparable to Figure 6.13b at a year between the periods of Figure 7.5b and c).

The warming has a strong seasonal dependence. In Figure 7.6a, the January surface air temperature is seen, averaged over all Januaries during 2040–2069. This may be compared to the annual average shown in Figure 7.5b for the same time period. The warming in northern hemisphere high latitudes is much larger in winter than in the annual average. Changes exceed 5 °C over large parts of the Arctic and reach even larger values in some locations. The warming exceeds 3 °C over a large part of Canada and northern Eurasia and exceeds 2 °C over most of the United States.

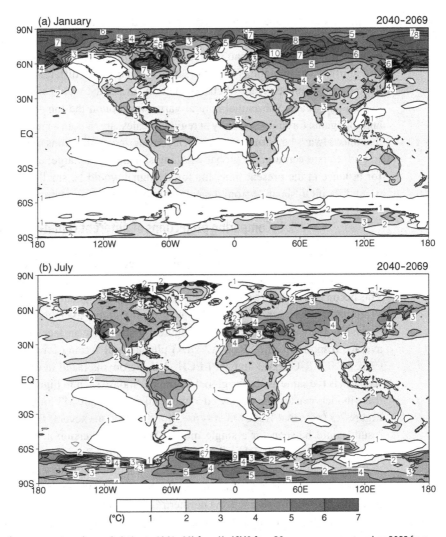

Fig. 7.6 Surface temperature change (relative to 1961–90) from HadCM3 for a 30-year average centered on 2055 from simulations with GHG and sulfate aerosol forcing for (a) January; (b) July. An ensemble average of three simulations is shown. Contour interval 1 °C.

The northern hemisphere summer case may be seen in Figure 7.6b for July, averaged over 30 years centered on 2055. The largest changes in Figure 7.6b occur in southern hemisphere high latitudes, i.e. in the winter hemisphere. The summertime warming over the northern hemisphere continents is still quite large. This includes strong effects in Europe and in the western part of the United States and Canada. One of the consequences of these warmer summer temperatures is increased evaporation, which tends to drive soil moisture down. Reduced summer soil moisture is simulated by a number of models, and while confidence in this prediction is not high owing to the complexity of land surface processes, it would have considerable potential impact on agriculture.

The 30-year averages in these figures are used to reduce the influence of natural climate variability, as simulated in the model. Most features at the large scale are due to the anthropogenic forcing. Regional patterns, however, would have more influence from interdecadal variability. Two simulations from the same model but with different initial conditions, producing a different sequence of chaotic natural variability, would show some differences at the regional scale. Substantial natural variations around the long-term change caused by anthropogenic forcing, especially at regional scales, are thus an aspect of the climate system that must always be borne in mind when considering applications of climate model results. In mid- to late-century, the global warming signal is quite large; earlier in the simulation, for instance at the present time, the forced signal would be smaller so it would be much more difficult to separate from the natural variability with certainty.

7.3.1 Comparing projections of different climate models

In discussing the global-average warming projected for the next century in Figure 7.4, comparing different models was discussed as one measure of the reliability of the results. This becomes even more relevant when considering the spatial dependence of the impacts. While the models agreed reasonably well on the global-average warming in the 2050s, regionally there are large variations. Figure 7.7 shows the surface air temperature annual-average response to the SRES A2 GHG plus aerosol forcing for three models – GFDL-CM2.0, NCAR-CCSM3 and MPI-ECHAM5 (acronyms listed in endnote 2). The forcing scenario is the same as for the global-average response seen in Figure 7.4. The results from these models can also be compared to the results from HadCM3 for the same time period in Figure 7.5 (note that HadCM3 was used for the previous several figures so that consistent results could be seen for a single model. This model version also has a relatively long published track record). Overall, the models in Figure 7.7 agree on the large-scale features of the warming, including poleward amplification, the continents warming before the ocean regions and the delayed warming in the North Atlantic and Southern Ocean. Focusing on a particular region, for instance, the northern Great Plains in the United States, one finds that one model predicts a 2 to 3 °C surface temperature change while the warming in the other two models is larger than 3 °C, even though this region is fairly large and the annual average value is shown. Other differences include the exact magnitude of the polar amplification, and the degree of warming in Australia and central Asia.

In Chapter 5, we noted that climate models do reasonably well at reproducing the large-scale features of precipitation in the current climate, but often have errors at smaller scales. It is thus worth asking how spatial patterns of precipitation compare among climate models. Precipitation patterns for northern hemisphere summer for three selected climate models are seen in Figure 7.8 for a 30-year average centered on 2085. The June to August (JJA) average is shown because including more months in the average helps to further reduce natural variability, which otherwise tends to be larger in precipitation than in surface temperature.

Large changes in tropical precipitation are prominent for each model. These are due to shifts in the tropical convection zones. The overall pattern of the tropical convection zones is not greatly altered – the Hadley circulation still rises in the tropics and sinks in the

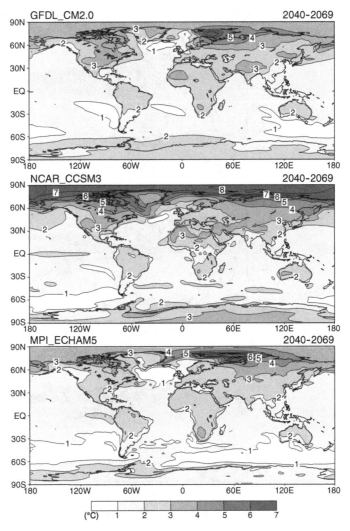

Fig. 7.7 Annual-average change in surface air temperature (relative to 1961–90) from three climate models (denoted by acronyms – see endnotes) for a 30-year average centered on 2055 from simulations with GHG and sulphate aerosol forcing. Contour interval 1 °C.

subtropics, and the Walker circulation still has preferred rising in the western Pacific – but the exact position, extent and intensity of the convection zones has been modified. While the tropical average precipitation increases, there are regions of substantial rainfall reduction, notably in some parts of the subtropics. Parts of the midlatitudes tend to have increased precipitation. Unfortunately, different models do not agree well on exactly where the wet and dry anomalies will occur when one examines the maps at the regional scale of a small country or state.

When a disagreement occurs among models, it usually has a scale that is typical of some set of physical processes. It may be that in one model there is a shift in the jet stream with

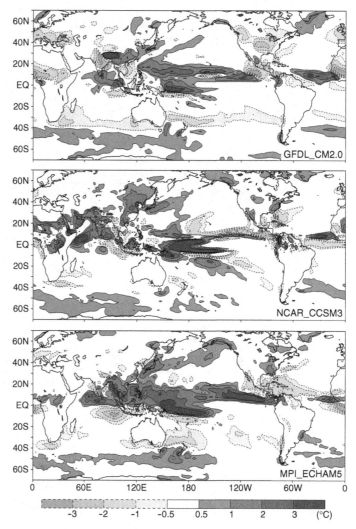

Fig. 7.8 Precipitation change for June–August average (relative to 1961–90) from three climate models (as in Figure 7.7) for a 30-year average centered on 2085 from simulations with SRES A2 GHG and sulfate aerosol forcing. Contour interval 1 mm day^{-1} (0.5 mm day^{-1} for the first interval).

attendant rearrangement of the precipitation patterns while this is smaller in another model. In one model a warming tendency in a particular continental region may be exacerbated by reduced soil moisture for a particular land surface type, whereas in another model sufficient soil moisture might be stored from the previous season to maintain evaporation and keep temperature increases modest. Two models might each have a moderate warming in a particular region if clouds were fixed, but in one model the warming might lead to a reduction in low-level stratus clouds and produce a feedback that increased the warming. Small differences in the representation of a particular process may thus lead to significant differences in the model response in particular regions.

There is thus an art to deciding the reliability of the model-predicted changes for particular aspects of the future climate. This is often phrased as the level of certainty (or uncertainty) regarding a particular result. The agreement of different models is one factor in assessing this. Note, however, that if models disagree on the particular distribution of a change, but tend to exhibit similar types of change, this is useful information. For instance, many models exhibit changes in the tropical convection zones. While they do not agree on where exactly drought or additional rain will occur, it is very reasonable to assess that regional distributions of tropical precipitation are likely to be sensitive to climate change. Similarly, from the results in Figure 7.7, one may infer that the warming is likely to be unevenly distributed, with some regions receiving greater impacts than others, even though we cannot yet say which regions.

Another factor in assessing reliability is the complexity of the feedbacks and processes involved, relative to the degree of testing that has been carried out on this component of climate models. For instance, regional impacts that involve particular changes in cloud cover would be viewed as less reliable than changes that involve primarily radiative processes. Changes involving predicted soil moisture response are assigned a lower level of reliability at this time because the processes are complex relative to the amount of scrutiny they have received in earlier stages of model development. Scientific programs are launched in each of the areas that are deemed to need more careful assessment, so it is hoped that the next generation of climate models will have smaller error bars associated with such processes.

7.3.2 Multi-model ensemble averages

Multi-model ensemble averages provide one commonly used way to condense information from many climate models that have run the same experiment. Each point in the map is an average over the response of multiple models. For instance, Figure 7.9 shows the average over 10 models of the precipitation change for 2070–99 relative to a base period average over 1961–90 for northern winter and summer seasons.[3] The June–August season precipitation map in Figure 7.9 may be compared to those for three of the individual models that go into this multi-model average in Figure 7.8. The multi-model average tends to have smoother, larger-scale features but smaller amplitude. Notice that the contour interval is much smaller in the multi-model average figure. This is because if an individual model produces a strong localized feature, such as a strong drought, in a particular location but the other models do not reproduce this in exactly the same place, the amplitude is reduced in the average. Because the models tend to agree more on the large scale than on smaller scales, the average tends to reflect this.

There are both advantages and drawbacks to multi-model ensemble averages. They permit many models to be presented in a single figure, draw attention to the larger-scale features of the change, and are less likely to be subject to the errors of one particular model. On the other hand, they tend to hide the discrepancies among the models. If used alone without further examination of the individual models, this could lead to excessive confidence in the pattern seen in the average. For instance, in the multi-model ensemble average in Figure 7.9b, the Sahel region in sub-Saharan Africa shows almost no change, but in Figure 7.8, some of the models produce a strong change in this region – but of differing sign. It is not that the

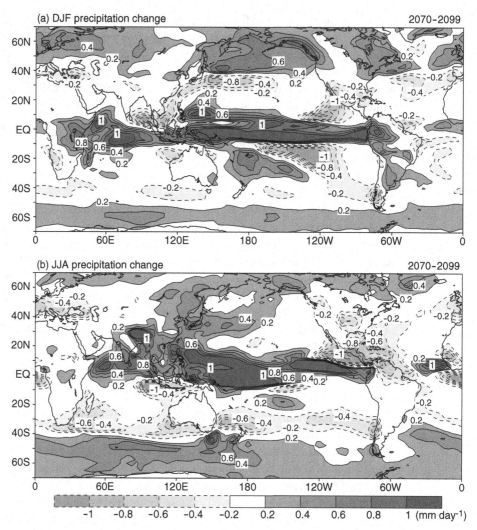

Precipitation change for a multi-model ensemble average for the 2070–99 average for (a) for December–February, and (b) June–August (relative to 1961–90) for a simulation with the SRES A2 GHG and sulfate aerosol forcing. Contour interval 0.2 mm day^{-1} (compared with 1 mm day^{-1} in Figure 7.8).

precipitation in that region is not sensitive to global warming in individual models, but rather that it is also very sensitive to differences among the models. Thus the small average change in this region is not a prediction with high confidence. Similarly in the southeast Asian monsoon region, the ensemble average is positive, but areas of negative precipitation change can be seen in individual models in Figure 7.8.

When drawing conclusions regarding a challenging variable such precipitation, climate scientists ideally take into account not only the multi-model ensemble average, but knowledge of the change exhibited in each of the individual models, as well as the fidelity of the simulation of the climatology in each of the models for the region under consideration. With this in mind a few general features may be noted. Winter time precipitation tends to increase in mid- and high latitudes. A factor favoring this increase is that warmer air tends to hold more moisture, which is thus available for poleward transport and precipitation in storms. In Figure 7.9, a poleward shift of the midlatitude storm tracks is also seen in the winter hemisphere, with a rainfall reduction at the subtropical edge. Details of this can be model-dependent. In the tropics, many regions that currently have high precipitation receive even more, while subtropical regions, which currently have little precipitation, tend to receive even less. This is partially explained by increased moisture associated with a warming, combined with the low-level flow: in regions of low-level convergence, moister air implies increased convergence of moisture favoring rainfall. In subtropical regions where the low-level flow associated with the Hadley and Walker circulations diverges, warmer air implies that moisture divergence tends to increase, decreasing the amount of moisture available locally for precipitation. However, changes in the circulation can overcome these effects on small scales, so there are many exceptions to these rough rules of thumb. The statistics of how often precipitation occurs and intensity of precipitation events also change, as further discussed in section 7.4.3.

An increase in precipitation in the equatorial Pacific is qualitatively reminiscent of the increase there during El Niño. As may be seen in Figure 7.6 and Figure 7.7, surface temperature warming occurs throughout the tropics, but there is a slight additional increase in the eastern equatorial Pacific. This occurs in a number of models, suggesting that dynamics similar to what occurs during El Niño may play a role in modifying the Pacific part of the tropical warming pattern. Some models, such as ECHAM4 and HadCM3, also indicate that the frequency and amplitude of ENSO may change in the warmer climate. Although there is not currently good agreement on what the changes may be, there is much evidence to suggest that ENSO behavior is sensitive to the climate state about which it oscillates, and so it is very plausible that there will be changes in ENSO.

7.3.3 Polar amplification of warming

Polar amplification of the warming is one of the main predicted features of global warming, so a brief discussion of mechanism and realism is in order. Paleoclimate records indicate that poleward amplification of global temperature change has been a feature of past warmer climates, thus increasing confidence in this prediction. One of the main contributing mechanisms is the snow/ice feedback as described in Chapter 6, which operates in these regions. The impacts are larger regionally than they are in the global average. Another contributing mechanism is the *lapse rate feedback*. The lapse rate (rate of temperature decrease with height) is larger at high latitudes than in the tropics. This affects the greenhouse feedback between the atmospheric temperature in the upper troposphere and the surface temperature.[4]

An additional factor is that in current climate there is normally a very strong inversion in winter time, with a cold atmospheric surface layer relative to upper levels. In the warmer climate, this is reduced. This contributes to the increased wintertime warming. A leading factor in the winter is that under normal conditions the energy balance depends strongly on release of heat stored in the ocean surface layer, while the atmosphere loses heat to space through infrared radiation. When the greenhouse effect is increased, infrared loss to space is slowed, and this is even more important in Arctic winter regions than in regions or seasons where incoming solar radiation is an important part of the energy budget. Other factors also contribute to or oppose polar amplification. While it is more complex than the simple explanation given here, it is a very reproducible feature among different climate models and may be taken to be a robust prediction. It is likely to be an important factor in ecosystem and societal impacts. Ecosystem impacts include such factors as melting of permafrost layers, changes in ocean ice cover and changes in the onset of the growing season.

7.3.4 Summary of spatial patterns of the response

For reference, some of the main points to be drawn from the model simulations in this section regarding spatial distribution of the warming are as follows.

- Polar amplification of the warming is a robust feature. It is due partly to the snow/ice feedback and partly to effects involving the difference in lapse rate between high latitudes and the tropics.
- In time-dependent runs, polar amplification is seen first in the northern hemisphere, while in the North Atlantic and Southern Ocean effects of circulation to the deep ocean slow the warming.
- Continents generally tend to warm before the oceans.
- There is a seasonal dependence to the response. For instance, winter warming in high latitudes is greater than in summer.
- The models tend to agree on continental scale and larger, but there are many differences at the regional scale. Regional-scale predictions (e.g. for California) tend to have higher levels of uncertainty, especially for some aspects (e.g. for precipitation). Much research is currently directed at this problem.
- Natural variability will tend to cause variations about the forced response, especially at the regional scale.
- Precipitation is increased (about 5–15%) on a global average, but regional aspects can be quite variable between models. There is reason to believe that regional changes are likely. Wintertime precipitation tends to increase.
- Summer soil moisture tends to decrease in large regions in the tropics and lower midlatitudes, which would have substantial implications for agriculture. Contributing factors include precipitation changes, greater evapotranspiration with increased temperature, and changes in snowmelt, but soil moisture models depend on such things as vegetation response, which are crudely modeled and have much regional dependence, so this effect is typically assigned a lower level of certainty.

7.4 Ice, sea level, extreme events

7.4.1 Sea ice and snow

In a warmer climate, both seasonal snow cover and sea ice will clearly have a smaller extent. Figure 7.10 quantifies this using the example of ECHAM5 simulated sea ice response.

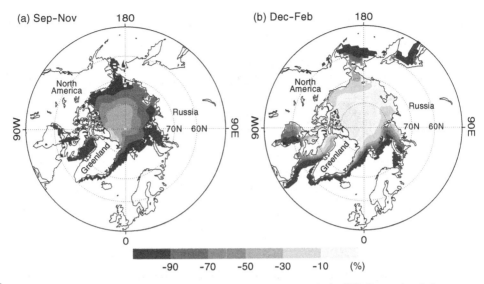

Fig. 7.10 Change in sea ice concentration simulated in ECHAM5 for a 2070–99 average in the SRES A2 scenario, relative to a base climatology for 1961–90, shown as a percent of the base climatology value. (a) September–November and (b) December–February. Shading levels show reductions exceeding 10%, 30%, 50%, 70% and 90% of the 1961–90 values. Changes are shown only for regions that had sea ice in the base climatology.

Here the change in ice coverage is shown for the late twenty-first century relative to a base climatology from recent climate. The change in ice concentration (fraction of the grid box covered by ice) is given as a percent of its value in the base period climate – for instance, the regions near the edge of the ice region, with values exceeding 70 and 90%, represent areas that had some ice cover in the base climatology that have much less or almost none in the warmer climate. In the northern hemisphere winter season, this retreat of the ice edge, often by a few hundred kilometers, is the clearest change. The ice concentration near the pole is not reduced (although the thickness is). In midwinter, the surface air temperature increases may be large, but the temperature in normal climate is far below freezing in the deep Arctic. The northern hemisphere fall season, September–November (SON), on the other hand, has large changes throughout the Arctic Ocean, as sea ice formation is delayed. Enormous regions that had ice cover in the base climatology are ice-free in 2070–99. Even near the pole, regions of 30 to 70% reduction in ice cover may be seen, implying large areas of open water at the North Pole in this simulation. A similar pattern applies in the summer season. The reduction in ice grows roughly proportional to the warming, so earlier in the century similar patterns of reduction are found but with smaller amplitude. The southern hemisphere has roughly comparable reductions in the area of sea ice coverage.

As discussed in Chapter 5, sea ice interacts strongly with the surface heat balance. Ice does not conduct heat very rapidly compared with the rate of transfer between ice-free ocean and atmosphere. Warmer atmospheric temperatures slow the rate of ice formation, and with less sea ice, more heat is tranferred from the ocean to the atmosphere. Large changes in this seasonal heat transfer thus contribute to high latitude warming of surface air temperature. Regional details of the sea ice reduction can differ from model to model, partly owing to

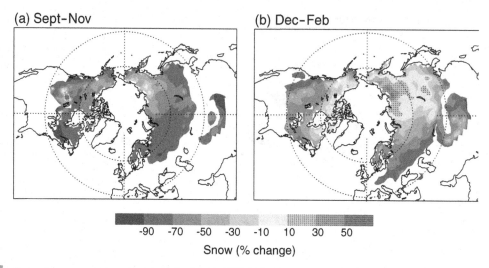

Fig. 7.11 Change in snow amount over land areas simulated in ECHAM5 for a 2070–99 average in the SRES A2 scenario, relative to a base climatology for 1961–90, shown as a percent of the base climatology value. (a) September–November and (b) December–February. Shading levels show reductions (negative values) exceeding 10%, 30%, 50%, 70% and 90% of the 1961–90 values, with stippling showing increases exceeding 10%, 30%, 50%. Changes are shown only for regions that had snow amounts exceeding 1 kg m^{-2} in the base climatology.

differences in some of the complex processes such as lead formation and ice dynamics outlined in Chapter 5, but there is substantial agreement on large reductions in the sea ice and on the seasonality of this reduction, as seen in the ECHAM5 example shown here.

Snow cover also experiences large reductions. An example is given in Figure 7.11, also from ECHAM5, for snow amount (mass of snow per unit area) in northern fall and winter. In winter (December–February), vast areas that had substantial snow mass in the base climatology, for instance much of the northern US and southern Canada, experience 50 to 90% reduction in the snow mass in the warmer climate in this simulation. Regions that had small amounts of snow in the base climatology are masked out in the plot so as not to be distracting, but the majority of these are essentially snow-free in the warmer climate. Melting snow can be a large source of soil moisture, so the snow reductions can also have considerable effect on surface water and energy balances in spring and summer seasons. Another measure of snow changes, relevant to the albedo feedback, is the area covered by snow. This tends to decrease most strongly in fall and spring associated with decreases in the length of the snow season. This can be seen strikingly in Figure 7.11, from the large areas with over 90% reduction of snow amount in the fall season. Note that in winter, there can actually be some areas with greater snow amount, owing to precipitation increases in areas that remain below freezing (in this example, in Siberia), but that in most places and seasons, the effect of warmer temperatures predominates.

7.4.2 Land ice

Land ice occurs in mountain glaciers, the parts of the polar ice caps that occur on land, and the vast *ice sheets* of Greenland and Antarctica. The retreat of mountain glaciers has been occurring over much of the globe, as is by now well known. These are among the indicators of the warming trend, and their retreat is expected to continue. The mountain glaciers contribute to near-term sea level rise while their long-term impact on sea level is limited by their mass. Changes in the Greenland and Antarctic ice sheets are of concern, although details of this are poorly constrained at this time. These contain enough ice to raise mean sea level by over 75 meters if it were all to melt, but this simple calculation is very unlikely to be relevant. Most of the Antarctic continent is cold enough to remain below freezing at the surface for anticipated warming values. In a warmer climate, snowfall at high latitudes is likely to increase, tending to increasing deposition onto much of Antarctica. Calculations of the *surface mass balance*, i.e. the net mass gained by precipitation onto the ice sheet, minus the ice mass lost by melting and sublimation, have been done by a number of models for both Antarctica and Greenland. Overall the surface mass balance is estimated to contribute a net increase in mass to Antarctica in the twenty-first century owing to precipitation increases, while for the Greenland ice sheet losses exceed gains. The net effect on sea level of land ice surface mass balance is estimated in IPCC (2007) at 0.04 m to 0.20 m through the twenty-first century (2090–99 relative to 1980–99, for the A1B scenario).[5] The mean contributions of Greenland, Antarctica and other land ice in this calculation are about 0.5, −0.7 (i.e. Antarctica gains mass) and 0.12 m, respectively. Note that this calculation does *not* take into account ice loss by the dynamical effect of ice flow down to the sea. The Greenland ice sheet is considered at risk for largely disappearing if high GHG levels are sustained over long periods of time. This could contribute up to 7 m of sea level rise over millennial time scales.

Ice sheet dynamics are complex and poorly understood. The rate of flow of the ice down hill toward the ocean depends on a number of factors. When parts of an ice sheet experience increased flow toward the ocean, "calving" of icebergs into the ocean occurs, with subsequent transport and melting. This in turn may affect the pressure on inland parts of the ice sheet and potentially the flow rate. Around parts of Antarctica, the ice tends to flow out into the surrounding ocean as ice shelves, part of which are floating. Warming can affect the stability of these. In 2002, an ice shelf known as Larsen B, an area of roughly 3000 square km or 7 billion tons of ice, shown in Figure 7.12, broke up in a period of months (from late January to early March). At the scale of the Antarctic continent this is small, but it is part of a series of retreats of ice shelves in the Antarctic peninsula over the last few decades. The loss of about 13 000 square km of ice shelf since 1974 has helped fuel concern regarding the rate of flow in parts of the Antarctic ice sheet. Radar monitoring of ice thickness and flow rates is expected to yield insight in coming decades. At this time, such events are useful as a cautionary illustration of the surprises that can arise in a complex system. Unlike many aspects of global warming, this was not predicted before it occurred.

Sea level rise in the global average has two main components: net melting of land ice, discussed above, and thermal expansion due to the warming of the ocean. Melting of sea ice has no effect on sea level because the floating ice already displaces the same volume of water.

In this century, sea level rise due to thermal expansion (mostly of the upper ocean) occurs at a modest rate. In GCMs in IPCC (2007), the range of sea level rise due to thermal expansion is about 0.13 to 0.32 m in the twenty-first century (1980–99 to 2090–99), based on the

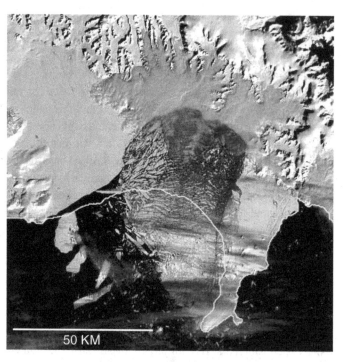

Fig. 7.12   Breakup of the Larsen B ice shelf in Antarctica, leaving thousands of icebergs adrift. The overlaid contour shows the ice edge prior to the collapse. Image courtesy of the National Snow and Ice Data Center, University of Colorado, Boulder, from the MODIS (Moderate Resolution Imaging Spectrometer) instrument on NASA's Terra satellite.[6]

A1B scenario. Similar results apply for the A2 scenario. In the short term, until 2020, the trend is about 5 to 20 mm per decade. Because mean sea level rise depends on warming the ocean and the flow of ice sheets, there is a substantial lag between the global mean surface warming and the eventual sea level rise. As discussed in Chapter 3, the contribution of the deep ocean depends on the surface warming being carried slowly through the deep ocean by the thermohaline circulation. This creates ongoing sea level rise for centuries even if greenhouse gases are stabilized. If GHG are stabilized at a level equivalent to 4 times preindustrial CO_2, the eventual rise from thermal expansion is estimated at 1 to 4 m (IPCC 2001, 2007).

When the contributions of land-ice surface mass balance and thermal expansion are combined, the range for the twenty-first century (under the A1B scenario) is estimated in IPCC (2007) at 0.21 to 0.48 m. To this must be added an uncertain amount due to ice sheet dynamical discharge. Even if greenhouse gas concentrations are stabilized at the end of the century, those concentrations imply a commitment to much more substantial sea level rise by both deep ocean warming and melting of the Greenland by sheet over subsequent centuries.

Regional variations in sea level also occur in climate models. These depend on changes in the ocean circulation, since if one region warms more than another there is a complex adjustment process involving ocean pressure gradients and currents. Changes in surface wind can likewise create changes in sea level from one part of an ocean basin to another. Such changes can be substantial, but there is little agreement on the spatial distribution of these among models.

7.4.3 Extreme events

As mean temperature rises, if the standard deviation of daily temperatures associated with weather variations remains similar, then the occurrence of temperature events currently considered extreme will become more common. The reasons for this are so simple that the effect is quite robust, despite it pertaining to uncommon events. Figure 7.13 illustrates this schematically. One can take a temperature that one currently considers extreme, i.e. occurs rarely for a given city, for instance 40 °C (104 °F). In Figure 7.13a, this small probability is seen as the small shaded area under the curve above 40 °C in the tail of the probability distribution. One then asks how often this temperature is exceeded in the warmer climate. Daily temperature probabilities often follow a roughly Gaussian distribution. When the distribution is shifted to center about a warmer mean, the probability of exceeding a threshold temperature increases quite quickly, as is shown in Figure 7.13b by the much larger shaded area now found above 40 °C. If 40 °C is the threshold for a heat wave that is considered a hardship, then there will be many more of these in the warmer climate. The same diagram could be applied to winter with a threshold of 0 °C to indicate why an increased number of thaws might be expected in a cold climate, such as a ski resort, where these events were formerly rare. On the beneficial side for those averse to cold, the occurrence of cold events below a certain threshold, for instance the number of frost days, which is important for agricultural activities, decreases in a corresponding manner.

The discussion above is for the case where the mean changes due to global warming while the variability about the mean, measured by the standard deviation, does not change much. This is roughly reasonable for temperature, since the variability is associated with movement of warm and cold air masses, and the equatorward temperature gradient tends

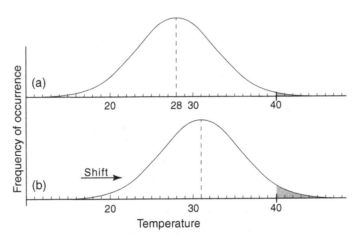

Fig. 7.13 Schematic of a simple way in which events currently considered extreme become more common with warming. (a) The probability density (or frequency of occurrence) of temperature values for a base climate period in some location where the mean is 28 °C. The shaded area gives the probability of temperatures warmer than 40 °C. (b) The probability distribution after the mean is shifted by global warming. The probability of temperatures warmer than 40 °C (area of shading) increases substantially.

not to change as much as the mean temperature. However, there are some effects that also change the standard deviation of temperatures. For instance, there is a tendency for the daily temperature range to decrease, as increased GHG slows nighttime cooling.

Because the atmosphere tends to hold more water vapor in a warmer climate, it is plausible that there may be changes in the probability of heavy rainfall events. However, current climate models do not reproduce the statistics of extreme events with high accuracy, so this remains in the higher uncertainty category at this time. A number of studies have indicated the possibility of increased intensity of tropical storms. Increases occur in both the sea surface temperature and the tropospheric temperature. The former favors and the latter disfavors storm intensification, so quantitative modeling is important. One recent study used a hurricane model nested in a large-scale environment estimated from climate models, using values approximating conditions 80 years into a 1% per year CO_2 increase run, i.e. shortly after CO_2 doubling (with the slowing of the warming by the oceans included).[7] This would roughly correspond to conditions typical of the 2050s relative to a 1961–90 average. They found that the fall of central pressure within the storm (a measure of intensity) increased by an average of about 14%, typical of an increase of about half a category on the Saffir–Simpson hurricane intensity scale. As a result there was a substantial increase in the number of category 5 storms.

7.5 Summary: the best-estimate prognosis

Table 7.1 indicates of some of the more certain effects that are predicted by climate models, adapted from the IPCC consensus reports. All models show substantial changes in climate, even though the changes vary from model to model on a sub-continental scale. The results from the models tend to have less agreement at smaller scales, so predictions for regions smaller than continental scale should be treated with great caution.

Figure 7.14 reinforces some of these in a schematic form. Changes in ENSO frequency and amplitude occur in some models. There is little agreement on the exact changes, but this indicates potential sensitivity. In a number of models an SST increase in the eastern Pacific cold tongue region occurs that looks somewhat like a sustained El Niño. This is best viewed as a change in the regional Pacific climate, although it appears to be affected by similar processes to the Bjerknes feedbacks discussed for El Niño.

Additional features to note include:

- Projected future temperature increases under all of the scenarios are considerably larger than the warming seen in the twentieth century.
- Temperature increases in southern Europe and central North America are greater than the global mean and are accompanied by reduced precipitation and soil moisture in summer.
- Changes in the day-to-day variability of weather are uncertain. However, episodes of high temperature will become more frequent in the future simply because of an increase in the mean temperature. There is some evidence of a general increase in convective precipitation.

Table 7.1 Summary of predicted climate changes, adapted from IPCC (1996, 2001, 2007).

Temperature	The lower atmosphere and Earth's surface warm.
	The stratosphere cools.
	The surface warming at high latitudes is greater than the global average in winter but smaller in summer. (In time-dependent simulations with a full ocean, there is less warming over the high latitude Southern Ocean).
	The surface warming is smaller in the tropics, but can be large relative to natural variability.
	For equilibrium response to doubled CO_2, the global average surface warming likely lies between $+2.0\,°C$ and $+4.5\,°C$, with a most likely value of $3.0\,°C$, based on models and fits to past variations.
	"Best-estimate" (IPCC 2007) temperature increase in 2090–99 relative to 1980–99 depends on future emissions. For A2 scenario $3.4\,°C$; B1 $1.8\,°C$; A1B $2.8\,°C$; A1FI $4.0\,°C$. Likely ranges estimated at 60% to 160% of these values (actual model ensemble ranges are smaller).
	Owing to the thermal inertia of the ocean, the temperature would increase beyond whatever time stabilization of greenhouse gases might be achieved.
Precipitation	The global average increases (as does average evaporation); the larger the warming, the larger the increase.
	Precipitation increases at high latitudes throughout the year; for equilibrium response to doubled CO_2, the average increase is 3 to 15%.
	The zonal mean value increases in the tropics although there are areas of decrease. Shifts in the main tropical rain bands differ from model to model, so there is little consistency between models in simulated regional changes.
Soil moisture	Increases in high latitudes in winter.
	Decreases over northern midlatitude continents in summer (growing season).
Snow and sea ice	The area of sea ice and seasonal snow cover diminish. Decrease in mountain glacier mass and extent.
Sea level	Sea level increases excluding rapid changes in ice flow for 2090–99 relative to 1980–99: for A2 0.23–0.51 m, B1 0.18–0.38 m.
	Even if greenhouse gases are stabilized, deep ocean warming creates ongoing sea level rise for centuries. If GHG are stabilized at a level equivalent to 4 times preindustrial CO_2, the eventual rise from thermal expansion is estimated at 1 to 4 m.
	Melting of the Greenland ice sheet will eventually add a substantial contribution, with additional contribution from the west Antarctic ice sheet, but uncertainties in ice sheet dynamics remain large.
	Substantial regional variations in sea level occur in climate models but there is little agreement on their spatial distribution.
Extreme events	Higher maximum temperatures and more hot days over land.
	Higher minimum temperatures over land, fewer frost days.
	More intense precipitation events.
	Increased typical cyclone intensity (considered likely; changes in frequency are considered uncertain).

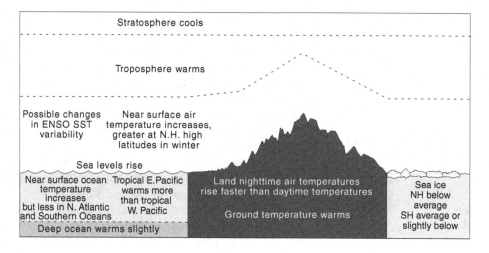

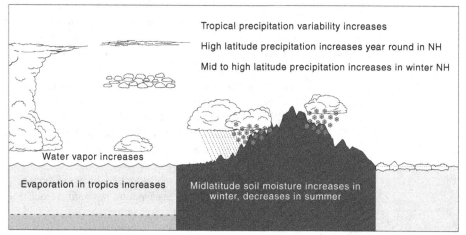

Fig. 7.14   Schematic summary of best-estimate climate changes due to greenhouse warming. NH, northern hemisphere; SH, southern hemisphere. Schematic adapted from IPCC (2001) and updated.

- Improved predictions of global climate change require better treatment of processes affecting the distribution and properties of cloud, ocean–atmosphere interaction, convection, sea ice and transfer of heat and moisture from the land surface. Increased model resolution will allow more realistic predictions of global-scale changes, and some improvement in the prediction of regional climate change.

7.6 Climate change observed to date

7.6.1 Temperature trends and natural variability: scale dependence

Because natural climate variability is so prevalent, what you expect to see in terms of trend relative to natural variations depends strongly on the spatial and time averages considered.

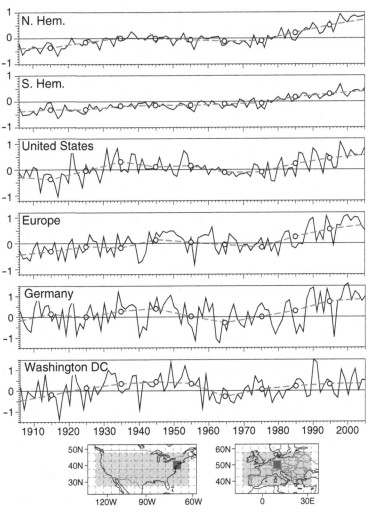

Surface temperature change relative to a 1960–90 base period for selected spatial averaging scales and regions. The time series marked US and Europe are for the 60 × 20 and 45 × 20 degree longitude by degree latitude averaging regions shown at the bottom; also indicated are the 5 × 5 degree regions labelled Germany and Washington DC. Annual and decadal averages are shown (decadal averages indicated by dots at the center of each decade). Data are from the University of East Anglia Climate Research Unit.[8]

Figure 7.15 provides an illustration of this, showing annual averages and decadal averages at the hemispheric scale, at a scale of 40 or 50 thousand km^2 (US and Europe) and at the scale of a 5×5 degree latitude–longitude box.

As one moves from larger to smaller averaging regions, the range of the natural variability increases. This holds for both the year-to-year variability and decade-to-decade variability. If one were to show temperatures for a particular month, for instance January values of each year for Washington, DC, the range would increase even more (the y-axis scale would increase from 1 °C to about 8 °C). At smaller scales, the variability is largely due to variations in energy transport as changes in the flow occur by random weather or climate variations. For instance, a slight increase in cold air transports from the north can easily have a large impact in a particular region, for a particular year. Because of energy conservation, as considered in Chapter 3, if one averages over many 5×5 degree boxes, the horizontal transports from one box to another must cancel. The energy transports through the edges of the larger averaging region still have effects, but as the area of the region increases, the relative importance of transports through the edges decreases. When one considers global averages, effects of horizontal transports must entirely cancel, and the only transports that can affect the surface temperature are exchanges with the deep ocean. Thus at the global scale, the vertical energy exchanges by radiation involved in the greenhouse effect are visible even on a decade-by-decade or year-by-year basis. At the local scale the greenhouse effect is still operating, but in a particular year, or even a particular decade, transports may cool one region and warm another. A consequence of this is that one can find regions or cities where, looking at the history of temperatures, the local residents might well ask, "What warming?" For instance, the time series for Washington DC does not have a significant trend. Even in a region such as Germany, where the decadal trend of the past 30 years is consistent with the larger-scale trend, particular cold years (such as 1996, seen as a downward spike in the figure) can be found because of the natural variability.

As one moves into the future in model simulations, this behavior continues: at the regional scale, natural variability continues to produce variations about an increasingly warm mean state. The impacts are thus best phrased in terms of shifts of the probability distribution, as discussed in Figure 7.13. For larger averaging regions, or longer time averages, the probability distribution shifts in the same way but has much smaller standard deviation, so the probability of being close to the mean shift is much higher.

7.6.2 Is the observed trend consistent with natural variability or anthropogenic forcing?

To determine if the observed trend is attributable to anthropogenic influences, one needs to know if such a trend can easily occur just by random sampling of the natural variability. If one had a very long time series of natural variability of surface temperature, one could find the probability that a given segment of 50 or 100 years would have an upward trend of the magnitude observed in the twentieth century. From the observed time series, we do not have multiple examples of 50- or 100-year trends to establish range for decadal and centennial scale natural variability, but we do have such information from the models. One method is to compare the range of variability range from an ensemble of model runs with natural forcing only to an ensemble of runs that also have the observed twentieth-century anthropogenic forcing (GHG + aerosol). Figure 7.16 shows an example of this. The range in the natural forcing runs comes both from specified forcings (volcanoes, changes in solar input, etc.) and climate variability (like El Niño or variations in the thermohaline circulation) that occurs even for constant radiative forcing. While the range of natural variability is substantial, the warming trend in the runs that include anthropogenic forcing emerges from this range late

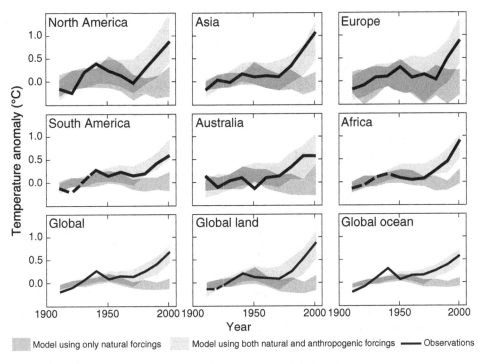

Fig. 7.16 Observed surface temperature decadal time series for several averaging regions (black line; dashed in decades with fewer observations), compared to the range of climate model response to estimated observed forcings for natural forcings only (medium shading), and natural forcings plus anthropogenic forcings (light shading). The shaded ranges are the 5 to 95% interval of the model ensemble, i.e. 10% of the models exceed this range in a given decade. Darkest shading where the two ranges overlap. After Hegerl *et al.* (2007).

in the twentieth century. This holds true separately for averages over land and over ocean regions, and for individual continents.

The terms *detection* and *attribution* are often used in the climate science community to describe different aspects of this problem. Roughly speaking, detection is used when a trend is found to be statistically unusual compared with natural climate variability, such as when the observed curve emerges from the range of variability in the simulations with natural forcings only. Attribution is used when you can provide concrete evidence that a trend is consistent with the response to anthropogenic forcings, such as the comparison to the anthropogenic-forcing runs, as well as being unlikely to be due to natural variability. Detailed statistical methods have been developed to make quantitative statements about attribution. Sophisticated techniques known as "fingerprinting" methods use weighted spatial averages chosen such that they are closely associated with the spatial pattern predicted for the warming (the "fingerprint"), while attempting to minimize the effect of spatial patterns of natural variability. However, the separation of the warming trend in models and observations from the natural variability range in Figure 7.16 conveys the essence of the argument that the observed warming is very likely attributable to human influence.

7.6.3 Sea ice, land ice, ocean heat storage and sea level rise

In section 6.8 and section 7.3, we discussed the role of ocean heat storage in slowing the warming initially; and in section 7.4, we discussed projections for future changes in ice and sea level. Here we consider what has been seen so far in observations. The time series of observations with high spatial coverage tends to be shorter for these quantities than for surface temperature. Thus at this point they should be viewed as corroborating evidence of the warming trend. It is also worth noting that these effects were predicted prior to the data sets being analyzed. Taken as a whole, they put together a picture of the warming trend generally consistent with what models simulate in the twentieth century and predict over the next decades.

Satellite observations of sea ice extent are available since 1978. Figure 7.17a shows anomalies of the annual average area covered by sea ice in the northern hemisphere over that time period. The decreasing trend amounts to about 3% of the total area per decade. Summer sea ice extent in the northern hemisphere has been decreasing by about 7% of the total area per decade. While it is difficult to do a formal statistical attribution on such a short time series, the decrease appears consistent with the warming in the northern polar regions (about 1 °C over the period of sea ice observations). Antarctic sea ice extent does not have a significant trend, consistent with small changes in the surface temperature in the southern polar regions.

Although it is often easy to see that a glacier has retreated, exact mass balances for glaciers are difficult to establish. The substantial uncertainties are due to difficulties in measuring thickness; it is typically much easier to measure changes in ice edge than ice volume.

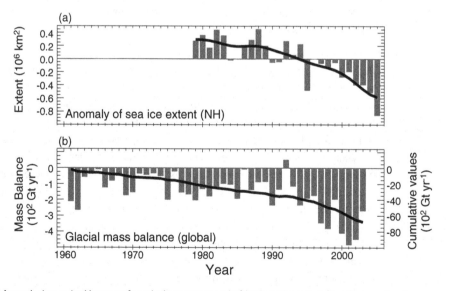

Fig. 7.17   (a) Anomaly time series (departure from the long-term mean) of Arctic sea ice extent. Bars show yearly values and the solid line shows a decadal average. (b) Global glacier mass balance. Yearly mass balance (left axis, negative denotes ice loss) is shown as bars; the solid line gives the cumulative global glacier mass balance (right axis). After Lemke *et al.* (2007).

Historical records exist for about 300 glaciers, supplemented by altimetry measurements in more recent times. Statistical techniques are then used to attempt to estimate regional and global mass balances. Figure 7.17b shows the results of such efforts. While the details should be viewed with considerable caution, the conclusion of cumulative ice loss is likely robust. As well as an indicator of warming, land ice has contributed to observed sea level rise. Glaciers are esimated to have contributed about 3 to 7 mm per decade, and the Greenland and Antarctic ice sheets about 4 to 7 mm per decade during 1961 to 2003.

In section 6.8, we discussed how ocean heat storage is important in delaying the warming associated with increasing greenhouse gases. Can one detect in observations the associated trend in ocean heat content? Figure 7.18 shows observational estimates of the change in heat content of the upper ocean. The change in heat content is given by (temperature change)$\times$(density)$\times$(heat capacity), vertically integrated through the upper ocean where most of the warming occurs (in this case through the upper 700 m), and integrated over the entire area of the ocean. The heat content change occurs in response to accumulated imbalance in surface heat flux or changes in the transport to and from the deep ocean. Natural variability of the heat content is substantially due to natural variations in the exchanges with the deep ocean, as occurs also in climate models. There are caveats because the observing system for subsurface temperature has imperfect coverage, depending largely on instruments lowered from ships until recent decades. Nonetheless, a slow upward trend is clearly visible among the natural variations. The size of this trend is approximately of the size predicted by climate models. For reference, it is simple to estimate the size of the heat content trend that would go with each $\mathrm{W\,m^{-2}}$ of surface heat flux imbalance. Multiplying by the surface area of the ocean ($3.6 \times 10^{22}\,\mathrm{m^2}$), $1\,\mathrm{W\,m^{-2}}$ converts to $1.1 \times 10^{22}\,\mathrm{J\,yr^{-1}}$. Thus even without a climate model, you can see that the magnitude of the trend is consistent with the magnitude that we have been discussing for the radiative imbalances.

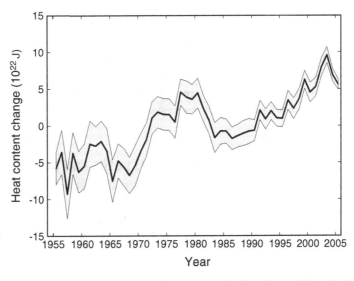

Fig. 7.18   Observed change in global annual ocean heat content for the ocean surface layer down to 700 m depth. After Bindoff *et al.* (2007); data from Levitus *et al.* (2005). The shaded area denotes an estimated 90% confidence interval.[9]

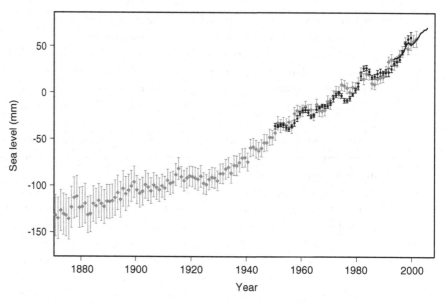

Fig. 7.19 Observed sea level change (relative to 1961–90) from coastal tide gauge measurements estimated by two methods (dots and diamonds; error bars denote 90% confidence interval). The black curve gives sea level change estimated by satellite altimetry. After Bindoff *et al.* (2007).[10]

The warming of the upper ocean seen in the heat content change also contributes to sea level rise. The contribution of thermal expansion to sea level rise for the period 1961 to 2003 was about 3 to 5 mm per decade.

Figure 7.19 shows estimates of the sea level change by three different methods. The longer time series are from coastal tide gauge measurements, with a curve for recent years added from satellite altimetry. Over the period 1961 to 2003, the trend in global mean sea level rise is estimated at 13 to 23 mm per decade (from the tide gauge data). This is higher than the sum of estimated trends from land ice and thermal expansion noted above, suggesting, for instance, a possible underestimate of the land ice contribution. Over the most recent decade, land ice and thermal expansion contribution estimates exhibit better agreement with sea level observations.

7.7 Emissions paths and their impacts

All of these climate changes depend on the future concentrations of greenhouse gases and other anthropogenic forcings. We thus return to the forcing scenarios introduced in section 7.1 to develop a sense of how different levels of greenhouse gases, determined by future socio-economic or political factors, and societal choices regarding environmental values and economic values, might affect the climate system response. Figure 7.20a shows the global average near-surface temperature change from an ensemble of climate models for

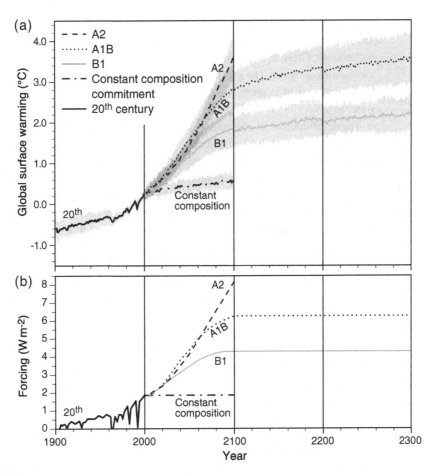

Fig. 7.20   (a) Global average change of near-surface temperature (relative to a 1980–99 base period) for selected scenarios; and (b) the radiative forcing of these scenarios. The response to estimated observed natural and anthropogenic forcing is shown in the twentieth century. An additional twenty-first century experiment, with the forcing constant at the year 2000 value, is shown. For the twenty-first century, the A2 scenario, examined in previous sections, is compared to the B1 scenario, which assumes that green technology and population reductions cap forcings by the end of the century; and to A1B which assumes a balance of green technology with fossil fuels. The A1B and B1 scenarios are continued beyond the year 2100 with forcing kept constant at the year 2100 value. Lines show the multi-model means, shading denotes the 1 standard deviation range. Discontinuities between different periods are simply due to differences in the number of models that have run a given scenario for each period and scenario. After IPCC (2007).[11]

several forcing scenarios. The global average radiative forcing associated with each scenario is given in Figure 7.20b. In the twenty-first century, the warming resulting from each of the A2, A1B, and B1 scenarios are compared.

Recall that in the A2 scenario it is assumed that only developed countries move towards lower emission energy sources, while the A1B and B1 scenarios assume a great reduction in regional income differences, and that the global population stops growing in mid-century and decreases thereafter. In the A1B scenario, a balance of fossil fuel and green technologies is assumed, whereas in the B1 scenario, there is an emphasis on global economic social and environmental sustainability, including increased economic equity and resource-efficient technology. These societal directions and choices lead to the considerable differences in the emissions pathways. The greenhouse gases and resulting radiative forcing increase slightly less at the end of century in A1B than in A2, and increase much less in B1 than in the other two.

These differences in emission path are directly reflected in the amplitude of the global warming over the twenty-first century, with larger warming in the A2 scenario than in the other two scenarios shown here. To be more precise, the global-average temperature response scales with the amplitude of the forcing. The larger the radiative forcing, the larger the temperature response. This relationship is to a good approximation linear in the range examined, consistent with the approximations discussed in Chapter 6. In turn, most physical climate effects are approximately proportional to the global-average temperature. Even large-scale measures of precipitation change increased approximately proportional to the temperature change. There has been considerable examination of whether certain parts of the physical climate system might exhibit "threshold" behavior. That is, are there certain threshold values of the forcing above which a relatively sudden, nonlinear, possibly irreversible change might occur? For instance, the thermohaline circulation under some conditions (that might be relevant to paleoclimate), such as large freshwater input into the sinking regions, can undergo a disproportionate decrease in strength in climate models. For the range of forcings considered in IPCC (2007), this does not happen in the parts of the physical climate system included in these models. This does not exclude the possibility that ice sheet dynamics or ecosystem response, which are not included in these models, may exhibit threshold-type behavior.

Also shown in Figure 7.20 are simulations in which the radiative forcing is set to a constant in a certain year, designed to show the global warming commitment (or to use a more precise term, the "constant composition commitment," since the atmospheric composition including GHG and aerosols is held constant). As discussed in section 6.8, this type of experiment with constant radiative forcing is used to see the rate of ongoing warming implied by the greenhouse gases that are already present in the atmosphere at a given time. One such experiment is shown for the case where greenhouse gases are held constant at today's values (i.e. human emissions suddenly cease); the other two experiments cap greenhouse gases at their values in 2100 for the A1B and B1 scenarios, respectively. Because the heat capacity of the ocean slows the warming, the surface temperature is not in equilibrium with the forcing at a given time. The warming continues, albeit at a lower rate of increase, into the future. The involvement of the deep ocean, which takes a very long time to equilibrate, results in an ongoing slow increase over centuries. The surface warming trend due to deep ocean lags, however, is substantially smaller than the rate of increase that occurs while the radiative forcing is increasing strongly with time. Thus reducing emission rates not only reduces the long-term climate change, but reduces the rate of increase, which can give human and natural systems a longer time scale to adapt.

To provide a sense of how the differing degrees of warming associated with these scenarios look geographically, Figure 7.21 shows the annual average surface temperature change for mid and late twenty-first century. The surface temperature maps are from a multi-model

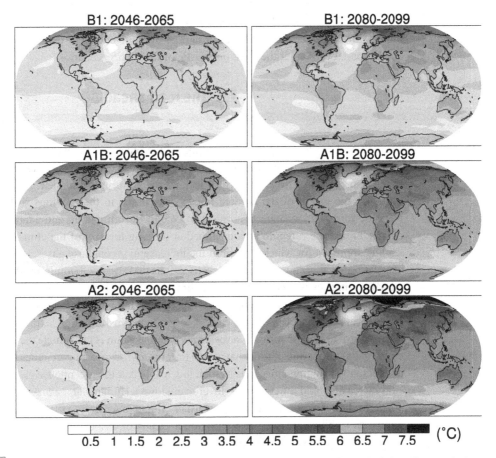

Fig. 7.21 Simulated annual-average surface air temperature change (relative to 1980–99) for the A2, A1B and B1 scenarios in mid and late twenty-first-century time periods: 2046–2065, and 2080–2099. A multi-model ensemble mean is shown, i.e. each point in the map is an average over the response of multiple models. Shading gives the temperature change in intervals of 0.5 °C. Adapted from Meehl *et al.* (2007).

ensemble mean; that is, the average is over all of the models that have done the simulation with the same scenario. This tends to capture the aspects that agree among the models, as discussed in section 7.3.2. Early in the century (not shown), the warmings tend to be similar among the different scenarios because the radiative forcings are relatively similar. By mid-century, the temperature change under the lower emission B1 scenario is visibly lower than the warming under the other two scenarios. For instance, a large part of the United States is about 1 °C warmer in the A1B scenario than in the B1 scenario (e.g. 2.5 °C versus 1.5 °C). In the late twenty-first century, temperature change in the A2 scenario exceeds 4.5 °C over much of the interior of North America. Under the B1 scenario, the same region has warmed by roughly half as much. This is consistent with the radiative forcing being only about half as large under the B1 scenario in the late twenty-first century.

Emissions pathways thus have a strong direct effect on the size of the anthropogenic warming and the rate at which the warming occurs. This in turn affects the extent and rate at which humans and ecosystems must adapt. The best estimates from climate models indicate that these changes will be substantial over the course of the next several decades and beyond, even under the most optimistic of the SRES scenarios.

7.8 The road ahead

The SRES scenarios used in the modeling of the physical climate system were excluded from explicitly considering accords to limit greenhouse gases. Societal choices to be made in the near future regarding action (or lack therof) to limit greenhouse gases will strongly influence the extent of the climate impacts discussed in the previous sections. Societal response to climate change falls into two categories: mitigation and adaptation. *Climate change mitigation* refers to actions aimed at limiting the size of the climate change – the leading approach for which is to limit the increase in greenhouse gases. *Adaptation*, when used for policy measures, usually refers to actions that attempt to minimize the impact of the climate change as it occurs. Of course, climate change adaptation can also contain starker aspects, since it can include substantial migration of human populations due to agricultural or sea level rise impacts, or in the case of ecosystem transformation, extinctions of species. We now turn to a set of *mitigation scenarios* where the aim is to estimate the greenhouse gas emissions as a function of time that would lead to *stabilization* of greenhouse gases – that is, of eventually bringing emissions to such low levels that the greenhouse gas concentrations stop increasing, and a new, stable level is achieved. Humans and ecosystems would then adapt, for better or for worse, to this warmer climate over time. Mitigation scenarios (also known as stabilization scenarios) provide a means of associating the expected levels of climate change once greenhouse gases are stabilized with choices made now or in the future regarding emissions reductions.

Figure 7.22 shows examples of such mitigation scenarios, divided into six categories that roughly span the range of stabilization outcomes. Each curve shows the center of a range of possible emissions scenarios that lead to eventual greenhouse gas concentrations in a specific range noted on the right hand side of the diagram. These scenarios take into account greenhouse gases other than carbon dioxide, but the focus is on CO_2 for two reasons: it is the largest contributor to greenhouse warming; and the costs of conversion from fossil fuel burning to alternative energy have been the sticking point in limiting climate change over the decades of discussion so far. The following main points can be noted:

- The highest emissions scenario categories, IV–VI, have the emissions continuing to increase over the first decades of this century in a manner roughly consistent with recent trends and with projections that do not include strong societal action.
- The different curves level off, and then begin to decrease at different times during the century.
- Recall that for a long-lived gas like carbon dioxide, a constant value of emissions on this plot yields an ongoing increase of the concentration. When emissions are increasing, the CO_2 concentration is increasing at an ever faster rate; when emissions level off, CO_2 concentration keeps going up at a fixed rate; when emissions are decreasing, concentration

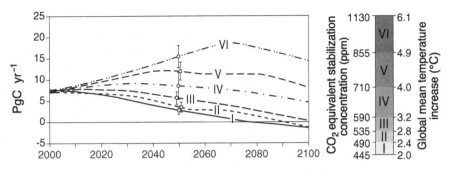

Fig. 7.22 Emissions pathways of carbon dioxide for six categories of mitigation scenario that lead to stabilization of greenhouse gases at different levels. The mitigation scenarios considered in Barker *et al.* (2007) include uncertainty estimates for the relationship between emissions in the concentrations of greenhouse gases that remain in the atmosphere. These are summarized here as error bars shown for the year 2050 (slightly offset for visibility). The center value of the range of scenarios in a given category is given for all years, for categories I–VI. The eventual value of greenhouse gas concentration when stabilization is achieved is given on the right hand bar as equivalent CO_2 (see text), with the range for each category shown as different shading, marked with the category number I–VI. The corresponding range of global average surface temperature for each category (using approximate best-estimate climate sensitivity of $3°C$ is also shown. Values condensed from Barker *et al.* (2007).

is still increasing, but less quickly. Stabilization of CO_2 concentrations is achieved only when emissions reach very low values (for some of these curves this does not happen within the century).

- The lower curves level off and then begin decreasing in the early decades – i.e. corresponding to societal action that begins now.
- By 2050, CO_2 emissions have been reduced relative to their values at the start of the century by 50% to 85% for category I, and 30% to 60% for category II. For category III, the range is from a 30% decrease to a 5% increase in 2050, while category VI permits a 90% to 140% increase.
- If emissions cannot be brought down quickly enough, the CO_2 overshoots the stabilization target. To bring the CO_2 concentrations back down to the target, negative emissions are then required – i.e. investment in methods for actively removing carbon dioxide from the atmosphere. This is assumed in the lowest two categories near the end of the century. The alternative would be to bring down emissions sooner, but this seems unlikely to be politically feasible given the track record so far.
- Other greenhouse gases increase and then are controlled along with CO_2 in these scenarios. These gases are converted to equivalent values of CO_2 in terms of radiative forcing impact over 100 years. The equivalent CO_2 concentration at stabilization is listed for each category. The categories correspond to greenhouse radiative forcing levels at stabilization of: 2.5–$3.0 \, W \, m^{-2}$ for category I through 6.0–$7.5 \, W \, m^{-2}$ for category VI. For comparison to scenarios examined earlier in the chapter, the radiative forcing for the A2 scenario over the last three decades of the century roughly compares to category VI (although recall that temperature has not risen to equilibrium in those transient runs). The average radiative forcing taken over the three decades at the end of the century for the

A1B scenario lies at the boundary between categories V and VI ($6\,\mathrm{W\,m^{-2}}$); that of B1 close to the boundary between III and IV ($4\,\mathrm{W\,m^{-2}}$).[12]

- A range of global-average surface temperature change (relative to preindustrial temperatures) corresponding to the new climate equilibrium for each CO_2 stabilization concentration is given based on an approximate best-estimate $2 \times CO_2$ climate sensitivity of $3\,^\circ\mathrm{C}$. Recall that the predicted global-average temperature change also has a range of uncertainty for a given greenhouse gas concentration. A rough estimate of this uncertainty would be to multiply the temperature axis by a factor of 0.7 to 1.4 for the range of equilibrium climate sensitivity seen in Table 7.1. The temperature evolution as a function of time would look generally similar to the idealized stabilization experiments seen in Figure 7.20 (adding about $0.6\,^\circ\mathrm{C}$, since changes in that figure were shown relative to 1980–1999).

Recalling that the climate impacts tend to roughly scale with the global average temperature, we have a qualitative sense of how the costs of adaptation would change among the different categories of emissions pathways. However, the state of the science is far from being able to provide quantitative dollar values for an economic cost–benefit analysis, especially considering that impacts on ecosystems are difficult to quantify both scientifically and economically, and that the costs are likely to be unevenly distributed among different regions and economic groups. One rule of thumb, espoused for instance by the Group of Eight industrialized economies (G8) in 2009, with a long history of prior discussion, is to aim to limit warming to $2\,^\circ\mathrm{C}$ above preindustrial temperatures. Emissions in 2050 provide a good indicator of whether this goal is likely to be met. The category I mitigation scenario range of 60% to 80% reduction of emissions by 2050 relative to 2000 values has thus been used in guiding target values in some attempts at climate change legislation.[13]

The technologies that are expected to come into play in reducing emissions include the following.

- Efficiency and conservation.
- Wind power (with geographical and energy storage constraints).
- Solar power (including solar thermal collectors and photovoltaic cells).
- Nuclear power (noting the environmental trade-off of nuclear waste storage instead of CO_2 emission).
- Hydroelectric power.
- Biofuels (including existing production of ethanol from sugarcane or other crops or crop by-products, and development of non-food sources such as perennial grasses or algae).
- Fossil fuel (primarily coal) power generation with carbon capture and storage (in which CO_2 is captured from the powerplant emissions stream, compressed, and then injected back into geological formations, typically coordinated with fossil fuel extraction).
- Ecosystem/agricultural management (including reduction of deforestation, and agricultural tillage, in which straw and other agricultural by-products are tilled into the soil to store carbon).

All of these technologies are in production in some form, but would have to be refined and scaled up enormously to meet the energy demand. It is encouraging that new investment in sustainable energy has increased rapidly over the period from 2002 to 2008 (from roughly 22 to 155 billion US dollars). However, the scale of the required infrastructure change is impressive. One convenient way of visualizing contributions to the change in energy supply, provided by S. Pacala and R. Socolow,[14] is as a "wedge" in which a low-emission technology grows from a small contribution today to a contribution that displaces $1\,\text{PgC}\,\text{yr}^{-1}$ of fossil fuel emissions 50 years from now. In the process, 25 PgC of emissions are prevented from going into the atmosphere. The following are examples of how much a particular technology would have to be scaled up over 50 years to provide a contribution of this size.

- Doubling the fuel efficiency of cars (assuming there are four times as many cars in 50 years, each traveling similar mileage to the average today).
- Cutting in half the average mileage each car travels (e.g. replacing trips by low-emission transportation, telecommuting, etc.).
- Energy-efficient buildings (reducing emissions associated with heating, cooling, lighting, refrigeration by 25% including in developing countries).
- Increasing efficiency of coal-based electricity generation from 32% to 60% (assuming twice current capacity in 50 years, and that efficiency increases to 40% would occur without carbon budget considerations).
- Wind power substituted for coal power, adding 2 million 1-megawatt-peak windmills (50 times current capacity).
- Photovoltaic power increased to about 700 times the current capacity to substitute for coal power (requires about 2–$3\,\text{m}^2$ of solar array per person).
- Nuclear power substituted for 700 GW of coal power (a doubling of current capacity).
- Biomass fuel production scaled to roughly 100 times the current Brazil or US ethanol production (requires about 1/6 of world cropland).
- Carbon capture and storage implemented for 800 GW of coal plants. In terms of storage, this requires that CO_2 injection increase to a factor of 100 times today's injection rates or the equivalent of 3500 times the injection by Norway's Sleipner project in the North Sea.
- Cease tropical deforestation completely plus double the current rate of new tree plantations.
- Conservation tillage applied to all cropland (10 times current).

The numbers in each of the above examples are rough, but they give a sense of the very sizable scale-up required for each sector. Using the fact that each is estimated to displace 1 PgC per year emissions in 50 years, we can estimate how many of these are required to move from an emissions path like category VI in Figure 7.22 to one that results in lower levels of CO_2. Carbon dioxide emissions in category VI increase by between 7 and $8\,\text{PgC}\,\text{yr}^{-1}$ over a 50-year period in the first half of the century. In other words, seven to eight of the above scale-ups of low emissions technology must be implemented just to keep emissions rates close to present values in the face of increasingly energy-intensive economies and population growth. Carbon dioxide would continue to increase in the atmosphere at the same rate as today. To actually slow the rate of increase of CO_2 concentrations would require additional wedges. Category I requires emissions to decrease by about 4 to $5\,\text{PgC}\,\text{yr}^{-1}$ in 50 years relative to today, or by about $12\,\text{PgC}\,\text{yr}^{-1}$ relative to the business-as-usual emissions in category VI. This implies that the equivalent of approximately 12 of the above items must be implemented over the coming five decades for the lowest category of stabilization target, roughly consistent with 2 °C warming above preindustrial values. Since only 11 items are listed, this implies doubling one of the items, or adding another technology at comparable scale. This underlines the likelihood that it is not a question of which technology should be scaled up, but rather that an 'all of the above' approach will be required if greenhouse gas concentrations are to be stabilized.

The prospects for reaching these lower stabilization levels may appear daunting. It is thus worth underlining that in the parts of the physical climate system currently included in climate models, the consequences of additional warming tend to increase gradually. This is not necessarily true for a number of subsystems relevant to human activities, including ice sheet flow, certain fisheries and agricultural impacts, and local ecosystems, where there is widespread concern that thresholds may be reached where the response to an additional increment of warming is not a further small increment of change, but a qualitative change to a new regime of behavior in that subsystem, such as the collapse of particular fisheries, or a substantial species loss in the ecosystem.[15] Nonetheless, if present societal choices should make achieving the lowest stabilization levels impractical, there would still remain ample reason to work instead toward the next higher categories of stabilization curve. Because there is substantial cost to changing our current infrastructure towards lower emissions technology, the guessing game society is engaged in is how much to invest now in converting towards lower emissions, compared with the costs down the road in terms of the impacts of different levels of climate change. Climate science will be an important tool whichever path is taken.

The student should now have an understanding of the processes in the physical climate system that lead to both climate variability and anthropogenic change, and of the strengths and limitations of the climate models that are used to predict these. This serves as a background for those who will work at the rapidly changing interface between climate change, human and ecosystem impacts, and adaptation strategies, and for all who must help to decide on mitigation actions.

Notes

1 Radiative forcing changes are estimated by Shine and Forster (1999) as follows: those due to greenhouse gas concentration changes from preindustrial to present in Figure 7.1 are estimated based on changes in concentrations of carbon dioxide, methane, nitrous oxide, CFC-11 and CFC-12 given in IPCC (1996), and the University of Reading radiation schemes and climatology. Those for the direct effects of sulphate aerosols are from the Langner and Rodhe (1991) "slow oxidation" case for the preindustrial to present day sulfate aerosol changes and the UK Meteorological Office Unified Model (Haywood *et al.* 1997).

2 Model data are from the Coupled Model Intercomparison Project 3 (CMIP3) archive hosted by PCMDI at the Lawrence Livermore National Laboratory – see Chapter 5 endnotes for a full acknowledgment. In Figure 7.4 seven models are selected for clarity, choosing the early contributors to the PCMDI CMIP3 archive (those shown contributed A2 runs earlier than May 2005). The model centers and names given by acronyms are as follows: CCCMA-CGCM3.1, Canadian Centre for Climate Modelling and Analysis Canadian Global Climate Model (version 3.1); CNRM-CM3, Centre National de Recherches Meteorologiques, France; CSIRO-Mk3.0, The

Australian Commonwealth Scientific and Industrial Research Organization (CSIRO) atmospheric research Mark-3.0 climate model; GFDL-CM2.0, 2.1, Geophysical Fluid Dynamics Laboratory (Princeton, USA); GISS-E-R, NASA Goddard Institute for Space Studies; NIES-MIROC-3.2 medres, National Institute for Environmental Studies, Japan, version 3.2, medium resolution; MPI-ECHAM5-OPYC, From the Max Planck Institute for Meteorology in Hamburg, Germany (known as ECHAM since the atmospheric component was originally developed by combining the ECMWF weather prediction model dynamical code with climate model parameterizations developed in Hamburg. This atmospheric model is coupled to an ocean GCM known as OPYC); MRI-CGCM2, Meteorological Research Institute, Japan (version 2, or 2.3.2 to be precise); NCAR-CCSM3 and NCAR-PCM1, National Center for Atmospheric Research (NCAR), Community Climate System Model (version 3.0) and Parallel Climate Model; UKMO-HadCM3, Hadley Centre for Climate Prediction, Met Office (United Kingdom) model, version 3.

3 The models included in the multi-model ensemble average of Figure 7.9 are UKMO-HadCM3, GFDL-CM2.0, GFDL-CM2.1, MRI-CGCM2, MPI-ECHAM5-OPYC, NCAR-CCSM3, NIES-MIROC-3.2, CSIRO-Mk3.0, CNRM-CM3, NCAR-PCM1.

4 More detail on the lapse rate feedback: recall from Chapter 3 that where moist convection occurs, the decrease of temperature with height tends to be smaller. Since radiative balance occurs at the top of the atmosphere, it is the upper tropospheric temperature which must warm sufficiently to restore balance in the infrared terms. If the lapse rate is large, then the surface warms more for a given upper tropospheric warming.

5 Ranges of land ice contributions to sea level rise are from Table 10.7, p. 820, of Meehl *et al.* (2007).

6 Areas of Larsen B and other ice-shelf ice loss from the National Snow and Ice Data center (http://nsidc.org).

7 Knutson and Tuleya (2004).

8 Temperature data set described in Brohan *et al.* (2006).

9 Estimates from other authors were compared to the Levitus *et al.* (2005) data in Bindoff *et al.* (2007).

10 Bindoff *et al.* (2007) sea level rise analysis includes reconstructed sea level fields relative to 1961–90 (updated from Church and White, 2006); coastal time gauge measurements relative to 1961–90 (from Holgate and Woodworth, 2004); satellite altimetry relative to 1993–2001 (Leuliette *et al.*, 2004). The number of coastal tide gauges decreases as one moves back in time to 140 in 1950, and just under 50 in 1900, primarily in the northern hemisphere.

11 The number of models in the multi-model ensemble mean and range in each century and scenario is as follows: 20th century 23; 21st century A2 17; A1B 21; B1 21; constant composition commitment 16; A1B commitment 22nd century 17, 23rd century 12; B1 commitment 22nd century 16, 23rd century 10. Linear trends from the corresponding control runs have been removed from these time series. IPCC (2007).

12 The values in Figure 7.22 should be taken as illustrating the approximate center of a category of emissions curves, rather than an exact emission curve. To reach the same stabilization value a curve that decreases less quickly early in the century must decrease more quickly later in the century. The bar at the right hand side should be viewed as a rough guide, since uncertainties in stabilization value and associated temperature cannot be expressed on a single scale. Barker *et al.* (2007) use $3\,°C$ as the approximate best-estimate doubled CO_2 sensitivity and $3.7\,W\,m^{-2}$ as the corresponding radiative forcing. These differ slightly from values from different estimates used in Chapter 6 but have approximately the same $0.8\,°C$ warming per $W\,m^{-2}$ radiative forcing.

13 Meinshausen *et al.* (2009) provide additional references and assess the probability of exceeding $2\,°C$ and the usefulness of 2050 emissions as an indicator. Examples of legislative emission targets include California legislation setting targets for reducing greenhouse emissions to 1990 levels by 2020 (AB32), and to 80% below 1990 levels by 2050 (CA Executive Order S-3-05), or the Waxman–Markey bill that passed the US House of Representatives in June 2009 (but with

unclear fate in the US Senate at this writing) which used reduction targets for carbon emissions from large sources at 17% below 2005 levels by 2020 and 83% below 2005 levels by 2050. The G8 declaration released at the L'Aquila, Italy summit, July 2009, citing the aim of not exceeding 2 °C temperature increase, supported a goal of developed countries reducing emissions by 80% by 2050, and a "willingness to share with all countries" a goal of at least 50% reduction – a goal that the developing world has yet to agree on. The European Union has in principle been in favor of the 2 °C objective for several years; the European Parliament used the 80% reduction by 2050 target in a non-binding resolution, February 2009, and reiterated it in European Presidency Conclusions in October 2009, just prior to the Copenhagen Conference that failed to produce a binding international climate accord.

14 Source for new investment in sustainable energy: UN Environmental Program "Global Trends in Sustainable Energy Investment 2009." The discussion of low emission technology scale-up follows Pacala and Socolow (2004), which provides a succinct overview of the contributions of various technologies. Their term, "wedges," for the growth in the sector is adopted for the very brief presentation here. Barker *et al.* (2007) provide more extensive coverage within the reach of a determined undergraduate.

15 Parry *et al.* (2007) provides an extensive summary of impacts and adaptation strategies, including effects on ecosystems. Their estimates of ecosystem impacts ranked by degrees above preindustrial of global-averaged surface warming include: for 1–2.5 °C, polar ecosystems increasingly damaged, 10–15% of species committed to extinction, coral reefs bleached, and major loss of habitat or species in regions such as South Africa, Queensland and the Amazon rainforest; for 2.5–3.5 °C, coral reefs overgrown by algae, major changes in polar systems, globally, 20–30% of species committed to extinction, over 15% of global ecosystems transformed; for 3.5–4.5 °C, over 40% of ecosystems transformed, extinction of 15–40% of the endemic species in global biodiversity hotspots. These potential impacts are not treated in the main text here because the author does not have the expertise for independent evaluation of these estimates, but they exemplify current work at the interface between climate change science and ecosystems impacts.

Glossary

This contains words or acronyms that are used several places in the text, or to review or provide definitions of mathematical terms, etc. It is not a good indicator of importance of a term. Definitions are informal, not exhaustive.

Absorption To convert the energy of incoming electromagnetic radiation into internal energy within a gas or surface. Note that absorption and re-emission is not the same as reflection.

Adiabatic Without exchange of energy or mass (between an air or water parcel and its surroundings).

Advection Process by which a wind or current carries properties (temperature, concentration, etc.) between regions that involves gradients of that property times the velocity in the direction of the gradient.

Aerosol Suspended particle.

Albedo Fraction of incident solar radiation that is reflected from a surface or from the Earth as a whole. When not obvious from the context, planetary albedo is used to denote the latter.

Anomaly Departure from normal climatological conditions. It is calculated by taking the difference between the value of a variable at a given time, such as pressure or temperature for a particular month, and subtracting the climatology of that variable for the same part of the seasonal cycle.

Anthropogenic Any phenomena created or influenced by humans.

C (1) Degrees centigrade. $1°C = 1 K = 1.4°F$, with K kelvin and °F Fahrenheit degrees. (2) Symbol for carbon.

Beta effect Effect of variations of the Coriolis force with latitude.

CFC Chlorofluorocarbon.

Chaos Irregular behavior of a system in which sensitive dependence on initial conditions (growth of small errors) limits predictability.

CO_2 Carbon dioxide.

Convection Overturning motions at small scale that transfer heat vertically. When condensation occurs it is termed moist convection, when it does not, dry convection. When used alone, whether it is moist or dry should be obvious from the context. Deep convection refers to moist convection through the depth of the troposphere.

Convection zones, tropical Regions of frequent deep convection in the tropics. More general and less unwieldy than Intertropical Convergence Zone.

Convergence Negative divergence, i.e. parcels tend to contract (see divergence).

Coriolis parameter Gives the strength as a function of latitude of the component of the Coriolis force that affects horizontal motions.

Correlation A measure (with values between -1 and 1) of the tendency of two variables to vary together (if negative then to vary with opposite sign). For a time series of N measurements of variables X and Y, correlation is computed as correlation $= \frac{1}{N} \left(\sum X'Y' \right) (sdev(X') \cdot sdev(Y'))^{-1}$, where X' denotes anomalies with the mean of X removed, $sdev(X')$ denotes standard deviation and $\sum$ denotes sum over the N values. For this estimate to be useful, N must be large.

Cryosphere The component of the climate system consisting of ice and snow.

Divergence (1) Short for horizontal divergence. Measures the rate at which a parcel would tend to expand in the horizontal owing to the flow field. (2) Measures the rate at which a parcel would tend to expand in 3D.

DJF Averaged over December, January and February.

Easterly Wind originating from the east.

ECMWF European Centre for Medium-range Weather Forecasting.

Emission (1) To produce and give off electromagnetic (here usually infrared) radiation. This occurs by changes in the excitation energy of electrons etc. in the molecules of a gas or surface. (2) To give off a substance (such as CO_2) into the atmosphere or ocean.

Emissivity (1) Measures the emission or absorption of (infrared) radiation by a unit of gas or a substance relative to blackbody emission. (2) As in (1) but for bulk emissivity of the whole depth of the atmosphere.

ENSO El Niño/Southern Oscillation.

Equation of state Gives the relation between density and temperature, pressure, salinity etc.

Evapotranspiration Evaporation as regulated by land vegetation.

Feedback For an initial change in a variable (or part of a system), the part of the response of other variables or other parts of the system that affects the initial change, for instance to amplify or decrease it (positive and negative feedback, respectively).

Flux Rate of transfer (of energy, momentum or mass).

Frequency (As used in power spectra) When expressed in cycles per year, the frequency is the inverse of the period of oscillation.

Gaussian (1) Common distribution of probability for a random variable. (2) A mathematical dependence similar to (1).

GCM General Circulation Model (or Global Climate Model). Most complex form of computer model for the large-scale atmosphere and or ocean.

Gradient Change in a variable per unit distance in the direction of fastest change (i.e. largest spatial derivative).

Greenhouse gases Trace gases that absorb infrared radiation, and thus affect Earth's energy budget, such as water vapor, carbon dioxide, methane, nitrous oxide and CFCs.

Hydrostatic balance Balance between gravity and vertical pressure gradient.

Index A variable used as an indicator of a phenomenon, such as the Southern Oscillation Index. Often a spatial average or measurement at a particular location are used. Index variables are chosen because they have measurements available and provide a reasonable summary of a phenomenon that might include many variables over large regions.

Infrared radiation Electromagnetic radiation of larger wavelength ($0.75\,\mu m$ to $1\,\mu m$) than visible light ($0.4\,\mu m$ to $0.75\,\mu m$). Emission increases with temperature.

Interannual Having time scales of years.

IPCC Intergovernmental Panel on Climate Change.

ITCZ Intertropical Convergence Zone.

Kelvin Unit for absolute temperature. $1\,K = 1\,°C$. Add 273.15 to temperature in $°C$ to get absolute temperature.

Lapse rate Rate of decrease of temperature with height.

Latent heat Heat released when water vapor condenses, or is removed for evaporation.

Longwave radiation Same as infrared radiation.

Meridional In the south–north direction.

Micron Micrometer (μm), $10^{-6}\,m$.

Midlatitudes The region between the subtropics and polar regions (roughly 30 and 60 degrees latitude).

Millibar (mb or mbar) Conventional unit for pressure in meteorology; $10^{-2}\,mb = 1\,Pa = 1\,N\,m^{-2} = 1\,(kg\,m\,s^{-2})\,m^{-2}$. Sea level pressure is slightly greater than $1\,bar = 1000\,mb = 1000\,hPa$ (hectopascal).

Monsoon A seasonal shift in wind patterns and convection zones and associated weather phenomena.

Nanometer $10^{-9}\,m$.

NINO-3 A common index for equatorial SST associated with ENSO. Average SST from $5°\,S$ to $5°\,N$, $150°W$ to $90°\,W$.

NOAA National Oceanic and Atmospheric Administration.

Noise Process in which there is a random element to the relation between the value at a given time and previous values.

Noise, white A process so random that the value at each time is uncorrelated to previous values.

Normalized Mathematical term indicating the amplitude of a quantity has been rescaled, for instance, by dividing by the standard deviation.

OLR Outgoing longwave radiation. Upward infrared radiation leaving the top of the atmosphere into space.

Parameterization Representation of the average effects of a small-scale process on the large scales, taking into account the dependence of the small scales on large-scale variables.

Parcel An imaginary small unit of air or water, used in deriving equations, or in considering stability to convection.

Power spectrum See endnote 11 from Chapter 1.

Residence time The time molecules of a given compound (or particle) stay in the atmosphere between production and removal or decay.

Salinity Concentration of salt.

Shoal To become shallower.

SOI Southern Oscillation Index. See endnote 9 from Chapter 1.

Southern Oscillation A historical term for the cyclic nature of surface pressure differences between the eastern and western Pacific along the equator and associated weather and climatic phenomena. Now used mostly in El Niño/Southern Oscillation (ENSO).

Spectrum (1) Short for power spectrum. (2) the range of wave lengths of light and other forms of electromagnetic radiation.

SST Sea surface temperature.

Standard deviation A measure of the typical amplitude of variations of a variable. For a time series of N measurements of a variable X, $sdev(X') = \left(\frac{1}{N-1} \sum (X')^2 \right)^{1/2}$.

Stratification Vertical distribution of density, which affects the work required to lift a parcel of air or water.

Teleconnections Remote effects of a phenomenon in one region upon climate variables in another region.

Thermal circulation Circulation driven by temperature differences.

Thermocline Subsurface layer in the ocean at about 100–200 m depth which separates the cold deep ocean from the relatively warm upper ocean.

TOGA Tropical Ocean–Global Atmosphere program.

Trade winds Easterly winds in the tropics.

Troposphere The lower part of the atmosphere (roughly below 10–12 km) with 80% of the atmospheric mass.

Upwelling Moving upward, especially subsurface water rising to the surface in the ocean.

Weather noise The effects of weather on climate time scales, with weather variability viewed as a random process.

Westerly Wind originating from the west.

Wind shear Change of wind per unit distance. If not specified, refers to vertical shear.

Wind stress Force per unit area of the wind on a surface in a direction parallel to the surface. Here usually wind stress at the surface of the ocean (stress can also apply at an imaginary surface in the atmosphere).

Zonal In the east–west direction.

References

Anderson, D. L. T. and J. P. McCreary, 1985: Slowly propagating disturbances in a coupled ocean–atmosphere model. *J. Atmos. Sci.*, **42**, 615–630.

Arakawa, A., A. Katayama and Y. Mintz, 1969: Numerical simulation of the general circulation of the atmosphere. In: *Proceedings of the WMO/IUGG Symposium on Numerical Weather Prediction, Tokyo, 1968*, session IV. Japan Meteorological Agency.

Arrhenius, S., 1896: On the influence of carbonic acid in the air upon the temperature of the ground. *Phil. Mag.* **41**, 237–276.

Barker T., I. Bashmakov, L. Bernstein *et al.*, 2007: Technical Summary. In: *Climate Change 2007: Mitigation. Contribution of Working Group III to the Fourth Assessment Report of the Intergovernmental Panel on Climate Change*, ed. B. Metz, O. R. Davidson, P. R. Bosch, R. Dave and L. A. Meyer. Cambridge and New York: Cambridge University Press.

Battisti, D. S., 1988: The dynamics and thermodynamics of a warming event in a coupled tropical atmosphere/ocean model. *J. Atmos. Sci.*, **45**, 2889–2919.

Battisti, D. S. and A. C. Hirst, 1989: Interannual variability in the tropical atmosphere/ocean system: influence of the basic state, ocean geometry and nonlinearity. *J. Atmos. Sci.*, **46**, 1687–1712.

Bell, G. D., 2008: Seasonal Forecasting of Tropical Cyclones: NOAA's seasonal hurricane outlooks. *Global Guide to Tropical Cyclone Forecasting*, ed. C. Guard. Australian Bureau of Meteorology.

Bender, C. M. and S. A. Orszag, 1978: *Advanced Mathematical Methods for Scientists and Engineers*. New York: McGraw-Hill.

Berger, A. and M. F. Loutre, 1991: Insolation values for the climate of the last 10 million years. *Quarternary Sci. Rev.*, **10**, 297–317.

Berlage, H. P., 1957: Fluctuations in the general atmospheric circulation of more than one year, their nature and prognostic value. *Roy. Neth. Meteor. Inst. Meded. Verh.*, **69**, 152.

Bindoff, N. L. and T. J. McDougall, 2000: Decadal changes along an Indian Ocean section at 32 degrees S and their interpretation. *J. Phys. Oceanog.*, **30**(6), 1207–1222.

Bindoff, N. L., Willebrand, J., Artale, V. *et al.*, 2007: Observations: oceanic climate change and sea level. In: *Climate Change 2007: The Physical Science Basis. Contribution of Working Group I to the Fourth Assessment Report of the Intergovernmental Panel on Climate Change*, ed. S. Solomon, D. Qin, M. Manning, *et al.* Cambridge and New York: Cambridge University Press.

Biasutti, M. and A. Giannini, 2006: Robust Sahel drying in response to late 20th century forcings. *Geophys. Res. Lett.*, **33**, L11706, doi:10.1029/2006GL026067.

Bitz, C. M., M. M. Holland, A. J. Weaver and M. Eby, 2001: Simulating the ice-thickness distribution in a coupled climate model. *J. Geophys. Res.*, **106**, 2441–2463.

Bjerknes, J., 1969: Atmospheric teleconnections from the equatorial Pacific. *Mon. Wea. Rev.*, **97**, 163–172.

Bohren, C. F. and B. A. Albrecht, 1998: *Atmospheric Thermodynamics*. New York: Oxford University Press.

Bourles, B., R. Lumpkin, M. J. McPhaden *et al.*, 2008: The PIRATA program: history, accomplishments, and future directions. *Bull. Met. Soc.*, **89**, 1111–1125.

Broecker, W. S., 1987: Unpleasant surprises in the greenhouse? *Nature*, **328**, 123–126.

Broecker, W.S., 1990: Salinity history of the northern Atlantic during the last deglaciation. *Paleoceanography*, **5**, 459–467.

Brohan, P., J. J. Kennedy, I. Haris, S. F. B. Tett and P. D. Jones, 2006: Uncertainty estimates in regional and global observed temperature changes: a new dataset from 1850. *J. Geophys. Res.*, **111**, D12106, doi:10.1029/2005JD006548.

Callendar, G. S., 1938: The artificial production of carbon dioxide and its influence on temperature. *Q. J. Roy. Meteorol. Soc.*, **64**, 223–237.

Cane, M. A. and E. S. Sarachik, 1983: Equatorial oceanography. *Rev. Geophys. Space Phys.*, **21**, 1137–1148.

Cane, M. A. and S. E. Zebiak, 1985: A theory for El Niño and the Southern Oscillation. *Science*, **228**, 1084–1087.

Cane, M. and S. E. Zebiak, 1987: Prediction of El Niño events in a physical model. In: *Atmospheric and Oceanic Variability*, ed. H. Cattle. Royal Meteorological Society Press, 153–182.

Cane, M., S. E. Zebiak and S. C. Dolan, 1986: Experimental forecasts of El Niño. *Nature*, **321**, 827–832.

Cattle, H. and J. Crossley, 1995: Modelling Arctic climate change. *Phil. Trans. R. Soc. London A.*, **352**, 201–213.

Cayan, D. R., and R. H. Webb, 1992: El Niño/Southern Oscillation and streamflow in the western United States. In: *El Niño – historical and paleoclimatic aspects of the Southern Oscillation*, ed. H. F. Diaz and V. Markgraf. Cambridge: Cambridge University Press.

Cess, R. D., G. L. Potter, J. P. Blanchet *et al.*, 1989: Interpretation of cloud-climate feedback as produced by 14 atmospheric general circulation models. *Science*, **245**, 513–516.

Cess, R. D., G. L. Potter, M.-H. Zhang *et al.*, 1991: Interpretation of snow-climate feedback as produced by 17 general circulation models. *Science*, **253**, 888–892.

Cess, R. D., M. H. Zhang, W. J. Ingram *et al.*, 1996: Cloud feedback in atmospheric general circulation models: an update. *J. Geophys. Res.*, **101**, 12791–12794.

Charney, J. G., A. Arakawa, D. J. Baker *et al.*, 1979: *Carbon Dioxide and Climate: A Scientific Assessment*. Washington, DC: National Academy of Sciences.

Chen, W. Y. and H. M. van den Dool, 1997: Asymmetric impact of tropical SST anomalies on atmospheric internal variability over the North Pacific. *J. Atmos. Sci.*, **54**, 725–740.

Cheney, R., L. Miller, R. Agreen *et al.*, 1994: TOPEX/POSEIDON: the 2-cm solution. *J. Geophys. Res.*, **97**, 24555–24563.

Church, J. A. and N. J. White, 2006: A 20th century acceleration in global sea-level rise. *Geophys. Res. Lett.*, **33**, L01602, doi:10.1029/2005GL024826.

Clerbaux, C. and D. Cunnold, 2006: Long-lived compounds. In: *Scientific Assessment of Ozone Depletion: 2006*. Geneva: World Meteorological Organization, pp. 1.1–1.61.

Colman, R., 2004: A comparison of climate feedbacks in general circulation models. *Clim. Dyn.*, **20**, 865–873, doi:10.1007/s00382-003-0310z.

Crowley, T. J., 1983: The geologic record of climate change. *Rev. Geophys. Space Phys.*, **21**, 828–877.

Crowley, T. J. and G. R. North, 1991: *Paleoclimatology*. Oxford: Oxford University Press.

Dansgaard, W., Johnsen, S. J., Clausen, H. B. *et al.*, 1993: Evidence for general instability of past climate from a 250-kyr ice-core record. *Nature*, **364**, 218–220.

Dargaville, R. J., S. C. Doney, and I. Y. Fung, 2003: Inter-annual variability in the inter-hemispheric atmospheric CO_2 gradient: contributions from transport and the seasonal rectifier. *Tellus B* **55**, 711–722, doi:10.1034/j.1600-0889.2003.00038.x.

Darnell, W. L., W. F. Staylor, S. K. Gupta, N. A. Ritchey and A. C. Wilber, 1992: Seasonal variation of surface radiation budget derived from international satellite cloud climatology project C1 data. *J. Geophys. Res.*, **97**, 15741–15760.

Denman, K. L., G. Brasseur, A. Chidthaisong, *et al.*, 2007: Couplings between changes in the climate system and biogeochemistry. In: *Climate Change 2007: The Physical Science Basis. Contribution of Working Group I to the Fourth Assessment Report*, ed. S. Solomon, D. Qin, M. Manning *et al.* Cambridge and New York: Cambridge University Press.

Dettinger, M. D., M. Ghil, C. M. Strong, W. Weibel and P. Yiou, 1995: Software expedites singular-spectrum analysis of noisy time series. *Eos Trans. Am. Geophys. Un.*, **76**, 12–21.

Dickinson, R. E., 1984: Modeling evapo-transpiration for three-dimensional global climate models. In: *Climate Processes and Climate Sensitivity*, ed. J. E. Hanson and T. Takahashi. Geophysical Monograph 29, Maurice Ewing Vol. 5, American Geophysical Union, 58–72.

Dickinson, R. E., A. Henderson-Sellers, P. J. Kennedy and M. Wilson, 1986: Biosphere–Atmosphere Transfer Scheme (BATS) for the NCAR Community Climate Model. *NCAR Tech. Note TN-275+STR*. Boulder, CO: NCAR.

Dines, W. H., 1917: The heat balance of the atmosphere. *Q. J. Roy. Meteorol. Soc.*, **XLIII**, 151–158.

Emanuel, K. A., 1988a: Toward a general theory of hurricanes. *Am. Sci.*, **76**, 317–379.

Emanuel, K. A., 1988b: The maximum intensity of hurricanes. *J. Atmos. Sci.*, **45**, 1143–1155.

EPICA community members, 2004: Eight glacial cycles from an Antarctic ice core. *Nature*, **429**, 623–628. doi:10.1038/nature02599.

Etheridge, D. M. *et al.*, 1998: Atmospheric methane between 1000 AD and present: evidence of anthropogenic emissions and climatic variability. *J. Geophys. Res.*, **103**, 15979–15994.

Falkowski, P., R. J. Scholes, E. Boyle *et al.*, 2000: The global carbon cycle: a test of our knowledge of Earth as a system. *Science*, **290**, 291–296, doi:10.1126/science.290.5490.291.

Fairbanks, R. G. 1989: 17,000-year glacio-eustatic sea level record: influence of glacial melting rates on the Younger Dryas event and deep-ocean circulation. *Nature*, **342**, 637–642.

Farman, J.C., B.G. Gardiner and J.D. Shanklin, 1985: Large losses of total ozone in Antarctica reveal seasonal ClO_x/NO_x interaction. *Nature*, **315**, 207–210.

Flato, G.M. and W.D. Hibler, III, 1992: Modeling pack ice as a cavitating fluid. *J. Phys. Oceanog.*, **22**, 626–651.

Flückiger, J., E. Monnin, B. Stauffer, J. Schwander and T.F. Stocker, 2002: High-resolution Holocene N_2O ice core record and its relationship with CH_4 and CO_2. *Glob. Biogeochem. Cyc.*, **16**, doi:1029/2001GB001417.

Foster, J., Liston, G., Koster, R. *et al.*, 1996: Snow cover and snow mass intercomparisons of general circulation models and remotely sensed datasets. *J. Clim.*, **9**, 409–426.

Fourier, J., 1827: Mémoire sur les temperatures du globe terrestre et des espaces planétaires, *Mem.de l'Academie Royale des sciences de l'Institute de France*, **7**, 569–604.

Folland C.K. and D.E. Parker, 1995: Correction of instrumental biases in historical sea surface temperature data. *Q. J. Roy. Meteorol. Soc.*, **121**, 319–367.

Folland, C.K., N.A. Rayner, S.J. Brown *et al.*, 2001: Global temperature change and its uncertainties since 1861. *Geophys. Res. Lett.*, **28**, 2621–2624.

Fröhlich, C., 2003: Long-term behaviour of space radiometers. *Metrologia*, **40**, 60–-65.

Fröhlich, C. and J. Lean, 1998, The sun's total irradiance: cycles, trends and related climate change uncertainties since 1978, *Geophys. Res. Lett.*, **25**, 4377–4380.

Fröhlich, C. and J. Lean, 2002: Solar irradiance variability and climate. *Astron. Nachrichten*, **323**, 203–212.

Giannini, A., R. Saravanan and P. Chang, 2003: Oceanic forcing of Sahel rainfall on interannual to interdecadal time scales. *Science*, **203**, 1027–1030.

Gill, A.E., 1980: Some simple solutions for heat induced tropical circulation. *Q. J. Roy. Meteorol. Soc.*, **106**, 447–462.

Gillette, D.A., W.D. Komhyr, L.S. Waterman, L.P. Steele and R.H. Gammon, 1987: The NOAA/GMCC continuous CO_2 record at the South Pole, 1975–1982, *J. Geophys. Res.*, **92**, 4231–4240.

Glantz, M.H., 1994: Drought, desertification, and food production. In: *Drought Follows the Plow*, ed. M.H. Glantz. Cambridge: Cambridge University Press, pp. 9–30.

Goldenberg, S.B. and L.J. Shapiro, 1996: Physical mechanisms for the association of El Niño and west African rainfall with Atlantic major hurricane activity. *J. Clim.*, **9**, 1169–1187.

Goody, R.M. and J.C.G. Walker, 1972: *Atmospheres*. Englewood Cliffs, NJ: Prentice-Hall.

Goody, R.M. and Y.L. Yung, 1989: *Atmospheric Radiation, Theoretical Basis*. 2nd edn. Oxford UK: Oxford University Press.

Gradstein, F.M., J.G. Ogg, A.G. Smith *et al.*, 2004: *A Geologic Timescale 2004*. Cambridge: Cambridge University Press.

Graedel, T.E. and P.J. Crutzen, 1993: *Atmospheric Change: An Earth System Perspective*. New York: W.H. Freeman.

Gray, W.M., 1984: Atlantic seasonal hurricane frequency: Part I: El Niño and 30-mb quasi-biennial oscillation influences. *Mon. Wea. Rev.*, **112**, 1649–1668.

GRIP Project Members, 1993. Climate instability during the last interglacial period recorded in the GRIP ice core. *Nature*, **364**, 203–207.

Halpern, D., 1987: Comparison of moored wind measurements from a spar and toroidal buoy in the eastern equatorial Pacific during February–March 1981. *J. Geophys. Res.*, **92**, 8197–8212.

Halpern, D., 1996: Visiting TOGA's past. *Bull. Am. Meteorol. Soc.*, **77**, 233–242.

Halpert, M. S. and C. F. Ropelewski, 1992: Surface temperature patterns associated with the Southern Oscillation. *J. Clim.*, **5**, 577–593.

Handel, M. D. and J. S. Risbey, 1992: An annotated bibliography on the greenhouse effect and climate change. *Climatic Change*, **21**, 97–255.

Hansen, J., G. Russell, A. Lacis *et al.*, 1985: Climate response times: dependence on climate sensitivity and ocean mixing. *Science*, **229**, 857–859, doi:10.1126/science.229.4716.857.

Hansen, J., I. Fung, A. Lacis *et al.*, 1988: Global climate changes as forecast by Goddard Institute for Space Studies three-dimensional model. *J. Geophys. Res.*, **93**D, 9341–9364.

Hansen, J., M. Sato, A. Lacis *et al.*, 1998: Climate forcings in the industrial era. *Proc. Natl Acad. Sci.*, **95**, 12753–12758.

Hansen, J., R. Ruedy, J. Glascoe and M. Sato, 1999: GISS analysis of surface temperature change. *J. Geophys. Res.*, **104**, 30997–31022, doi:10.1029/1999JD900835.

Hartmann, D. L., 1994: *Global Physical Climatology*. San Diego: Academic Press.

Hayes, S. P., L. J. Mangum, J. Picaut *et al.*, 1991: TOGA-TAO: a moored array for real-time measurements in the tropical Pacific Ocean. *Bull. Am. Meteorol. Soc.*, **72**, 339–347.

Haywood, J. M., R. J. Stouffer, R. T. Wetherald, S. Manabe, V. Ramaswamy, 1997: Transient response of a coupled model to estimated changes in greenhouse gas and sulfate concentrations. *Geophys. Res. Lett.*, **24**, 1335–1338.

Hegerl, G. C., H. von Storch, K. Hasselmann *et al.*, 1996: Detecting greenhouse-gas-induced climate change with an optimal fingerprint method. *J. Clim.*, **9**, 2281–2306.

Hegerl, G. C., F. W. Zwiers *et al.*, 2007: Understanding and attributing climate change. In: *Climate Change 2007: The Physical Science Basis. Contribution of Working Group I to the Fourth Assessment Report of the Intergovernmental Panel on Climate Change*, ed. Solomon, S., D. Qin, M. Manning *et al.* Cambridge and New York: Cambridge University Press.

Held, I. M., 2005: The gap between simulation and understanding in climate modeling. *Bull. Am. Meteorol. Soc.*, **86**, 1609–1614.

Held, I. M. and B. J. Soden, 2000: Water vapor feedback and global warming. *Annu. Rev. Energy Environ.*, **25**, 441–475.

Held, I. M., T. L. Delworth, J. Lu, K. L. Findell and T. R. Knutson, 2005: Simulation of Sahel drought in the 20th and 21st centuries. *Proc. Natl. Acad. Sci.*, **102**(50), 17891–17896.

Hibler, W. D. III, 1979: A dynamic thermodynamic sea ice model. *J. Phys. Oceanogr.*, **9**, 815–846.

Hibler, W. D., III and S. J. Johnsen, 1979: The 20-yr cycle in Greenland ice core records. *Nature*, **280**, 481–483.

Holgate, S. J. and P. L. Woodworth, 2004: Evidence for enhanced coastal sea level rise during the 1990s. *Geophys. Res. Lett.*, **31**, L07305, doi:10.1029/2004GL019626.

Hulme, M., 1994: Validation of large-scale precipitation fields in General Circulation Models. In: *Global Precipitation and Climate Change*, eds. M. Desbois and F. Desalmand. NATO ASI Series, Vol. I 26. Berlin: Springer-Verlag, pp. 387–406.

Hunke, E. C. and J. K. Dukowicz, 1997: An elastic-viscous-plastic model for sea ice dynamics. *J. Phys. Oceanog.*, **27**, 1849–1867.

Hurrell, J. W., 1995: Decadal trends in the North Atlantic oscillation: Regional temperatures and precipitation. *Science*, **269**, 676–679.

Hurrell, J. W., 1996: Influence of variations in extratropical wintertime teleconnections on Northern Hemisphere temperatures. *Geophys. Res. Lett.*, **23**, 665–668.

Indermühle, A, T. F. Stocker, F. Joos *et al.*, 1999: Holocene carbon-cycle dynamics based on CO_2 trapped in ice at Taylor Dome, Antarctica. *Nature*, **398**, 121–126.

IPCC, 1990: *Climate Change: The IPCC Scientific Assessment of Climate Change. Report Prepared for IPCC by Working Group I*, ed. J. T. Houghton, G. J. Jenkins and J. J. Ephraums. Cambridge and New York: Cambridge University Press.

IPCC, 1992: *Climate Change 1992: The Supplementary Report to the IPCC Scientific Assessment. Report Prepared for IPCC by Working Group I*, ed. J. T. Houghton, B. A. Callander and S. K. Varney. Cambridge and New York: Cambridge University Press.

IPCC, 1996: *Climate Change 1995: The Science of Climate Change. Contribution of Working Group I to the Second Assessment Report of the Intergovernmental Panel on Climate Change*, eds. J. T. Houghton, L. G. Meira Filho, B. A. Callander *et al.* Cambridge and New York: Cambridge University Press.

IPCC, 2001: *Climate Change 2001: The Scientific basis. Contribution of Working Group I to the Third Assessment Report of the Intergovernmental Panel on Climate Change*, eds. J. T. Houghton, Y. Ding, D. J. Griggs *et al.* Cambridge and New York: Cambridge University Press.

IPCC, 2007: *Climate Change 2007: The Physical Science Basis. Contribution of Working Group I to the Fourth Assessment Report of the Intergovernmental Panel on Climate Change*, eds. S. Solomon, D. Qin, M. Manning *et al.* Cambridge and New York: Cambridge University Press.

Ji, M., A. Leetmaa and V. E. Kousky, 1996: Coupled model predictions of ENSO during the 1980s and the 1990s at the National Centers for Environmental Prediction. *J. Clim.*, **9**, 3105–3120.

Jin, F.-F. and Neelin, 1993: Modes of interannual tropical ocean–atmosphere interaction – a unified view. Part I: Numerical results. *J. Atmos. Sci.*, **50**, 3477–3503.

Johnsen, S. J., H. B. Clausen, W. Dansgaard *et al.*, 1992: Irregular glacial interstadials recorded in a new Greenland ice core. *Nature*, **359**, 311–313.

Jones, P. D. and Moberg, A., 2003: Hemispheric and large-scale surface air temperature variations: an extensive revision and an update to 2001. *J. Clim.*, **16**, 206–223.

Jones, P. D., M. New, D. E. Parker, S. Martin and I. G. Rigor, 1999: Surface air temperature and its changes over the past 150 years. *Rev. Geophys.*, **37**, 173–199.

Jones P. D., T. J. Osborn, K. R. Briffa *et al.*, 2001: Adjusting for sampling density in grid box land and ocean surface temperature time series. *J. Geophys. Res.*, **106**, 3371–3380.

Jouzel, J., 2004: *EPICA Dome C Ice Cores Deuterium Data*. IGBP PAGES, World Data Center for Paleoclimatology, Data Contribution Series # 2004-038. NOAA/NGDC Paleoclimatology Program, Boulder CO, doi:10.3334/CDIAC/cli.007.

Jouzel, J., B. Stauffer and J. P. Steffensen, 1992. Irregular glacial interstadials recorded in a new Greenland ice core. *Nature*, **359**, 311–313.

Jouzel J., F. Vimeux, N. Caillon *et al.*, 2003: Magnitude of isotope/temperature scaling for interpretation of central Antarctic ice cores. *J. Geophys. Res.*, **108** (D12), 4361, doi:10.1029/2002JD002677.

Kalnay, E., M. Kanamitsu, R. Kistler *et al.*, 1996: The NCEP/NCAR 40-year reanalysis project. *Bull. Am. Meteor. Soc.*, **77**, 437–471.

Kasahara, A. and Washington, W. M., 1967: NCAR global general circulation model of the atmosphere. *Mon. Wea. Rev.* **95**, 389–402.

Katayama, A., 1972: Calculation of radiative transfer. In: *Proceedings of the WMO/IUGG Symposium on Numerical Weather Prediction, Tokyo, 1968*, pp. IV-8-7–IV-8-10. Japan Meteorological Agency.

Keeling, C. D. and T. P. Whorf, 2005: Atmospheric CO records from sites in the SIO air sampling network. In: *Trends Online: A Compendium of Data on Global Change.* Oak Ridge, TN: Carbon Dioxide Information Analysis Center, Oak Ridge National Laboratory.

Keeling, C. D., R. B. Bacastow, A. E. Bainbridge *et al.*, 1976: Atmospheric carbon dioxide variations at Mauna Loa Observatory, Hawaii. *Tellus*, **28**, 538–551.

Keeling, C. D., S. C. Piper, R. B. Bacastow *et al.*, 2001: Exchanges of atmospheric CO_2 and $13CO_2$ with the terrestrial biosphere and oceans from 1978 to 2000. I. Global aspects. *SIO Reference Series*, No. 01-06. San Diego: Scripps Institution of Oceanography.

Kiehl, J. T., 1992: Atmospheric general circulation modeling. In: *Climate System Modeling.* ed. K. E. Trenberth. Cambridge: Cambridge University Press.

Kiehl, J. T. and K. E. Trenberth, 1997: Earth's annual global mean energy budget. *Bull. Am. Meteor. Soc.*, **78**, 197–207.

Kistler, R., E. Kalnay, W. Collins *et al.*, 2001: The NCEP-NCAR 50-year reanalysis: monthly means CD-ROM and documentation. *Bull. Am. Meteor. Soc.*, **82**, 247–267.

Knutson, T. R. and R. E. Tuleya, 2004: Impact of CO_2-induced warming on simulated hurricane intensity and precipitation: sensitivity to the choice of climate model and convective parameterization. *J. Clim.*, **17**, 3477–3494.

Langner, J. and H. Rodhe, 1991: A global three-dimensional model of the tropospheric sulfur cycle. *J. Atmos. Chem.*, **13**, 225–263.

Latif, M., K. Sperber, J. Arblaster *et al.*, 2001: ENSIP: The El Niño simulation intercomparison project. *Clim. Dynam.*, **18**, 255–276.

Legler, D. M. and J. J. O'Brien, 1984: *Atlas of tropical Pacific wind stress climatology 1971–1980.* Tallahassee, FL: Department of Meteorology, Florida State University.

Lemke, P., J. Ren, R. B. Alley *et al.*, 2007: Observations: changes in snow, ice and frozen ground. In: *Climate Change 2007: The Physical Science Basis. Contribution of Working Group I to the Fourth Assessment Report of the Intergouvernmental Panel on Climate Change*, ed. S. Solomon, D. Qin, M. Manning *et al.* Cambridge: Cambridge University Press.

Le Quéré, C., M. R. Raupach, J. G. Canadell *et al.*, 2009: Trends in the sources and sinks of carbon dioxide. *Nature Geosci.*, **2**, 831–836, doi:10.1038/ngeo689.

Leuliette, E. W., R. S. Nerem and G. T. Mitchum, 2004: Calibration of TOPEX/Poseidon and Jason altimeter data to construct a continuous record of mean sea level change. *Mar. Geodesy*, **27**(1–2), 79–94.

Levitus, S., J. Antonov and T. Boyer, 2005: Warming of the world ocean, 1955–2003. *Geophys. Res. Lett.*, **32**, L02604, doi:10.1029/2004GL021592.

Lorius, C., J. Jouzel, D. Raynaud, J. Hansen and H. Le Treut, 1990: The ice-core record: climate sensitivity and future greenhouse warming. *Nature*, **347**, 139–145.

Madden, R. and P. R. Julian, 1972: Description of global-scale circulation cells in the tropics with a 40–50 day period. *J. Atmos. Sci.*, **29**, 1109–1123.

Manabe, S. and R. T. Wetherald, 1975: The effects of doubling the CO_2 concentration on the climate of a general circulation model. *J. Atmos. Sci.*, **32**, 3–15.

Manabe, S., J. Smagorinsky and R. F. Strickler, 1965: Simulated climatology of a general circulation model with a hydrologic cycle. *Mon. Wea. Rev.*, **93**, 769–798.

Manabe, S., R. J. Stouffer, M. J. Spelman and K. Bryan, 1991: Transient responses of a coupled ocean-atmosphere model to gradual changes of atmospheric CO_2. I. Annual mean response. *J. Clim.*, **4**, 785–818.

Mason, S. J., L. Goddard, N. E. Graham *et al.*, 1999: The IRI seasonal climate prediction system and the 1997/98 El Niño event. *Bull. Am. Meteorol. Soc.*, **80**, 1853–1873.

Masters, G. M., 1998: *Introduction to Environmental Engineering and Science*, 2nd edn. Upper Saddle River, NJ: Prentice Hall.

McCabe, G. J. Jr. and Dettinger, M. D., 1999: Decadal variability in the relations between ENSO and precipitation in the western United States. *Int. J. Climatol.*, **19**, 1399–1410.

McCreary, J. P., 1985: Modeling equatorial ocean circulation. *Ann. Rev. Fluid Mech.*, **17**, 359–409.

McPhaden, M., A. J. Busalacchi, R. Cheney *et al.*, 1998: The Tropical Ocean–Global Atmosphere observing system: a decade of progress. *J. Geophys. Res.*, **103**, 14169–14240.

Mechoso, C. R., A. W. Robertson, N. Barth *et al.*, 1995: The seasonal cycle over the tropical Pacific in coupled ocean-atmosphere general circulation models. *Mon. Wea. Rev.*, **123**, 2825–2838.

Meehl, G. A., T. F. Stocker, W. D. Collins *et al.*, 2007: Global climate projections. In: *Climate Change 2007: The Physical Science Basis. Contribution of Working Group I to the Fourth Assessment Report of the Intergovernmental Panel on Climate Change*, ed. S. Solomon, D. Qin, M. Manning *et al.* Cambridge and New York: Cambridge University Press.

Meinshausen, M., N. Meinshausen, W. Hare *et al.*, 2009: Greenhouse-gas emission targets for limiting global warming to 2 °C. *Nature*, **458**, 1158–1162, doi:10.1038/nature08017.

Mitchell, J. F. B. and T. C. Johns, 1997: On modification of global warming by sulfate aerosols. *J. Clim.*, **10**, 245–267.

Molina, M. J. and F. S. Rowland, 1974: Stratospheric sink for chlorofluormethanes: Chlorine atom catalyzed destruction of ozone. *Nature*, **249**, 810–812.

Monnin, E., A. Indermühle, A. Dallenbach *et al.*, 2001: Atmospheric CO_2 concentrations over the Last Glacial termination. *Science*, **291**, 112–114.

Myhre, G., E. J. Highwood, K. P. Shine and F. Stordal, 1998: New estimates of radiative forcing due to well mixed greenhouse gases. *Geophys. Res. Lett.*, **25**, 2715–2718.

Neelin, J. D., M. Latif, M. A. Allaart *et al.*, 1992: Tropical air–sea interaction in general circulation models. *Clim. Dynam.*, **7**, 73–104.

Neelin, J. D., D. S. Battisti, A. C. Hirst *et al.*, 1998: ENSO theory. *J. Geophys. Res.*, **103**, 14261–14290.

Nicholson, S. E., C. J. Tucker and M. B. Ba, 1998: Desertification, drought, and surface vegetation: an example from the West African Sahel. *Bull. Am. Meteor. Soc.*, **79**, 815–829.

Niiler, P. P. and J. D. Paduan, 1995: Wind-driven motions in the northeast Pacific as measured by Lagrangian drifters. *J. Phys. Oceanog.*, **25**, 2819–2830.

Niiler, P. P., A. Sybrandy, K. Bi, P. Poulain and D. Bitterman, 1995: Measurements of the water-following capability of holey-sock and TRISTAR drifters. *Deep-Sea Res.*, **42**, 1951–1964.

Oberhuber, J. M., 1993: Simulation of the Atlantic circulation with a coupled sea–ice–mixed-layer isopycnal general circulation model. *J. Phys. Oceanog.*, **23**, 808–829.

Olson, J. S., J. A. Watts and L. J. Allison, 1983: *Carbon in Live Vegetation of Major World Ecosystems*. DOE/NBB Report No. TR004. Oak Ridge, TN: Oak Ridge National Laboratory.

Parkinson, K., Y. Vinnikov and D. J. Cavalieri, 2006: Evaluation of the simulation of the annual cycle of Arctic and Antarctic sea ice coverages by 11 major global climate models. *J. Geophys. Res.*, **111**, C07012, doi:10.1029/2005JC003408.

Parry, M. L., O. F. Canziani, J. P. Palutikof *et al.*, 2007: Technical Summary. In: *Climate Change 2007: Impacts, Adaptation and Vulnerability. Contribution of Working Group II to the Fourth Assessment Report of the Intergovernmental Panel on Climate Change*, ed. M. L. Parry, O. F. Canziani, J. P. Palutikof *et al.* Cambridge: Cambridge University Press, Cambridge, 23–78.

Pacala, S. and R. Socolow, 2004: Stabilization wedges: Solving the climate problem for the next 50 years with current technologies. *Science*, **305**, 968–972, doi: 10.1126/science.1100103

Petit, J. R., J. Jouzel, D. Raynaud *et al.*, 1999: Climate and atmospheric history of the past 420,000 years from the Vostok ice core, Antarctica. *Nature*, **399**, 429–436.

Philander, S. G. H., 1985: El Niño and La Niña. *J. Atmos. Sci.*, **42**, 2652–2662.

Philander, S. G. H., 1990: *El Niño, La Niña, and the Southern Oscillation*. San Diego: Academic Press.

Philander, S. G. H., R. C. Pacanowski, N. C. Lau and M. J. Nath, 1992: Simulation of ENSO with a global atmospheric GCM coupled to a high-resolution, tropical Pacific ocean GCM. *J. Clim.*, **5**, 308–329.

Piechota, T. C. and J. A. Dracup, 1996: Drought and regional hydrologic variations in the United States: associations with the El Niño/Southern Oscillation. *Water Resources Research*, **32**(5), 1359–1373.

Pond, S. and G. L. Pickard, 1997: *Introductory Dynamic Oceanography*, 2nd edn. Oxford UK: Butterworth-Heinemann.

Press, W. H., S. A. Teukolsky, W. T. Vetterling and B. P. Flannery *et al.*, 1992: *Numerical Recipes in FORTRAN: The Art of Scientific Computing*, 2nd edn. Cambridge: Cambridge University Press.

Qu, X. and A. Hall, 2006: Assessing snow albedo feedback in simulated climate change. *J. Clim.* **19**, 2617–2630.

Randall, D. A., R. D. Cess, J. P. Blanchett *et al.*, 1994: Analysis of snow feedbacks in 14 general circulation models. *J. Geophys. Res.*, **99**, 20757–20771.

Randall, C. E., G. L. Manney, D. R. Allen *et al.*, 2005: Reconstruction and simulation of stratospheric ozone distributions during the 2002 Austral winter. *J. Atmos. Sci.*, **62**, 748–764.

Rasmusson, E. M. and T. H. Carpenter, 1982: Variations in tropical sea surface temperature and surface wind fields associated with the Southern Oscillation/El Niño. *Mon. Wea. Rev.*, **110**, 354–384.

Raynaud, D., J. Jouzel, J. M. Barnola *et al.*, 1993: The ice record of greenhouse gases. *Science*, **259**, 926–934.

Rayner, N. A., P. Brohan, D. E. Parker *et al.*, 2006: Improved analyses of changes and uncertainties in marine temperature measured in situ since the mid-nineteenth century: the HadSST2 dataset. *J. Clim.*, **19**, 446–469.

Revelle, R. and H. E. Suess, 1957: Carbon dioxide exchange between atmosphere and ocean and the question of an increase of atmospheric CO_2 during the past decades. *Tellus*, **9**, 18–27.

Reynolds, R. W., 1988: A real-time global sea surface temperature analysis. *J. Clim.*, **1**, 75–86.

Reynolds, R. W. and T. M. Smith, 1994: Improved global sea surface temperature analyses using optimum interpolation. *J. Clim.*, **7**, 929–948.

Reynolds, R. W. and T. M. Smith, 1999: Improved global sea surface temperature analyses using optimum interpolation. *J. Clim.*, **7**, 929–948.

Reynolds, R. W., N. A. Rayner, T. M. Smith, D. C. Stokes and W. Wang, 2002: An improved in situ and satellite SST analysis for climate. *J. Clim.*, **15**, 1609–1625.

Ropelewski, C. F. and M. S. Halpert, 1987: Global and regional scale precipitation associated with El Niño/Southern Oscillation. *Mon. Wea. Rev.*, **115**, 1606–1626.

Ropelewski C. and M. Halpert, 1989: Precipitation patterns associated with the high index phase of the Southern Oscillation. *J. Clim.*, **2**, 268—284.

Rowell, D. P., C. K. Folland, K. Maskell and M. N. Ward, 1995: Variability of summer rainfall over tropical north Africa (1906–92): Observations and modelling. *Q. J. Roy. Meteorol. Soc.*, **121**, 669–704.

Royer, D. L., 2006: CO_2-forced climate thresholds during the Phanerozoic. *Geochim. Cosmochim. Acta*, **70**, 5665–5675.

Ruddiman, W. F., 2001: *Earth's Climate: Past and Future*. New York: W. H. Freeman.

Sabine, C. L., M. Heimann, P. Artaxo *et al.*, 2004: Current status and past trends of the global carbon cycle. In: *The Global Carbon Cycle: Integrating Humans, Climate, and the Natural World*, ed. C. B. Field and M. R. Raupach. SCOPE 62. Washington DC: Island Press.

Sarmiento J. L. and N. Gruber, 2002: Sinks for anthropogenic carbon. *Physics Today*, **55**, 30–36.

Sarmiento J. L. and N. Gruber, 2006: *Ocean Biogeochemical Dynamics*. Princeton, NJ: Princeton University press.

Schmitz, W. J. Jr, 1995: On the interbasin scale thermohaline circulation. *Rev. Geophys.*, **33**, 151–173.

Schmitz, W. J., 1996a: *On the World Ocean Circulation*. Vol. I, *Some Global Features/ North Atlantic Circulation*. Woods Hole Oceanographic Institution Series 96–03. Woods Hole, MA: Woods Hole Oceanographic Institution.

Schmitz, W. J., 1996b: *On the World Ocean Circulation*. Vol. II, *The Pacific and Indian Oceans/A Global Update*. Woods Hole Oceanographic Institution Series; 96-03. Woods Hole, MA: Woods Hole Oceanographic Institution.

Schopf, P. S. and M. J. Suarez, 1988: Vacillations in a coupled ocean–atmosphere model. *J. Atmos. Sci.*, **45**, 549–566.

Schonher, T. and S. E. Nicholson, 1989: The relationship between California rainfall and ENSO events. *J. Clim.*, **2**, 1258–1269.

Semtner, A. J. Jr, 1976: A model for the thermodynamic growth of sea ice in numerical investigations of climate. *J. Phys. Oceanog.*, **6**, 379–389.

Siegenthaler, U., T. F. Stocker, E. Monnin *et al.*, 2005: Stable carbon cycle–climate relationship during the late Pleistocene. *Science*, **310**, 1313–1317.

Shine, K. P. and P. M. de F. Forster, 1999: The effect of human activity on radiative forcing of climate change: a review of recent developments. *Global Planet. Change*, **20**, 205–225.

Slater, J. F., J. E. Dibb, B. D. Keim, *et al.*, 2001: Relationships between synoptic-scale transport and interannual variability of inorganic cations in surface snow at Summit, Greenland: 1992–1996. *J. Geophys. Res.*, **106**D, 20897–20912.

Soden, B. J. and I. M. Held, 2006: An assessment of climate feedbacks in coupled ocean-atmosphere models. *J. Clim.*, **19**, 3354–3360.

Steele, M., J. Zhang, D. Rothrock *et al.*, 1997: The force balance of sea ice in a numerical model of the Arctic Ocean. *J. Geophys. Res.*, **102**C, 21061–21079.

Stockdale, T. N., A. J. Busalacchi, D. E. Harrison and R. Seager, 1998: Oceanic modelling for ENSO. *J. Geophys. Res.*, **103**, 14325–14355.

Straus, D. M. and J. Shukla, 1997: Variations of midlatitude transient dynamics associated with ENSO. *J. Atmos. Sci.*, **54**, 770–790.

Suarez, M. J. and P. S. Schopf, 1988: A delayed action oscillator for ENSO. *J. Atmos. Sci.*, **45**, 3283–3287.

Sverdrup, H. U., M. W. Johnson and R. H. Fleming, 1942: *The Oceans: Their Physics, Chemistry and General Biology*. Englewood Cliffs, NJ: Prentice-Hall.

Tang, B. H. and J. D. Neelin, 2004: ENSO influence on Atlantic hurricanes via tropospheric warming, *Geophys. Res. Lett.*, **31**, L24, 204, doi:10.1029/2004GL021072.

Tarasov, L. and W. R. Peltier, 2005. Arctic freshwater forcing of the Younger Dryas cold reversal. *Nature*, **435**, 662–665.

Thompson, D. W. J. and J. M. Wallace, 1998: The Arctic Oscillation signature in the wintertime geopotential height and temperature fields. *Geophys. Res. Lett.*, **25**, 1297–1300.

Trenberth, K. E., ed., 1992: *Climate System Modeling*. Cambridge: Cambridge University Press.

Trenberth, K. E. and D. J. Shea, 1987: On the evolution of the southern oscillation. *Mon. Wea. Rev.* **115**, 3078–3096.

Trenberth, K. E., G. W. Branstator, D. Karoly *et al.*, 1998: Progress during TOGA in understanding and modeling global teleconnections associated with tropical sea surface temperatures. *J. Geophys. Res.*, **103**, 14291–14324.

Turco, R. P., 1997: *Earth Under Siege: From Air Pollution to Global Change*. Oxford, UK: Oxford University Press.

Tyndall, J., 1861: On the absorption and radiation of heat by gases and vapours, and on the physical connection of radiation, absorption, and conduction. *Phil. Mag.*, **22**, 169–194.

United Nations Environment Programme (UNEP), 2000: *Montreal Protocol on Substances that Deplete the Ozone Layer* as adjusted and or amended in London, 1990; Copenhagen, 1992; Vienna, 1995; Montreal, 1997; and Beijing, 1999. Geneva, Switzerland: World Meteorological Organization.

US Government Printing Office, 1976: *US Standard Atmosphere, 1976.* NOAA-S/T76-1562. Washington, DC: US Government Printing Office.

Visbeck, M. H., J. W. Hurrell, L. Polvani and H. M. Cullen, 2001: The North Atlantic Oscillation: Past, present, and future. *Proc. Natl Acad. Sci.* **98**, 12876–12877, doi:10.1073/pnas.231391598.

Walker, G. T., 1923: Correlation in seasonal variations of weather, VIII: A preliminary study of world weather. *Mem. Indian Meteor. Dep.*, **24**, 75–131.

Walker, G. T. and Bliss, E. W., 1932: World weather. V. *Mem. Roy. Meteorol. Soc.*, **4**, 53–84.

Wallace, J. M., T. P. Mitchell, E. M. Rasmusson *et al.*, 1998: On the structure and evolution of ENSO-related climate variability in the tropical Pacific: Lessons from TOGA. *J. Geophys. Res.*, **103**, 14169–14240.

Wang, H., J.-K. Schemm, A. Kumar *et al.*, 2009: A statistical forecast model for Atlantic seasonal hurricane activity based on the NCEP dynamical seasonal forecast. *J. Clim.*, **22**, 4481–4500.

Washington, W. M. and C. L. Parkinson, 2005: *An Introduction to Three-dimensional Climate Modeling*, 2nd edn. University Science Books.

Webster, P. J., 1983: Large-scale structure of the tropical atmosphere. In: *Large-scale Dynamical Processes in the Atmosphere*, ed. B. J. Hoskins and R. P. Pearce. New York: Academic Press, pp. 235–275.

Welander, P., 1955: Studies of the general development of motion in a two-dimensional ideal fluid. *Tellus* **7**, 141–156.

Winton, M., 1997: The effect of cold climate upon North Atlantic Deep Water formation in a simple ocean–atmosphere model. *J. Clim.*, **10**, 37–51.

Winton, M., 2000: A reformulated three-layer sea ice model. *J. Atmos. Ocean. Tech.*, **17**, 525–531.

Worley, S. J., S. D. Woodruff, R. W. Reynolds, S. J. Lubker and N. Lott, 2005: ICOADS Release 2.1 data and products. *Int. J. Climatol.*, **25**, 823–842.

Wyrtki, K., 1975: El Niño – the dynamic response of the equatorial Pacific ocean to atmospheric forcing. *J. Phys. Oceanog.*, **5**, 572–584.

Xie, P. and P. A. Arkin, 1996: Analyses of global monthly precipitation using gauge observations, satellite estimates, and numerical model predictions. *J. Clim.*, **9**, 840–858.

Xie, P. and P. A. Arkin, 1997: Global precipitation: A 17-year monthly analysis based on gauge observations, satellite estimates and numerical model outputs. *Bull. Amer. Met. Soc.*, **78**, 2539–2558.

Xie, S.-P., A. Kubokawa and K. Hanawa, 1989: Oscillations with two feedback processes in a coupled ocean–atmosphere model. *J. Clim.*, **2**, 946–964.

Xue, Y. and J. Shukla, 1993: The influence of land surface properties on Sahel climate. Part I: Desertification. *J. Clim.*, **6**, 2232–2245.

Zachos, J., M. Pagani, L. Sloan, E. Thomas and K. Billups, 2001: Trends, rhythms, and aberrations in global climate 65 Ma to present. *Science*, **292**, 686–693.

Zebiak, S. E. and M. A. Cane, 1987: A model El Niño Southern Oscillation. *Mon. Wea. Rev.*, **115**, 2262–2278.

Zeng, N., J. D. Neelin, K.-M. Lau and C. J. Tucker, 1999: Enhancement of interdecadal climate variability in the Sahel by vegetation interaction. *Science*, **286**, 1537–1540.

Index

absorptivity, 43, 196, 200
adiabatic, 82
  lapse rate, 82
    dry, 82, 83, 99, 168
    moist, 94
aerosol model, 157
aerosols, 46, 222, 226
albedo, 46
annular modes, 141
anthropogenic
  climate change, 1
  forcing, 248
  perturbation, 65
atmospheric boundary layer, 47, 82
atmospheric chemistry model, 157
atmospheric circulation, 50–58, 79–80
atmospheric window, 42

Beta effect, 74, 99
biogeochemistry, 34, 177
biogeochemistry model, 157
biophysical land surface models, 154
biosphere, 2, 34, 37
Bjerknes hypothesis, 18, 103
blackbody radiation, 41

carbon cycle, 8, 157
CFC, 7
climate, definition of, 1
climate change, 5, 7–13, 24–28
  anthropogenic, 1
  mitigation, 256
  observed to date, 246–252
climate drift, 178
climate dynamics, 1
climate feedback parameter, 199, 201, 202
climate feedbacks, 201, 204
  cloud, 208
  lapse rate, 211
  snow/ice, 207
  water vapor, 205
climate model hierarchy, 175
climate modeling, 2, 145, 157–158
  time scales, 162, 163
climate sensitivity, 205

climate sensitivity parameter, 204
climate system modeling, 157–158
climatology, 2
cloud fraction, 208
cloud top feedback, 210
clouds
  effect on climate change, 39
computational cost, 150
computational instability, 151
conditional instability, 170
conservation of mass, 85, 119
  applied to moisture, 89
  applied to salinity, 91
convection, 35
  dry, 47, 83, 168–169
  moist, 35, 90, 94, 148, 169–171
convective adjustment, 171
convective heating, 47, 90
Coriolis force, 54, 72–74
Coriolis parameter, 74
cryosphere, 34, 37

data assimilation, 126
delayed oscillator model, 123
discretization, 145
doubled-CO_2 response, 205
drag coefficient, 154
dynamical climate system, 3

Earth system model, 2, 157
eddy transports, 41, 53
El Niño, 4, 13, 103–136
  dynamics of transition phases, 116
  extreme phases, 106
  first forecast, 22
  forecast limits, 127–131
  history, 14, 18
  observations, 19–22
  prediction, 125–127
  remote impacts, 131–136
El Niño/Southern Oscillation
  see ENSO
electromagnetic spectrum, 41
emission temperature, 198
emissions scenarios, 221–225, 252

Printed in the United States
By Bookmasters